The New York Times
Book of
Science Literacy

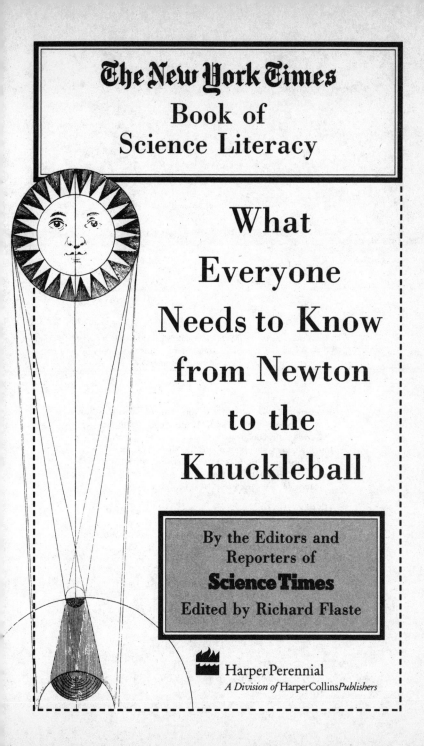

The New York Times
Book of Science Literacy

What Everyone Needs to Know from Newton to the Knuckleball

By the Editors and
Reporters of
Science Times
Edited by Richard Flaste

HarperPerennial
A Division of HarperCollinsPublishers

92 93 94 95 96 CW 10 9 8 7 6 5 4 3 2

ACKNOWLEDGMENTS

In addition to those writers specifically listed as contributors and designated by initials at the end of each piece, *The New York Times'* science editors have contributed profoundly to these articles. At the time this book was created, the editing staff of "Science Times" was headed by science and health editor Philip M. Boffey, and included his deputy editor, Cory Dean, along with William Dicke and Claiborne Ray. Holcomb B. Noble, a former deputy editor for the section, and Erik Eckholm, a deputy and then editor before moving on to other assignments, also were deeply involved in many of the articles here. The art director for "Science Times," Nancy Sterngold, has lent her intelligence to all aspects of the production for many years. Sydney Bodnick has been an invaluable regular on the copyediting staff each Monday when "Science Times" goes to press. Brenda Nicholson Parrino, the Science Department's news assistant, has been at the heart of the operation since the beginning.

Steve Wasserman, the editorial director of Times Books, and Paul Golob, associate editor, were major influences on the book's shape and tone.

And special thanks to Jordan Flaste, who provided skillful, flawless research assistance.

"Science Times" was created in 1979 by A. M. Rosenthal, who, when he was *The New York Times'* executive editor, boldly committed the paper to an era of pioneering science coverage. For most of its years, the inspired guiding spirit of "Science Times" was Arthur Gelb, the former managing editor of the paper.

CONTRIBUTORS

Lawrence K. Altman [LKA]
Natalie Angier [NA]
Sandra Blakeslee [SB]
William J. Broad [WJB]
Jane E. Brody [JEB]
Malcolm W. Browne [MWB]
Alan Cowell [AC]
Cory Dean [CD]
James Gleick [JG]
Daniel Goleman [DG]
Philip J. Hilts [PJH]
Gina Kolata [GK]
John R. Luoma [JRL]
Andrew Pollack [AP]
Elisabeth Rosenthal [ER]
Harold M. Schmeck, Jr. [HMS]
Peter Steinfels [PS]
William K. Stevens [WKS]
Walter Sullivan [WS]
Matthew L. Wald [MLW]
John Noble Wilford [JNW]

CONTENTS

The Search for Origins

Understanding Human Behavior

The Pursuit of Health

Our Troubled Environment

The Promise of Technology

Curiouser and Curiouser

The New York Times
Book of
Science Literacy

INTRODUCTION
by Richard Flaste

Quick, what is the second law of thermodynamics? Lots of people know the answer (in perhaps its most simplified expression it means that energy disperses), but many more—some of them highly educated, well-read people—don't. This question, and others like it, are often used to define who is scientifically literate and who is not, an expression of the belief that science literacy is a matter of memorizing vocabulary words, principles, and times and places of important events. Foolishness, really. Science itself is so vast, ranging from cosmology to molecular biology, that many scientists in one field would have to declare themselves illiterate in another if this testing were the only way to go about certifying competence. I believe that the best road to science literacy is one that offers a broad familiarity with the day's major scientific issues and the key concepts that underlie them in plain language (let the specialized vocabulary come along when it will). And that is the path taken by this volume, a book that is meant to be an introduction to a great many areas of science—as well as a good read for those more scientifically advanced among us, who simply want to brush up on the news. This volume is, in those respects, a direct extension of "Science Times," the weekly section of *The New York Times* from which all of the articles were drawn. Like "Science Times" itself, it declares that science need be neither intimidating nor arduous; it can, in fact, be irresistibly alluring. We are talking here, after all, about us—our bodies, our world, our universe.

Gaining a better understanding of science is more than a matter of self-improvement. There are many who feel that America itself is threatened by an ignorance and loathing of science. Teachers usually get the blame. That is, if science is being taught at all: of the country's 24,000 high schools, 7,000 offer no physics and, incredibly, 2,000 don't even offer biology. When Jon D. Miller, director of the Public Opinion Laboratory at Northern Illinois University, interviewed 2,000 adults recently, he found that superstition was often stronger than scientific thinking (two out of five interviewees thought that "some numbers are especially lucky for some people") and that science was deeply distrusted, with more than a third of the respondents agreeing with the statement "Science tends to break down people's ideas of right and wrong."

Miller's survey, moreover, was conducted during a time when fear was growing over whether electromagnetic fields created by telephone wires and toasters were a hazard to people (see page 282 of this volume); when bizarre materials called superconductors had been created, which could change our experience of the world (page 298); and when nutrition science was managing to enlighten many of us on some issues and confuse us on others in a period of swirling controversy (page 223 for a discussion of cholesterol). These are complex issues that do not yield to superstition and ignorance. People who think that science is something you believe in or not (as in, do you believe in evolution?) have handicapped themselves. People who are comfortable believing in communication from the dead or in ESP, for instance, have willfully deprived themselves of the wholesome skepticism necessary to address the real world. They have opted for ignorance.

In the quest for a truer view, this society's most readily available purveyor of information, journalism, can be both a boon and a burden. Journalism, after all, is a primary source of our sense of the world as tumultuous, fearful, and puzzling. Too often the news media present science in ways that can be alarming or just confusing. Frequently journalists are simply wrong in their understanding of what the research is saying. Sometimes they take small conclusions from small studies and treat them as if these weak numbers had the power to overthrow everything we've ever learned, rather than seeing such results as just one more station on the road to a far-off destination.

When I've talked to scientists over the years, some of them treat the journalism that describes their fields as if it were about something else entirely; they say it does not reflect the world they know because it is too superficial and (perhaps just as important) because it employs a completely different language, presenting the scientific material according to journalism's own conventions. But none of this negates journalism's great potency. A journalistic television program, book, or newspaper article, done well, is able to introduce the public to an area of scientific investigation succinctly and clearly. Journalism must distill an issue and then convey it as appealingly as possible without hyping it. The journalist is trained to do this in the same rigorous way that a scientist is trained to create an experiment. A good newspaper story must not only possess clarity, but should help the reader to conceptualize the issues at stake. What is the mainstream thought on a particular issue? What are the dissenting voices? How credible is either side? In this way, journalism can be a more effective medium than the professional literature, which often obscures what it has to say through poor writing, formulaic presentation, and general obfuscation that manages only rarely to hint at the true adventure of serendipity and misfortune that is the scientist's journey.

This book, the product of some of the nation's finest science journalists, lives by the rules of the trade: jargon is rarely permissible, and unfamiliar concepts should be defined in ordinary language, often using metaphor to replace mathematics. Each article (by universally accepted journalistic mandate) must be interesting from the very first paragraph. In journalism, no topic is too difficult; a failure to make even the most arcane subject comprehensible is a failure of the communicator and not the reader. Any reporter who makes himself seem clever at the reader's expense has failed; no reader should ever come away from a science article feeling stupid. Readers should be uplifted, informed, and emboldened to investigate further.

As you travel along with the writers and editors of "Science Times," stopping to peer inside a dinosaur's egg (page 71) or glimpse the baffling stuff called antimatter (page 24), it is important, even indispensable, to have some overarching view of what science is and how it works. All of the fields of science have some crucial aspects of their inquiry in common; once you know that, any one of the fields is more approachable, more deeply satisfying.

I say that with the fervor of a convert. For me, as for so many people I know, science in high school and the early years of college was the bitterest fruit in the groves of academe. Too much rote. Too much specialized vocabulary that would be of no use to me later. And painful memories: one martinet of a tenth-grade chemistry teacher, in particular, very nearly extinguished in me the flame of scientific curiosity forever. As I began college, I was powerfully determined to study literature and politics, and only when required would I take a course in science. Even then, it would be a course for nonmajors, certainly, and I would explain to anyone who wanted to listen that science wasn't my thing. When C. P. Snow wrote his famous essay about the two cultures (science and the humanities) and how far apart from each other the people in those cultures were, that idea suited me just fine. I was on the side of literature and art and thought. The humanities, that was where I wanted to hang out. Science was a place full of aliens.

The irony is that one day I would find myself earning a living in the scientists' world. By the time I had become the editor in charge of "Science Times," arriving at that assignment from elsewhere at the paper, science for me had already begun to shed its hard repellent shell and reveal itself as an exhilarating way of seeing the world. Despite the catastrophes of high school and the aversion that followed, I was finally able to make myself comfortable in the environs of science, like a traveler who decides that a once forbidding land is now the only place he wants to call his home. And it was understanding what scientists did and how they thought that accomplished this miraculous transformation.

Scientific thinking is, on one level, quite easy. Robert Pollack, a former Columbia University dean and a biologist, stresses that "science is essentially a structure for asking questions and every five-year-old is a natural scientist because every five-year-old is curious about the world." This is undoubtedly so, as far as it goes, but it is perhaps misleading also, because, as Dr. Pollack would certainly agree, science is so much more than that. It requires habits of thought that allow you to be so skeptical that you will even be willing to challenge your own deepest assumptions. It means approaching specific problems in accepted, time-tested ways.

Once you get the knack of scientific thinking it becomes possible to do all sorts of things that once might have seemed beyond the pale, like reading the findings in a professional journal or challenging a newspaper article because it offers too little data to allow you to come to an informed conclusion or because it is too credulous of what the scientists themselves have to say.

One soon gains an abiding appreciation of that old paradox about learning: the more we learn the more ignorant we seem. No sooner do we land a man on the moon than we realize that we don't even know where the solar system ends (page 38); we manage to learn the defects that cause a few genetic diseases and then discover that we are barely out of our infancy in understanding the genetic structure of a human being (page 169). And then there is AIDS! That scourge struck at a particularly smug time. When it arrived, scientists were even allowing themselves some optimism about cancer; they had improved the remedies and felt they were getting closer to the true cause on a genetic level. But AIDS was a humbling thing, a disease whose origins, biology, prevention, and future course were largely outside of science's knowledge. Only slowly (although some would say that actually the work progressed with lightning speed if you look at comparable scientific developments) have scientists begun to discern the nature of the disease. Two articles in this volume (pages 190 and 195) hint—and can do no more—at some of the issues surrounding AIDS.

What is the scientific process? First, it is a human process, at its most engaging, an all-consuming effort to get inside a problem. Joshua Lederberg, the former president of Rockefeller University and a Nobel laureate, once said that in his genetics work he made himself think, "What would it be like if I were one of the chemical pieces in a bacterial chromosome?" The biologist and philosopher P. D. Medawar, perhaps the best ever to write on this topic, explained the scientific process this way: "Scientists are building explanatory structures, telling stories which are scrupulously tested to see if they are stories about real life." Getting more specific about the quest for useful findings, he said:

> Most of the day-to-day business of science consists not of hunting for facts . . . but of testing hypotheses. . . . A well-designed and technically successful experiment will yield results of two different kinds: the experimental results may square with the hypothesis, or they may be inconsistent with it. If the results square with the implications of the hypothesis, then the scientist takes heart and begins to hope that he is thinking on the right lines; he will then, if he has any sense, expose his hypothesis to still more exacting experimental tests . . . but no matter how often the hypothesis is confirmed—no matter how many apples fall downwards instead of upwards—the hypothesis embodying the Newtonian gravitation scheme cannot be said to have been proved to be true. Any hypothesis . . . may conceivably be supplanted by a different hypothesis later on.

All sciences strive for rigorous approaches to standardize and systematize what they learn so that science can be distinguished from superstition. If a pencil falls off the desk and a door swings open at the same time, are they related events? Maybe, but probably not. Nevertheless, coincidences are seductive (page 343). That's why a ballplayer might insist on repeatedly wearing the same shirt he wore when he hit a home run. It is, he says, his lucky shirt. But science and luck are often enemies. It is too easy to link things together through intuition and coincidence, come to the wrong conclusion, and be no better off than you were before. Science seeks to discover which linkages are legitimate and which aren't. That's what an experiment is: if I ask a group of people to take aspirin regularly to see if it diminishes the number of heart attacks in the group, and subsequently the heart attacks do seem fewer, how can I be relatively sure the aspirin is linked to the benefit? Maybe this was just a fortunate group. The answer, in part, resides in the dreaded field of statistics.

My own experience of science was changed permanently by a college course, taken relatively late in life, in experimental statistics. Like many other students, I approached the subject with great trepidation. But it turns out that statistics is mostly just arithmetic, most of which involves simply pushing the buttons of a calculator. By far the most important part of statistics is the logical mindset it forces you into, and that's primarily what I've retained over the years. Perhaps better than any other discipline, statistics teaches skepticism. You learn that very little seems absolutely certain and that much of science phrases its questions in terms of probability, not certainty. How likely is it that a particular variable (the aspirin) is causing something we are observing (fewer heart attacks than expected)? In an experiment a scientist arranges data in such a way as to yield an answer about likelihood. The statistical analysis of his data may show that, because the heart attack rate was so much lower than expected for this group, the aspirin can be thanked at a "confidence level" of 95 percent. The other 5 percent is to allow for the possibility that chance caused the result, and that the aspirin had nothing to do with it. At 95 percent (and usually no less), the scientist says his findings are "statistically significant," with a "P value" of .05. The scientist might then come to a conclusion and say that he has found a link between aspirin and fewer heart attacks. (Although he still hasn't proven anything—the correlation still might have been caused by chance or some other variable he had not thought to consider.)

The catch here, of course, is that statistics can be massaged to give you the answer you want. Maybe the scientist has stacked the deck somehow, intentionally or inadvertently. It is always important to know how well a scientist has taken into account all the possible variables that might provide a given result. For instance, if you read that Asians suffer fewer cases of breast cancer than Westerners and that this observation seems linked to the fact that they eat less saturated fat, you have to ask yourself: so what? Asians and Westerners lead very different lives, and this is true often even when the Asians move to the West. Could it be that some other factor is really the critical one, like the exercise gained from riding bicycles instead of driving cars? Could it be that total calories leading to greater accumulation of body fat are more the cause of breast cancer than the type of fat people eat? And, regardless of what the statistics say, have the researchers looked at enough subjects and gathered enough data to persuade us? Or have they looked at a great many people in making their point but actually asked their question so inefficiently that the large number of subjects is just a smokescreen for a bad experiment?

I remember once when a reporter filed a story on a particularly alarming environmental issue, I asked him whether the science seemed good. He told me he wasn't equipped to know; he was just reporting what the scientists said. Such willful ignorance made my blood boil. All of us are equipped to challenge with sensible skepticism anything that scientists tell us as we try to reach conclusions that satisfy us. The paleontologist and science writer Stephen Jay Gould recalls how he would plummet into despair when his nonscientist father felt buffaloed by science: "My heart would sink whenever my father attributed the carelessness of scholars to his own ignorance based on lack of professional training. I could never get him to understand that advanced degrees and letters after a name guarantee no new level of wisdom and that, in the end, there is no substitute for old-fashioned, careful reading."

One of the most serious questions we have to ask ourselves when we try to be intelligently skeptical about many of today's scientific findings has to do with the nature of the danger—so much of what we read, after all, is about danger—that is being portrayed. What does science know about the true risk of a particular peril? A heart attack is much more likely for the fat fifty-year-old man than the lean one, but it is also true that it is a rare occurrence for both of them. So while it is probably a good idea not to be a fat fifty-year-old, it is mischievous and misery-making to overstate the danger.

Are comparisons of risks meaningful? Look at these figures: the average annual risk of death from smoking a pack of cigarettes a day has been calculated at 1 in 300, from a motor vehicle accident it's 1 in 4,100, and from drinking water containing the EPA limit of trichloroethylene it is 1 in 500 million. Risk comparison is one of those areas that inflame great debate. Is the foregoing really a sophisticated and insidious way of falsely reassuring us about a possible carcinogen in the water? Or is it a useful elucidation of real differences in risks in a world full of them, large and small? I tend to take the latter position; but either, I think, is tenable in certain circumstances. The important thing is to approach all assertions of risk thoughtfully. Drive carefully and don't smoke.

In all my years as a science editor, the trap I worried about most was the one set by scientists who saw themselves as advocates for one point of view or another. This is generally not a problem when the point of view is one that is well established with overwhelming evidence, as is the case with the danger of cigarette smoking. It is a problem in a great many other cases where debate is still the order of the day. Then the scientist has stepped out of his role and chosen to become a spokesman for, in effect, an interest group. Of course, journalists are guilty here, too. For years, we have turned to scientists for authoritative answers, and many of them have done their job well, telling us what the evidence said and suggesting just how great the uncertainty might be. But we have often been too reverential (like that credulous reporter who just reported what the "experts" said), and we have, I fear, encouraged many of them to step into a kind of messianic role in which they overstate what their findings say—about the environment, or nutrition, or a drug treatment. We tend to accept it; then when contrary evidence comes along, we assume that science in general is full of baloney.

Sometimes, scientifically trained people even view it as their duty to simplify and preach. In the spring of 1990, for instance, thirty-eight different professional organizations joined in a government Panel's announcement that every American above two years of age should eat less fat and more spinach. They decided to speak as one to diminish the confusion about nutrition these days, even though a handful of conflicting scientific findings give us all very good reason to feel confused. A *Times* editorial chastised the professional groups, saying that "eminent authorities can err en masse, especially when they coalesce and suppress confusion. . . . The cholesterol panel may be right, but its methods are those of puffery, not science."

The selections in this book have been drawn from "Science Times," January 1987 to May 1990, a time frame that in some respects limited the scope of the selections but also ensured their freshness. Many of the selections were edited slightly, especially to remove references to time or to events that were no longer relevant, or, where feasible, to accomplish minor updating; in some cases longer original articles were shortened to better serve the purposes of this volume.

It is probably a good idea to read this volume in small segments, allowing an article to introduce you to a field of recent investigation, whetting your appetite for it. There is so much touched on here that if you try to take it all in at once, the result, I fear, will be numbing rather than stimulating.

The goal of this book is to bring to you the major scientific topics of the day in a way that only journalism can. In many of the longer articles the writer will guide you so that you can actually see the science at work and be steered away from the puffery. The articles have been chosen not because they represent everything that was ever important to science—often they just take the smallest piece of a big question—but because they are about developments that, for the most part, were in the news. For one reason or another, they were on our minds. They percolated to the surface in "Science Times" because scientists were arguing about them, pursuing them, and the rest of us were fascinated by them. They were part of the great debate about ourselves and our universe. Even the piece about the knuckleball's wild movements (page 364) qualifies here. And the article on why the night sky is dark (page 29) exhibits yet another cherished aspect of science: the indulgence of pure wonder. Each week as the editors create a new "Science Times" they are working to concoct a kind of smorgasbord of the latest knowledge, allowing readers to select the morsels that interest them most. So it is with this book. Written by science journalists who have devoted their lives to making these diverse worlds of exploration accessible to anyone who really cares, the articles here touch every molecule of our humanity.

The Cosmos
Around Us

THE GENIUS OF NEWTON

FEWER than seven hundred copies of the book are sold each year, and even specialists in the subjects it covers rarely read more than excerpts from the six-hundred-page work. But in 1987, on the three hundredth anniversary of its publication, scientists on both sides of the Atlantic gathered to acclaim it anew as the greatest and most influential scientific book ever written. The Latin title of this formidable tome is *Philosophiae Naturalis Principia Mathematica* (*Mathematical Principles of Natural Philosophy*), generally shortened to the *Principia*. Its author was Sir Isaac Newton, the English mathematician whose genius laid the foundation of modern science and revolutionized the worldviews of statesmen and philosophers.

The Smithsonian Institution's National Museum of American History in Washington, D.C., presented a six-month exhibit devoted to the *Principia,* and in Great Britain four new postage stamps were issued celebrating the 1687 publication of the first edition of the book. Physicists, historians, and scholars held symposiums at the University of Maryland, the New York Academy of Sciences, and elsewhere in the United States and Britain to reexamine Newton's pervasive legacy.

Among the leading physicists participating in the symposiums on Newton's *Principia* was Dr. Subrahmanyan Chandrasekhar of the University of Chicago, whose own achievements include a mathematical description of space-time around black holes and the discovery of a law describing the limits of the masses of dwarf stars. "It is important for all of us to look again at the works of this amazing man," he said in an interview. "It is fashionable today to think of Einstein as the epitome of scientific genius, and compared with us ordinary mortals, Einstein was indeed a giant. But compared with Newton, Einstein runs a very distant second. In his *Principia,* Newton created the science of dynamics at a single stroke. The *Principia* underlies nearly every aspect of modern science. Perhaps most astonishing, Newton wrote the entire book in only eighteen months. Can you imagine Einstein having done such a thing? I cannot."

Many of Newton's speculations seem as pointed and fresh today as they did in his own time, scientists say. In the powerful analyses *Principia* provides of the movements of comets, for instance, Newton even hinted at the existence of a cloud of comets far outside the orbits of planets, from which individual comets would swoop in to approach the sun only rarely. Evidence of the existence of the Oort cloud (named for the Dutch astronomer Jan Hendrik Oort) has accumulated in recent years and appears to corroborate Newton's surmise.

Modern astronomers often use enormous reflecting telescopes of a type invented by Newton. In principle these telescopes are nearly identical to the two-inch-diameter instrument devised by Newton to study the heavens three centuries ago.

In the *Principia,* Newton introduced the notion of gravity as a universal force and showed how to use this idea in calculating the motions of planets, satellites, and comets. These insights led Newton's friend and colleague Edmond Halley to conclude in 1705 that the bright comet of 1682 was the same one that had appeared in 1531 and 1607. Halley's correct prediction that the comet would reappear in 1758–59 brought him posthumous renown. The *Principia* also explains how to gauge the tidal effects of the moon and sun, how to measure the precise shape of the earth (it is flattened at the poles), how to interpolate points along a complex trajectory, and how to calculate the exact path of a comet from only three observations.

Newton's complex computations, based on an imaginative blend of geometry and algebraic relationships, relied heavily on a new branch of mathematics invented independently by Newton and his German rival, Gottfried Wilhelm Leibniz. The calculus, a mathematical method vital to planetary astronomy, to most fields of physics and chemistry, and even to sociology and the kind of actuarial analysis used by insurance companies to set rates, made it possible to deal mathematically with continuous dynamic change.

The intellectual power of Newton's demonstrations is still a source of wonder to physicists. But remarkable though Newton's achievements were, they might have been overlooked if he had failed to write the *Principia.* "We know from his papers he had performed prodigies in a number of fields," said Dr. Richard S. Westfall, a participant in several symposiums and author of *Never at Rest,* a biography of Newton. "As we also know, he had completed nothing." Had Newton died before writing the *Principia,* he said, "we would at most mention him in brief paragraphs lamenting his failure to reach fulfillment."

Newton was finally impelled to write the book by Halley's encouragement and support. Newton started work on the gigantic project in 1684 and completed the three "books" and the hundreds of propositions, theorems, and proofs of the *Principia* in just eighteen months, drawing largely on work he had done twenty years earlier and set aside in favor of other projects, especially alchemy. The first edition was published in 1687, and Newton lived to publish two revised editions. According to Dr. Paul Theerman, curator of the Smithsonian's *Principia* exhibition, real fame came to Newton only after the mathematician moved to London from Cambridge and assumed public office as Master of the Mint, supervising the production of Britain's coins and devising safeguards against counterfeiting. "At that point," he said, "Newton and his book suddenly became fashionable."

Among the books in the Smithsonian collection are various early-eighteenth-century popularizations of the *Principia,* including one with the title *Sir Isaac Newton's Theory of Light and Colours and His Principle of Attraction, Made Familiar to the Ladies in Several Entertainments.*

"Despite his rather uncongenial personality," Dr. Chandrasekhar said, "Newton became one of London's major social lions. People used to stand for hours to catch a glimpse of his carriage being driven to the mint where he worked." The great scientist's neurotic suspicions of associates reached such a peak in 1693–94 that some contemporary accounts describe him as insane. In that period he accused his friend the philosopher John Locke of having tried to "embroil" him with women and thereby destroy him. The son of a twice-widowed mother, Newton spent much of his childhood in solitude. He was apparently unpopular (and academically unsuccessful) in his early school years; according to one account, he began studying in earnest to avenge himself academically on a class bully who had beaten him physically.

In recent years some historians have ascribed Newton's lifelong psychological problems in part to poisoning by various heavy metals, especially mercury. He spent many years devoted to alchemical research, during which he was persistently exposed to toxic vapors. Newton's teachings, which seemed to contemporaries to explain the material universe, were quickly adopted by the European movement known as the Enlightenment, one of many driving forces behind both the French and American revolutions.

Newton's theory of universal gravitation, powerful though it was in explaining the movements of planets and comets, as well as of the apple that legend says inspired him by falling on his head, has not survived unscathed. Einstein's general theory of relativity showed that gravity is a consequence of the curvature that space undergoes around a massive object and that Newton's analysis was only approximately correct. "But remember," Dr. Chandrasekhar said, "Newton himself tried to correct the impression that he had explained gravity. He made it clear that he was only attempting to calculate what gravity does—not to explain the underlying nature of gravity. That question remains a profound puzzle even today."

How would Newton feel about the course science has taken since his death? "I think he would have been troubled by the development of quantum theory," Dr. Chandrasekhar said, "since so much in quantum physics is indeterminate and acausal. But he would have been far less surprised by today's science than would any of his contemporaries. He would have been much more disturbed, I think, by today's religious evangelism."

Despite Newton's important role in the evolution of modern civilization, the great book by which his work came to be known is unlikely to become a best-seller ever again. The only edition of the *Principia* in print in the United States is published in small numbers by the University of California Press. It is a 1929 modern-English version of the 1729 translation Andrew Motte made from the Latin. "We know it is a terribly difficult book," said David C. Bratt, a spokesman for the University of California Press. "But this is Isaac Newton. You can't treat Newton as you would anyone else." [MWB]

Q&A

Q. *Why is the distance between stars measured in light-years? And how long, in miles, is a light-year?*

A. The stars are spaced so far apart that any measurement of the distance between them in familiar units would be unwieldy, according to Dr. Joe Patterson, professor of astronomy at Columbia University. So scientists created a more manageable astronomical unit called the light-year, which is the distance it takes light to travel in one year at a speed of about 186,200 miles per second. One light-year is about 6 trillion miles. The closest star to Earth, Alpha Centauri, is 4.2 light-years away. This is the average distance between most stars.

THE HUNT FOR GRAVITY WAVES

SCIENTISTS are planning the most ambitious effort yet to find and exploit one of the most elusive of the predicted phenomena of nature: gravitational waves. As long ago as 1915, Albert Einstein predicted in his general theory of relativity that the violent birth and death of stars in the universe would give off gravitational waves, bathing the earth in a unique kind of radiation. But so far, after decades of searching with increasingly sensitive detectors, scientists have not found any. Even so, agreement that they exist has grown steadily in recent years, and now the Bush administration has called for construction of a large observatory, with outposts on both coasts, to detect them. The facility, known as the laser interferometer gravitational wave observatory, could be completed by 1995 at a cost of $192 million. The California Institute of Technology (Caltech) is joined in the new observatory project by the Massachusetts Institute of Technology (MIT), which has also worked long and hard at developing the art of laser interferometry.

The discovery of gravitational waves would rank as one of the most important observational feats in modern physics and astronomy. It would provide a new confirmation of Einstein's general theory of relativity, a foundation of modern science whose validity is difficult to prove experimentally. And it would provide a powerful new tool for astronomy. Gravitational waves might allow scientists to get their first clear view of the collapse of exploding stars, the formation of mysterious black holes, and the residue of the Big Bang that is believed to have started the universe.

Einstein's general theory of relativity changed the concept of space from an empty void into a curving fabric of space and time. When stars collapse or make other violent motions, the theory suggests, ripples in the space-time fabric move off in all directions at the speed of light. These waves are not the mechanism that transmits the force of gravity between objects, but are distortions in that force. They are a phenomenon that can accompany gravity, much as calm water in a small lake can suddenly be churned into waves by a storm. In theory, gravity waves from distant cosmic events should move objects on Earth an infinitesimal amount, measured in distances that are much smaller than the nucleus of an atom. The new observatory would have to be sensitive enough to measure these waves and help pinpoint their cosmic origin.

"It opens a totally new window on the universe," says Dr. Rochus (Robbie) E. Vogt, director of the project and a physics professor at Caltech in Pasadena. "A very large part of astronomic objects are obscured. Gravitational waves penetrate things with impunity. So with gravitational waves you can look behind obscurations, behind the dust, into the cores of stars, where matter behaves in very strange ways."

The risk, skeptics say, is that the apparatus will not be sensitive enough to detect the waves. "It's not guaranteed that we'll get them," Dr. Vogt concedes. "But the fact that I and other people are willing to gamble our careers on it shows that we are confident technology is now ready."

A few physicists say detectors should be improved before a giant project is started. "It's an interesting field and a lot of progress has been made," says Dr. Richard L. Garwin, a pioneer of gravity wave research, who is now a scientist at the International Business Machines Corporation (IBM). "But there's a lot more that can be done to improve the apparatus."

Gravity is one of the central mysteries of science. It is the force that draws together all masses and particles of matter in the universe, but exactly how it works is still an enigma. Attempts at detection of gravity waves started in 1957 when an American physicist, Dr. Joseph Weber, set up a suspended aluminum cylinder to try to register the faint waves, measuring how much the cylinder moved with electrodes strapped to its surface. In 1969 he announced success, causing a commotion. But no physicist could reproduce his work.

The first indirect evidence of gravity waves came after a 1974 discovery by Dr. Joseph H. Taylor, Jr., and Dr. Russell A. Hulse, then at the University of Massachusetts at Amherst, of a pair of stars spiraling toward each other. They calculated that the slight decay in the stars' orbits agreed very closely with the loss of energy predicted for gravitational radiation. Since then the field has grown dramatically, with scores of physicists around the world joining the search. Detection devices of the cylinder type have proliferated, the best in theory being able to measure fairly small gravitational strains. The newest technique is to bounce a laser beam back and forth between suspended weights. In theory a gravity wave will affect the weights and throw the beam ever so slightly out of phase, allowing detection.

The largest such laser interferometer is at Caltech. It has two arms at right angles, each 130 feet long. Its sensitivity is great enough to register the pressure of light from a small camera's flashbulb, says Dr. Ronald W. P. Drever, its creator, who said it was fundamentally a development tool in the search for ways to improve the sensitivity of such devices.

The proposed observatory would be huge, with two arms each 2.5 miles long, increasing its sensitivity. At the ends of the arms as well as at their vertex, test masses fitted with mirrors would be suspended from wires. Superstable laser beams would flash back and forth between the test masses to measure the effect of gravitational waves on them. If everything works properly, the device would be one hundred to one thousand times as sensitive as the Caltech interferometer. A key expense in its construction is the long, four-foot-wide evacuated pipe stretching down the length of the arms, a device meant to keep the apparatus free of any disturbance from minute particles. The identical stations on the East and West coasts are needed so independent measurements can be constantly compared to help eliminate signals stemming from local seismic disturbances. Having at least two instruments in operation is considered the first step in getting a fix on the cosmic origin of the waves. "Two stations give you a circle in the sky," says Dr. Drever. "You need a third station to really locate it." Two European collaborators are at work on a plan to build other stations, probably in Germany and Italy.

Dr. Richard A. Isaacson, program manager for gravitational physics at the National Science Foundation, says the first-year construction budget request for the observatory was $47 million, with the total cost coming to $192 million. "This is the real world and big projects often run into snags," he concedes.

If built, the observatory would allow scientists to peer into the heart of celestial physics, observing gravitational waves from some of the most spectacular events in the universe, including supernovas, or exploding stars. Their waves would shed light on the earliest moments of supernova violence, when most of the explosive process is shielded from view by debris. Supernovas are rare, occurring in the Milky Way galaxy once every twenty or thirty years. But the new observatory's sensitivity should allow it to sense gravity waves from supernovas in distant galaxies, which might be detected as often as once a month.

The device should also be able to catch the first direct glimpses of black holes, dense stars with such strong gravitational fields that light cannot escape. Black holes have long been postulated but have yet to be detected unambiguously. "That's my own feeling of where the pay dirt will be," says Dr. Rainer Weiss, a physicist at MIT who is working on the project. "If they're there, we'll find black holes by the way they radiate." Another target is the violent final stages in the life of a pair of neutron stars as they spiral together after long orbiting one another, an event impossible to witness with optical telescopes. Still another target is the unknown.

"Our mysterious object we cannot predict," says Dr. Vogt of Caltech. "I know that there will be one. Ultimately, gravitational wave astronomy will just become a regular component of looking at the universe." Gravity wave enthusiasts are already thinking beyond the new observatory, hoping to create a system of three artificial satellites that would be even more sensitive.

"On the ground you are limited by the seismic background noise of the earth itself," says James E. Faller, a physicist at the Joint Institute for Laboratory Astrophysics at the University of Colorado at Boulder. "Space provides two things, a larger arena to work in and quietness."

Known as the laser gravitational wave observatory in space, the trio of satellites would use laser beams and weights to detect very-low-frequency gravity waves from the host of double stars that dot the galaxy. The three satellites would be 6.25 million miles apart, operating perhaps as soon as early next century. So far, no cost estimates have been drawn up.

"There's been a renaissance of gravitational physics in the last ten or fifteen years," says Dr. Faller. "The exquisite sensitivity that's now available to the physics community has put extraordinary things in our reach. Finally, we're able to do things. Suddenly it's become a real field." [WJB]

ORBITAL CHAOS

Astronomers trying to explain the peculiar geology of Miranda, a moon of Uranus, believe they have found an answer: it must have spent millions of years in a chaotic orbit, tumbling and lurching erratically instead of rotating at predictable intervals. Orbital chaos was once thought to be a dynamical impossibility, but astronomers now think it answers a variety of questions about the solar system. Jack Wisdom of the Massachusetts Institute of Technology recently showed that a moon of Saturn, Hyperion, is now tumbling chaotically and that chaotic orbits explain how asteroids reach Earth in the form of meteors.

Miranda has puzzled scientists since Voyager satellite pictures brought it into view—a small moon, three hundred miles in diameter, with rifts and ovals that suggested a complicated geological history. But a body that small should not have had enough internal heat to power such geological churning. "So that was a mystery," said Richard Greenberg of the Lunar and Planetary Laboratory at the University of Arizona in Tucson. "It shouldn't have been enough to drive any sort of geology, let alone the sort of large-scale exotic geology we see on Miranda."

Dr. Greenberg and Robert Marcialis, also at the Lunar and Planetary Laboratory, report in the journal *Nature* that the necessary heat could have been generated by tidal stresses if the moon was once irregularly shaped and orbiting chaotically. Furthermore, they calculate that many other satellites of the outer planets may have gone through similar processes before ending up in their present, predictable orbits.

THE MYSTERY OF
ANTIMATTER

HALF a century after the discovery of the baffling, rare stuff known as antimatter, scientists are beginning to collect it in minuscule amounts, creating a wave of excitement around the world as they explore its nature and applications and redouble their efforts to find it in the heavens. Breakthroughs in making and storing this enigmatic form of matter are enabling scientists to study more of it than ever before—yet still in amounts so small that the assembled mass would be invisible to the naked eye. At the edge of space, occasional rays of antimatter are being recorded by instruments that are lofted by balloons in an attempt to resolve some of the riddles that surround antimatter's origin and that of the universe.

Right now this work is pure research, but scientists say that in the future antimatter could be used in rocket engines, weapons, power generation, and other applications. Antimatter, with its *Alice in Wonderland* quality, is unlike anything else. Fundamentally, it is the mirror image of matter but with an opposite electrical charge. It cannot exist in the presence of matter; they annihilate each other. Those who muse about it like to dwell on the possibilities it presents of shadowy galaxies, stars, and planets that are, in effect, an anti-universe.

In the 1990s the antimatter search is to expand as an $80 million instrument is mounted on the side of America's orbiting space station to sweep the sky with unprecedented sensitivity for traces of the exotic material. While no evidence exists that the cosmos has clumps of antimatter, a few astronomers are nonetheless searching for stars and even whole galaxies made of it. Their search is driven in part by their belief that nature has a deep and abiding symmetry in which, metaphorically speaking, every plus has a minus, every yin a yang. All the matter in the universe should thus have an antimatter counterpart.

"There's no proof it isn't out there, and some theories suggest it's waiting to be found," said Dr. Mark E. Wiedenbeck, an astrophysicist at the University of Chicago. Dr. Robert L. Forward, senior scientist at the research laboratories of the Hughes Aircraft Company, said: "The real question in everybody's mind is why the universe seems to be made up of matter when, on a cosmic scale, antimatter is just as easy to make. It's one of the outstanding big mysteries." The search for symmetry has been fueled by recent experiments with atom smashers in which subatomic particles collide, creating the extremely high temperatures and pressures believed to have existed at the birth of the universe. The collisions produce tiny but intense fireballs of energy that can condense into showers of subatomic particles in which every electron has an antielectron, every proton an antiproton, and so on.

The antimatter particles discovered so far have the same mass and other measurable qualities as comparable particles of matter. When matter and antimatter particles collide, they are transformed into pure energy, mainly in the form of gamma rays. The process releases all the latent energy the particles contain and is many times more efficient than the nuclear reactions that power atomic and hydrogen bombs.

On earth, made exclusively of matter, a tiny particle of antimatter created in a laboratory usually lasts only a few millionths of a second before it is annihilated in a collision with matter. But in theory, whole islands of antimatter could be floating in the universe, cut off from matter by the empty void of space. If a large chunk of antimatter fell to earth, the planet would be vaporized in a blinding flash of energy. "An antigalaxy would look just exactly like any other galaxy," says Dr. Floyd W. Stecker, an astrophysicist at the National Aeronautics and Space Administration's Goddard Space Flight Center in Greenbelt, Maryland. "For the universe as a whole, it could turn out there's just as much antimatter as matter."

The prophet of antimatter was Dr. Paul A. M. Dirac, a British physicist who in 1928, at the age of twenty-six, predicted the existence of antiparticles on the basis of his strong belief that there should be balance in mathematics and nature. While pondering an equation describing the behavior of electrons, he realized it had positive and negative solutions. If the universe was symmetrical—and his deep sense of symmetry told him that it was—ordinary electrons, which have a negative charge, would have strange counterparts of the opposite charge. No one had observed such particles. But in 1932 Dr. Carl Anderson, a twenty-six-year-old physicist at the California Institute of Technology, came across odd streaks amid particle tracks in a cloud chamber. The paths looked exactly like those of electrons except they curved in the wrong direction. The anti-electron, or positron, had been discovered.

In 1933, Dirac won the Nobel Prize in Physics for the accuracy of his intuition. In his acceptance speech, he elaborated on his vision and predicted the discovery of antiprotons and other types of antimatter particles, as well as antimatter stars. "We must regard it rather as an accident that the Earth (and presumably the whole solar system), contains a preponderance of negative electrons and positive protons," he told the Stockholm audience. "It is quite possible that for some of the stars it is the other way about." Another of Dirac's predictions was confirmed in 1955 when Emilio Segre and Owen Chamberlain, physicists working with an atom smasher at the University of California at Berkeley, slammed protons into special targets and afterward located a few traces of antiprotons. Later it turned out that the flash of energy created in many high-energy collisions condenses into a matching pair of matter-antimatter particles. The principle of such creation is described by Einstein's famous equation $E = mc^2$—energy equals mass times the square of the speed of light.

For a long time antimatter on earth was so rare it appeared mainly in science fiction. In the *Star Trek* series, for example, the engines of the starship *Enterprise* were powered by antimatter. All that changed in 1982 as advances in magnets and other seemingly unrelated fields allowed physicists for the first time to make, capture, store, and use antimatter for their experiments. The breakthrough was made at the European Laboratory for Particle Physics, known as CERN, near Geneva. The key stride was a method to store antiprotons in a large circular tube from which all air had been removed. The antiparticles, whirling through the tube at nearly the speed of light, were held in a powerful magnetic field that kept them from striking the sides. Physicists used the antiparticles in atom smashers to probe the atom. Not long ago they extended their success by trapping antiprotons in magnetic "bottles" and holding them motionless, allowing intensified studies of their characteristics.

Scientists say strides in antimatter production and storage might eventually allow annihilation energy to be used for rockets and weapons, although such developments are far off. "Even with adequate funding, it will take literally decades of hard work before mirror matter is available in quantity," said Dr. Forward of Hughes Aircraft, which is exploring the use of antimatter in rocket propulsion. Paradoxically, islands of antimatter may already exist somewhere in the heavens. In the early days of antimatter research, astrophysicists quickly realized that the discoveries on earth had implications for the cosmos. According to the widely accepted Big Bang theory, the universe started as an infinitely hot and dense dot of energy that exploded with a dazzling brilliance and created the matter all about us. Astrophysicists reasoned that the laboratory findings of particle-antiparticle symmetry meant the Big Bang could have created an equal amount of antimatter.

Astronomers began to hunt for it. Optical telescopes were no help in this search since antimatter stars would appear identical to those made of matter. Astronomers instead looked for subtler clues. In our Milky Way galaxy, they reasoned, collisions involving gas, dust, stars, and planets made of antimatter would produce telltale bursts of gamma rays. The gamma rays would be unable to penetrate the earth's atmosphere, but satellites in orbit about the earth could detect them. But no significant gamma rays were found. Astronomers also looked beyond the Milky Way to a few known cases where distant galaxies were in collision. Again, there appeared to be no large annihilations.

The big surprise came in 1979—not from orbiting satellites but from a balloon-borne experiment launched from the National Scientific Balloon Facility in Palestine, Texas, and carried to the edge of space. It found antimatter itself above the earth. Astrophysicists had long suspected that the protons that form many of the cosmic rays streaking through the galaxy occasionally created antiprotons in the same way atom smashers do, that is, by collisions with bits of matter. They predicted they would find one antiproton cosmic ray in every twenty thousand cosmic rays. But the scientists in charge of the balloon experiment, one of the first to actually record cosmic antimatter on the edge of space, discovered some three times more antiprotons than they had expected. "It was a puzzlement," says Dr. Robert L. Golden of New Mexico State University, who headed the group. "Today people have explanations that range from the mundane to the very exotic."

A relatively simple explanation is that intergalactic space simply has more gaseous matter than expected, which triggers more collisions and therefore more production of antimatter cosmic rays. A more exotic explanation is that some of the excess rays are actually "primary antiprotons," not created ones, speeding through space from distant antimatter galaxies. "If some other galaxy is made of antimatter, then the cosmic rays it accelerates would be anticosmic rays," says Dr. Martin H. Israel, an astrophysicist at Washington University in St. Louis. "It's possible that some would leak into our galaxy." Despite such riddles, many theorists assume the majority of the cosmos is made of matter, although they admit the evidence is paltry.

To better understand the extent of antimatter in the cosmos, NASA is planning a major experiment for America's space station that would collect bits of antimatter speeding past earth. This Particle Astrophysics Magnet Facility, known as Astromag, is to cost some $80 million and be launched by the space shuttle sometime this decade. Thirty feet long and thirteen feet wide, it will take up more than half the shuttle's payload bay. Once in space, Astromag will be bolted to the superstructure of the space station.

Most of Astromag's bulk will be taken up by its superconducting magnets, cooled by liquid helium to four degrees above absolute zero. At that temperature, magnets can become very powerful and superconducting, meaning electricity circulates forever amid their coils in the next best thing to perpetual motion. Cosmic particles that speed into Astromag will have their paths bent by this powerful magnetic field, the curvature revealing the nature of the particle. Detectors would radio this information to earth for analysis.

According to scientists, the most spectacular find would be composite antiparticles such as anti-helium, anti-oxygen, or anti-iron. Unlike elementary particles, these cannot be created by collisions of matter in space and would therefore be either relics of the Big Bang or lone travelers from antimatter stars or galaxies. "Even one anti-helium particle would be clear proof that there are large aggregations of antimatter out there," says Dr. Israel, chairman of a NASA panel on Astromag. "That gets very exciting."   [WJB]

HOW OLD IS THE SUN?

Refined estimates of the age of meteorites and new theories on stellar evolution have led an astronomer at Yale University to conclude that the sun is about 200 million years younger than had been assumed: approximately 4.49 billion years old, instead of 4.7 billion years. Previous estimates of the sun's age were calculated on the assumption that meteorites formed somewhat later. But observations of nearby star formation, using infrared and radio telescopes, indicate that meteorites and planets form before the star is fully developed. As the protostar becomes denser, its increasing luminosity blows away the surrounding dust and gas, leaving behind meteorites and planets, and eventually its core reaches a mass that triggers the nuclear fusion that is the characteristic energy process of stars.

In a report recently in the *Astrophysical Journal,* David B. Guenther, the Yale astronomer, said: "The well-determined age of the meteorites, the theories describing the origin of the solar system—improved to fit recent infrared and radio observations of star-forming regions—and solar evolutionary calculations have all combined to permit, for the first time, a relatively precise determination of the sun's age."

More precise radioactive dating methods have fixed the age of the oldest meteorites orbiting the sun at about 4.53 billion years. Calculations that take into account the new theories of stellar evolution would put the date of the sun's birth, when the nuclear conversion of hydrogen into helium began through fusion in its core, at 4.49 billion years, plus or minus 40 million years. If the sun is indeed 200 million years younger than thought, Dr. Guenther said, this would still have little effect on predictions of the middle-aged sun's future. The sun has burned about half its hydrogen supply, astronomers estimate, and that means it will begin to die in another 5 billion years or so.

WHY IS THE NIGHT SKY DARK?

WHILE some astrophysicists work on high-velocity field dwarfs or magnetohydrodynamic solitons, others have been brooding lately about an older problem: why is the night sky dark? It is the puzzle that will not go away—the dark-sky paradox, also known as Olbers's paradox, for Wilhelm Olbers, a nineteenth-century astronomer who resurrected it a century after Edmond Halley had mentioned a certain "metaphysical paradox" to the Royal Society: "That if the number of Fixt stars were more than finite, the whole superficies of their apparent Sphere would be luminous."

The nub of the problem was the relatively new idea that the universe was unbounded and eternal. In an infinite universe, with stars and galaxies scattered in every direction, any line of sight from an earthbound observer was bound to intersect a star. Stargazers should not see dark spaces between stars; the spaces should be filled with more stars, just as someone in a deep forest sees nothing but trees in the spaces between trees. The sky should blaze with a uniform brightness.

That lights grow dim with distance was no answer. Twice as much distance to a star does mean one-fourth the apparent brightness; but it also means eight times the number of stars. Faraway stars lose brightness, true, but there should be enough of them to compensate. In the generations since, science has solved, debated, and solved Olbers's paradox many times over. Yet now a group of researchers, offering new calculations in the *Astrophysical Journal,* contends that modern astrophysics has brought its own brand of confusion. "There can hardly be another astronomical subject that is so fundamental in nature yet so widely misunderstood," the researchers write, adding that the situation is "scandalous." Of ten recent explanations in textbooks, they say, just three "may loosely be termed right."

Some astronomers proposed long ago that starlight must be absorbed by dust in the vast reaches of interstellar space. That proposal foundered when calculations revealed that, if so much starlight were being absorbed, the dust would grow so hot that it would re-radiate the energy and light up the night sky. Indeed, the flow of energy offers another way of stating the paradox, seemingly different but mathematically equivalent. Stars send off radiation in the form of photons, and that energy has to go somewhere. A more or less straightforward calculation suggests that all space should eventually grow as hot as the surface of a star.

"Photons move around, but they're not lost," says James E. Peebles of Princeton University. "If the universe is infinite and eternal, one would have an infinite accumulation of energy density. Which is absurd." Some theorists, up to the present day, have cited the expansion of the universe, after its explosive creation in the Big Bang, as the reason for the darkness. Since expansion means that stars are all receding—and the more distant the star, the greater its apparent speed—less of the light can reach us.

Olbers and Halley, of course, understood the paradox in the context of Euclidean geometry and Newtonian physics—just three dimensions, no Big Bang, no expansion, no relativity, no topological curvature, and none of the other peculiar gifts of Einstein's revolution. Those ideas created new interest in the paradox in the 1950s, but scientists soon found that they brought new difficulties as well. "The interest lapsed because of the complexity of the new cosmological models, trying to understand what happens to the radiation in the framework of expanding curved space," says Edward R. Harrison of the University of Massachusetts, an astronomer who has surveyed the history of the paradox and found fifteen distinct solutions offered over the centuries. "This was all very complicated. It was quite a problem to see clearly that the riddle is not affected by whether space is curved or not. It doesn't matter if the universe is finite or infinite, and it doesn't really matter much if it's expanding or static."

The astrophysicists offering new calculations—Paul S. Wesson of the University of Waterloo in Ontario and K. Valle and Rolf Stabell of the University of Oslo's Institute for Theoretical Astrophysics—find that, at most, the expansion of the universe cuts the amount of light reaching us by a factor of two. That is well short of the trillion or so needed to account for the actual darkness.

It was Dr. Harrison who first put forward what now seems the most persuasive key to Olbers's paradox: the galaxies' finite age. No one knows exactly how long ago was the cosmic explosion from which the universe sprang—it might have been 10 billion years, or 14 billion, or 18 billion—but whatever the age, the explosion puts an inviolable limit on the range of our vision. Any stars more than 18 billion light-years away are invisible because their light has not yet had time to reach us. It is as if the visible universe were enclosed by a spherical boundary, the "particle horizon" or "event horizon," moving outward by one light-year each year.

The astrophysicists from Ontario and Oslo say that their calculations confirm that view beyond doubt, the finite age of the galaxies definitively restricting "the number of photons that have been emitted into intergalactic space." Yet the darkness is not absolute. For the last twenty years, astronomers have known that the apparent emptiness of space is filled with a low-level background radiation, the remnant of the Big Bang. In a sense, the background radiation, too, has become part of Olbers's paradox—a light from all directions, uniform in intensity, though far murkier than the flaming surface of stars.

Halley's contemporaries, including Newton, could not have guessed at the Big Bang, but they knew perfectly well that light travels at a finite speed. They might have realized that a less-than-eternal universe answers Olbers's paradox. But it is hard to grasp the idea that looking far away in distance means looking far back in time. Furthermore, the age of the universe—six thousand years, according to biblical calculations—was a delicate issue for Newtonian astronomers. "The idea that you could look back to God on the job was a bit too much in Victorian circles, and it delayed a solution," Dr. Harrison observes. "You look out to the dark gaps between the stars, and you look right back to the creation." [JG]

Q&A

Q. *What star or standard is used when rating the apparent brightness of stars?*

A. Ancient astronomers first relied on their own eyes in comparing the brightness of stars and classified them in six orders of magnitude, each about two and a half times as bright as the next. According to Owen Gingerich, a professor of astronomy at Harvard University, the Alexandrian astronomer Ptolemy's "rough casting of stars into magnitude bins is the first record we have of stars being classified this way.

"In a modern edition of the *Almagest,* where the star catalogue appears, Ptolemy listed fourteen or fifteen that he considered to be of the first magnitude," or the brightest stars in the heavens, Mr. Gingerich says. The stars Ptolemy named or described in the second century A.D. were Arcturus, Vega, Capella, Regulus, Spica, Fomalhaut, Betelgeuse, Rigel, Achernar, Sirius, Procyon, Canopus, Alpha Centauri, Aldebaran, and Denebola.

"Missing from his list are such first-magnitude stars as Antares and Altair, but he included such stars as Fomalhaut and Denebola, which are not on the list when measured accurately," Mr. Gingerich notes. "There are stars visible in the Southern Hemisphere that are also of the first magnitude, but that fell south of where Ptolemy was observing and recording."

David Crawford, an astronomer at the Kitt Peak National Observatory in Tucson, Arizona, explains that modern magnitude classifications are tied to the historic scale, but that "now the zero point is set with an average of several thousand stars.

"Since we can measure them carefully now, the brightest ones are brighter than zero, and in a dark sky, which there isn't much of any more, the faintest ones we can see are about 6.5," he says. "The most-used modern scale, that of UBV magnitude, stands for ultraviolet blue visual. We measure them with filters that find those regions of color. The eye is not sensitive to ultraviolet, for example."

REACHING FOR THE STARS

THE story of the universe can be read in the many languages of radiated energy, the cosmic babel carrying across time and space the narrative that presumably began with the primeval atom that exploded 10 billion to 20 billion years ago in the Big Bang. With the new technologies of space-age astronomy, notably instruments like the Hubble Space Telescope deployed in the spring of 1990 high above the earth and plagued with technical problems, scientists are hoping to decipher more of these languages and see that each one tells a different part of the story of the universe's origin, evolution, and structure. No single instrument on the ground or in space, not even a perfectly functioning Hubble, can read more than part of the story. The Hubble will be joined by a fleet of other large telescopes to be launched this decade to decipher radiated messages that cannot be read from earth.

The most ambitious attempts to intercept these messages are the four Great Observatories planned by the National Aeronautics and Space Administration (NASA), starting with the Hubble, which is primarily for visible and ultraviolet observations. The others are the Gamma Ray Observatory, the Advanced X-Ray Astrophysics Facility, and the Space Infrared Telescope Facility. Several smaller telescopes will also be dispatched soon. "Never before have we launched so many science spacecraft with such frequency," says Dr. Lennard Fisk, NASA's administrator for science. "We will be awash in new data and new discoveries."

The observations of these space telescopes are expected to revolutionize astronomy by examining one by one all the radiations racing through the heavens. For these languages of cosmic discovery are actually different wavelengths of energy that span the electromagnetic spectrum, from gamma rays, X rays, and ultraviolet rays through visible light to infrared and radio waves. These are to eager and clever astrophysicists what ancient hieroglyphics are to archaeologists. Only by observations across the spectrum have astronomers learned much of what is now known of the birth and death of stars, the resounding collisions of atoms and galaxies, the violent whoosh of matter falling into gravitational holes, and the annihilation of matter meeting antimatter. They are even seeing the faint afterglow of the Big Bang itself, the theorized moment of creation. "It's one universe and everybody is using different instruments to look at the same universe," says Dr. John N. Bahcall, an astrophysicist at the Institute for Advanced Study in Princeton, New Jersey, one of the prime movers in the development of the Hubble telescope.

The space telescopes will not put out of business such venerable mountaintop observatories as Palomar and Kitt Peak. Indeed, there is a building boom in ground-based telescopes, with the largest of them all, the 396-inch Keck Telescope in Hawaii—four times as powerful as Palomar's 200-inch instrument in California—already poised to gather new data. Dr. Maarten Schmidt, an astronomer at the California Institute of Technology, says space telescopes are bound to make discoveries that will send scientists to ground-based telescopes for follow-up investigations, and vice versa.

Except for distant radio waves, the other wavelengths of energy are blocked out completely or partly by earth's thick atmosphere. Gamma rays, which are screened out completely, bear witness to the universe's violent nature. They are the product of collisions and massive collapses. When matter and antimatter, electrons and their antiparticles, positrons, meet and annihilate, gamma rays result. The collision of atoms in interstellar gas and the explosion of stars as supernovas, the collapse of aging stars into dense neutron stars or into black holes with a gravitational grip so strong light cannot escape them—these catastrophes send out waves of gamma rays. Some of the first discoveries of cosmic gamma rays were made in the early 1970s by Defense Department satellites sent aloft to monitor compliance with the nuclear test–ban treaty. Their instruments kept picking up short bursts of intense gamma rays. They emanated from points in space that appear to have no counterpart emission in other wavelengths. Astrophysicists still cannot explain the gamma-ray bursters.

The ubiquity of X rays was another surprise of space-age exploration. In 1962, an X-ray detector sent up by a rocket that rose briefly above the atmosphere discovered that the radiation filled the entire sky and came from a single point in the constellation Cygnus. Scientists think the X rays are produced by tremendous heat from gases of one star being sucked into a black hole, a theorized collapsed star so massive that its gravity pulls in everything around and prevents the emission of visible light. Dr. George Field, an astronomer at the Harvard-Smithsonian Center for Astrophysics, says the X-ray energies from similar cascades of matter might reveal the forces creating the recently observed "great walls," interconnected clusters of galaxies. In the 1970s, satellites detected thousands of X-ray hot spots in the universe, places where temperatures may exceed 1 million degrees Fahrenheit. The radiations are associated not only with suspected black holes but also with the violent centers of galaxies, intergalactic gases, and quasars, brilliant but enigmatic objects near the edge of the universe.

Dr. Charles J. Pellerin, Jr., NASA's director of astrophysics, says preparations are on schedule to build the $1.2 billion Advanced X-Ray Astrophysics Facility. Launching by a shuttle is planned in 1997. Since X rays will not reflect from mirrors if they arrive straight on, the telescope's six highly polished glass mirrors with gold coatings will be configured to detect the incoming radiation as it grazes the surfaces at shallow angles. The spacecraft will be capable of making images of the X rays, and its instruments should be one hundred times more powerful than those of the Einstein satellite. Astronomers are hoping, therefore, to learn from the X rays something about the dark matter that must pervade the universe. As they figure it, only 10 percent of the mass of the universe has been observed or inferred; there must be something else out there, invisible so far, to account for the gravitational forces shaping galaxies and clusters of galaxies.

The hottest stars radiate most of their energy in ultraviolet, and only by examining them in these wavelengths can astronomers understand their composition and the processes making them so hot.

Ultraviolet observations are one of the Hubble's chief advantages over ground-based telescopes. But Hubble's main reason for being is to look at the universe in visible light unfiltered and undistorted by earth's atmosphere. This most familiar form of light is emitted from atoms that have been energized through a collision with other atoms. Contrary to claims of some overzealous supporters, Hubble's 94.5-inch mirror would not necessarily be able to collect light from any deeper in the universe than can some of the best ground-based telescopes, even if the flaw in one of its mirrors is repaired. But it may see those most distant objects with ten times greater clarity and detect especially faint objects that have gone unobserved.

Eventually, Hubble's sharper vision might enable astronomers to pick out individual stars in clusters that show up as no more than a hazy blur through ground telescopes. Hubble scientists say the telescope's resolving power can separate individual beams of a car's headlights three thousand miles away. This capability, astronomers say, will be the key to observing beacon stars for measuring distances in the universe. From such measurements scientists expect to determine the size and age of the expanding universe.

The much larger mirrors of ground-based telescopes, particularly new ones under construction, will remain superior collectors of light and will be the instruments of choice in making broad surveys of the sky. Searching for quasars, for example, requires night after night of steady observation, looking among hundreds of thousands of faint lights for one shining object. Hubble will also be able to make some observations of infrared radiation, but astronomers will have to wait until the end of the decade for a major advance in that area.

Infrared emissions are a form of radiated heat and are especially informative to astronomers in examining interstellar dust clouds and searching for stars at birth and for planetary bodies around other stars. Although young stars shine mostly in ultraviolet light, their nurseries are obscured by dust that absorbs the ultraviolet and blocks astronomers' view of the birth process. But the heat created in the dust by the ultraviolet absorption emits infrared radiation, which can give scientists an indirect view of star birth. Preliminary studies have been completed for building the Space Infrared Telescope Facility, estimated to cost $800 million. Its super-cooled instruments are expected to have a sensitivity one thousand times greater than that of the 1983 satellite. Dr. Pellerin says he plans to seek authorization for the project in 1993, looking to a shuttle launching in 1999.

Radio waves are still observed almost exclusively by large disk antennas on the ground. In the 1960s astronomers detected a uniform glow of radio energy pervading the universe that is the faint echo of the Big Bang, radiations from the first few hundred thousand years after the creation of the universe. By extending the human view of the heavens to all the wavelengths of radiated energy, the prospects for discovery and surprise fill astronomers with excitement. No longer will they be so starved for information. "We're working the whole spectrum from radio to gamma," says Dr. Field of the Harvard-Smithsonian Observatory. "The game is to fit together and make sense of the views from all the wavelengths." [JNW]

Q&A

Q. *How do we know how far away the stars are?*

A. "This is something that takes up weeks in an introductory astronomy class," says Andrew Fraknoi, an astronomer who is the executive officer of the Astronomical Society of the Pacific. However, he says, modern measurements depend on a relatively simple initial principle. "It involves triangulation, measuring how an object in the distance shifts when seen from one angle and then another angle." Then simple geometry is used to compute the actual distance.

The process works something like this: Observe something in the distance from two different angles. For example, hold your finger to your nose and blink one eye, then the other. The finger appears to shift quite a bit. Now hold the finger at arm's length and do the same thing. The finger still shifts, but it shifts less. The difference allows the calculation to be made. The bigger the baseline, the easier it is to obtain a measurement. "What is the biggest distance apart that two viewing points could get on earth?" Dr. Fraknoi asks. "The answer turns out to be the earth's positions on opposite sides of the solar system as we travel around the sun."

"The problem is, the shift is unbelievably small, for even the nearest star, even when we take the 186-million-mile baseline that we get because the sun is about 93 million miles away." Not until 1838 was someone actually able to measure that small a shift, one and a half seconds of arc; a second of arc is one 3,600th of a degree. It is relatively easy to measure an object in the solar system, like the moon, "but that is our cosmic back-yard," Dr. Fraknoi says, "and stars are outrageously far away. The closest is 25 thousand billion miles away, not what we call spitting distance."

As the distances get larger, the shifts get even smaller. Only about one thousand of the nearest stars have been measured using this method.

OUTER LIMITS

FOUR billion miles away, farther than any man-made object has ever traveled, *Pioneer 10* has yet to find and cross the outer boundary of the solar system. Each day its instruments extend the known dimensions of the sun's sphere of influence with no end in sight. The findings have forced scientists to revise some ideas about the shape and dynamics of the heliosphere, the region where the sun's gases and magnetic field create an enveloping shield against most incoming interstellar gases and cosmic radiation.

In the summer of 1988 scientists and engineers at the Ames Research Center in Mountain View, California, marked the fifth anniversary of *Pioneer*'s passage beyond the orbit of the most distant planet by reporting that the five-hundred-pound craft, cruising at 28,400 miles an hour, continued to transmit faint reports from the cold and virtually empty space near the edge of the solar system. They were especially heartened by *Pioneer*'s good health.

Launched in 1972, it was the first spacecraft to traverse the asteroid belt beyond Mars and conduct a reconnaissance of Jupiter, which was supposed to be the limit of its voyage of discovery. Now *Pioneer* is more than four billion miles from the sun, almost forty-five times the distance from the earth to the sun and so far that it can no longer use the dim sun for guidance. It is so far beyond Pluto and Neptune that radio signals, traveling at the speed of light, take twelve and a half hours to make the round trip between spacecraft and earth. Robert W. Jackson, the deputy flight director, said that the craft was "going fine" and that if its power supply is carefully guarded it should continue transmitting scientific data for at least four years, possibly much longer. That could be enough time for *Pioneer* to detect and report on conditions where the heliosphere ends and true interstellar space begins.

When it reaches the boundary, *Pioneer* should observe distinct changes in the environment through which it is traveling. The outer flow of the gases from the sun will disappear and be replaced by interstellar gases, and the craft will detect an increase in cosmic radiation. The boundary region will likely be extensive, but scientists concede they are not sure where it will begin. Years ago they had predicted the boundary would be just beyond Jupiter's orbit. But *Pioneer* is now six times beyond that point, and it continues to report a rush of solar wind all around it. The wind is a million-mile-an-hour flow of charged gases that boil off the sun's surface and create a tenuous atmosphere throughout the solar system. The pressure of the wind deflects most interstellar particles.

"Every year we've been extending our projections of how long it will last," said Dr. John A. Simpson, a professor of physics at the University of Chicago and member of the *Pioneer* science team. "In the next three years we have a strong chance of encountering the boundary," said Dr. James A. Van Allen, a physicist at the University of Iowa and another *Pioneer* scientist, speaking in 1988.

Dr. Van Allen and Dr. Darrell Judge, a *Pioneer* scientist from the University of Southern California, based the three-year estimate on years of observations of solar wind activity, much of it made by *Pioneer 10* itself, suggesting that the size of the heliosphere may fluctuate markedly over an eleven-year cycle—in a sense, breathing in and breathing out. They also believe the heliosphere has a spherical shape rather than the streamlined teardrop shape that had been assumed. "It's quite a dynamic situation," Dr. Van Allen said. "Strangely enough, when solar activity is at a minimum in the cycle, the pressure of the solar wind is greater, and this expands the heliosphere. At solar maximum, there's more turbulence, but not the same momentum, and so the heliosphere contracts." The solar maximum is the time when the sun erupts with sunspots, the violent magnetic storms on the solar surface. This last occurred in 1980 and is due to recur in 1991.

If Dr. Van Allen and Dr. Judge are right, then the heliosphere could contract enough for *Pioneer* to get across the boundary before its nuclear electric generators run out of fuel and the radio goes dead. Other scientists suspect that the frontier may be much farther out. "We're still a considerable distance from any boundary," Dr. Simpson said. "But *Pioneer* may very well get at least close enough for us to extrapolate its dimensions properly." Scientists said *Pioneer's* instruments should be able to detect unmistakable evidence of the boundary crossing, though it may not be a sudden and sharp event. The density of solar wind would diminish, and the cosmic ray intensity should become stronger and constant with no more fluctuations caused by the solar wind's sweeping action.

In one model, the encounter between the supersonic solar wind and the inrushing interstellar gas could create a strong shock wave, perhaps "a transition shock." Another model postulates a more placid boundary. The heliosphere may just peter out, with neutral atoms in the interstellar gas exchanging electrons with the solar wind and thereby dissipating its energy.

In recent years, scientists have also followed *Pioneer's* extended mission for evidence of a tenth planet and for gravity waves such as those theorized by Einstein. The spacecraft was leading the search in conjunction with its twin, *Pioneer 11,* which was journeying in another direction toward the edge of the solar system. According to Einstein's general theory of relativity, cataclysms in the universe, like exploding stars and colliding galaxies, should generate gravity waves with extremely long wavelengths. They have never been detected on earth, and the *Pioneer*s have had no luck, either. The two craft are thought to be so stable and so far from other gravitational perturbations that they should be sensitive to the effects of such gravity waves.

For the same reason, astronomers have examined *Pioneer's* trajectory for any minute course changes that might come from the gravitational tugging of the long-sought Planet X beyond the known planets. Dr. John Anderson, a scientist at the Jet Propulsion Laboratory in Pasadena, California, reported that *Pioneer 10* had found absolutely no evidence of any uncharted planets. But Dr. Anderson said scientists had not given up on the possibility of finding such a planet, whose existence is suggested by the strong but unexplained gravitational forces that continue to perturb the orbits of Neptune and Uranus. If the source of these forces is a large planet, he said, it must be too far away to be observed yet by *Pioneer.*

As *Pioneer 10* cruises far away from the sun and planets, its photometer, a light-sensing device, is now able to observe and measure cosmic light, the accumulated light of all extragalactic sources. Gary N. Toller, an astronomer from the Applied Research Corporation, a contractor for the National Aeronautics and Space Administration, said measurements of the unadulterated cosmic light could tell how bright the galaxies were in the distant past and thus could provide clues to conditions early in the universe when stars began clumping into galaxies.

For after *Pioneer* finally leaves the solar system, it is destined to wander silently out among the stars. In 26,135 years, according to an analysis of its trajectory, *Pioneer 10* should reach its next celestial destination, the vicinity of the star Proxima Centauri.   [JNW]

Q&A

Q. *Have astronomers been able to use the location, speed, and direction of the observable galaxies to identify the probable location of the Big Bang?*

A. The Big Bang theory, widely accepted by astronomers and physicists, holds that all matter and energy in the universe originated in one primordial explosion billions of years ago. But scientists do not identify a location for the Big Bang "because the Big Bang created space and time as well as the matter and energy within space and time," explains David J. Helfand, chairman of the department of astronomy at Columbia University. "There was not a preexisting space in which the universe exploded."

Further, cosmologists do not calculate the "center" of the universe by measuring the speed and direction of celestial bodies. Such a calculation would assume the universe is three-dimensional, he says, and "the universe is not likely to have three-dimensional space." He suggests thinking of the earth in its galaxy as one of a number of dots painted on the surface of a balloon. "If you blow up the balloon it looks to you like everything is moving away from you. But it looks that way to everybody."

OUR LUMPY UNIVERSE

AS Hamlet would say to an astrophysicist these days, without drawing an argument, there are more things in the heavens "than are dreamt of in your philosophy." For astrophysicists who look beyond individual galaxies of stars to clusters of galaxies and, at an even larger scale, to aggregates of clusters called superclusters are intrigued and mystified by what they see. To their bewilderment the universe seems to be organized in even more vast chains of galaxies, structures so large and complex that they defy understanding in terms of current cosmological theory.

The picture of the universe's structure became more confused with the discovery in 1987 of what appeared to be an immense conglomeration of galaxies, a feature that is being called a supercluster complex. The complex, which includes our own Milky Way galaxy, is one hundred times more massive than any previously known structure and encompasses millions of galaxies, stretching across 10 percent of the observable universe. The discoverer, R. Brent Tully of the University of Hawaii's Institute of Astronomy, said the observation of such an enormous structure posed "major challenges to conventional theories of galaxy formation." Dr. Tully said at the time that other observations suggested the existence of at least four other supercluster complexes "of comparable magnitude," indicating that "this must be a general property of the universe."

Astrophysicists were cautious in their initial reactions, neither rejecting nor accepting Dr. Tully's interpretation of his discovery, and prepared to debate the issue for years. Simon White, a theoretical astrophysicist at the University of Arizona, said, "I think he's got enough here to force people to really try and confirm or deny it with new observations and analysis." Dr. Tully said he felt "quite comfortable" with his interpretations, calling them "not certain, but highly probable." He based his conclusions on a new supercomputer analysis of the positions of galaxy clusters.

Jeremiah P. Ostriker, a Princeton University astrophysicist, said: "If this is right, it's extremely important because it's hard to produce this kind of structure in current theory. It could mean there's some vital missing ingredient in standard theories of the development of the universe."

Searching for the missing ingredient, should that be necessary, would add to the ferment in cosmology today. Efforts to conceptualize the history and structure of the universe were already running into trouble because of the growing realization that the universe was not as uniform as had been assumed.

Astronomers first glimpsed form and movement in the universe in the 1920s. The familiar constellations are groupings of stars only in the eye of earthly beholders. In reality, Edwin P. Hubble found, billions of stars tend to occur in "island universes," and our Milky Way is only one of a multitude of these galaxies. Analysis of changes in light from the distant stars and galaxies—the displacements toward the red end of the spectrum, or redshift—showed that the universe was expanding in all directions. Everywhere astronomers looked, the galaxies appeared to be distributed uniformly and without any discernible pattern of larger structure. In the 1930s Fritz Zwicky of the California Institute of Technology saw some evidence for galactic clustering. But even the evidence that solidified support for the Big Bang, the prevailing theory that the universe exploded into being 10 billion to 20 billion years ago as a superdense fireball, did nothing to change scientists' image of a smooth, uniform universe.

In 1964 radio astronomers at Bell Laboratories detected microwave background radiation throughout the universe, the leftover radio noise of the explosive moment of creation. The smoothness of the background radiation signals indicated that the universe began with a uniform distribution of matter. Astrophysicists say the new discoveries of larger-scale structure raise questions about developments after the Big Bang—what events and forces gave such a rich texture to the universe? According to the Big Bang theory, the initial expansion was rapid, which would account for the universe's general smoothness.

But there must have been some wrinkles in the early distribution of matter. Either some clumping of matter set in motion gravitational forces to form the galaxies and later the clusters of galaxies, or perhaps some large-scale variations in the density of matter survived after the early, rapid expansion of the universe slowed down, and galaxies could have formed around the edges of these vast areas and be strung out in chains. While cosmologists were still debating these alternative theories, in recent years astronomers with improved telescopes and better computers for building and testing theoretical models began seeing that clusters of galaxies clumped into superclusters. They also observed groups of galaxies occurring in flat sheets and in chains or filaments stretching across millions and millions of light-years. They found some galaxies that appeared to surround a dark void, like dots painted on a toy balloon.

Such a variety of observed shapes may be a problem of perspective. Like the blind men examining the elephant, astronomers using different observational techniques, X-ray or radio or optical telescopes, may be describing different aspects of the same structures because of their particular points of view. But there was no doubt that they were seeing surprisingly large structures. More than ever, by the 1980s, cosmologists strongly suspected that there was much more to the universe than met the eye, and this has sorely tested their genius for conceiving cosmic theories. One of their most popular new concepts is that of the missing mass.

Various calculations, based on an assumption that the universe is closed in a balance between energy and matter, suggests that as much as 90 percent of the universe's mass has gone undetected. This dark matter, composed of as-yet-unknown elementary particles, is assumed to exert the dominant gravity in the universe, causing the clumping of galaxies. "The hypothesis works very well to explain the formation of individual galaxies and groups of galaxies," says Dr. White of the University of Arizona. "But superclusters cause us problems."

Dr. Ostriker, who has worked on these problems for years, conceded, "As observers find structure larger and larger, it gets harder and harder to explain things theoretically." Telescopes with new electronic cameras enable astronomers to examine many galaxies in a single field of view, whereas it used to take them a lifetime to study a few hundred. Large dish radio antennas, as well as optical telescopes, give them glimpses of the galactic clusters shaped like bubbles. X-ray and optical telescopes provide the raw material for plotting clusters and superclusters on computer-generated maps of the sky.

Dr. Tully was searching for the edge of the Local Supercluster, the region including the Milky Way, when he became aware that the structure was much larger than previously thought. He then looked at the "rich" clusters farther out in space, calculating their motion and plotting their distances from each other. Using a supercomputer, Dr. Tully constructed maps of the distribution of the clusters of galaxies as they would appear to an observer at various points in outer space. He compared their distance from each other with random spacings between clusters. He concluded that the clusters were related to each other because the probability of their occurring by chance was statistically slight.

As a result, Dr. Tully decided that about sixty of these clusters were concentrated in a single supercluster complex. He calls it the Pisces-Cetus Supercluster Complex, after the constellations in which it is found. "It was supposed that on a large scale, things in the universe are smoother and homogeneous," Dr. Tully said. "My findings show that is not the case. It's lumpy. That was unanticipated."

Dr. Tully suggested that the theory of cosmic strings "might provide an explanation" for what he is seeing. According to this theory, anomalies in space, called topological defects, were created in the first fraction of a second after the Big Bang. These defects could provide the focuses for the accumulation of matter. But he acknowledged that if such significant concentrations of matter existed at the moment after the Big Bang, "they should have given rise to irregularities in the relic background radiation."

The cluster Dr. Tully has identified is about 1 billion light-years long and 150 million light-years across. Since the universe is thought to be at least 10 billion years old, a distance of 1 billion light-years represents as much as 10 percent of the expanding universe's reach.   [JNW]

THE GREAT RED SPOT RIDDLE

A spinning tub of water has convinced experts on chaotic systems that they have figured out one of the Solar System's most baffling landmarks, the Great Red Spot of Jupiter. Using an apparatus that resembles a high-tech dishpan, a group at the University of Texas managed to create a miniature Red Spot of their own—a vigorous whirlpool sitting happily amid the turbulent flow. Like its cosmic counterpart, the laboratory spot seems to last indefinitely.

The Great Red Spot, a gigantic oval large enough to swallow up the earth several times over, began puzzling astronomers three centuries ago. The *Voyager* flybys of the late 1970s produced spectacular pictures that led researchers to describe the spot as a kind of giant hurricane in the planet's atmosphere. But earthly storms fade rapidly, dissipating their energy into the surrounding atmosphere. The Great Red Spot seems immortal.

The Texas group—Harry L. Swinney, Joel Sommeria, and Steven D. Meyers at the university's Center for Nonlinear Dynamics in Austin—began trying to re-create the spot in 1985. They based their experiment on a computer simulation by Philip S. Marcus, now at the University of California at Berkeley. Dr. Marcus contended that, according to his calculations, the turbulent chaos of a rapidly rotating fluid could spontaneously create a single coherent vortex, "like a Phoenix out of the ashes." The Texas researchers say they have confirmed his predictions; as they spin their tank four times a second, they see many tiny vortices evolve and combine until a single oval whirlpool remains.

"We think we've succeeded in modeling the basic dynamical properties of the Red Spot," said Mr. Meyers of the Texas group, which reported its findings in *Nature*. "It's easily created in the laboratory under a wide range of conditions." The researchers hope their findings will shed light on the interplay of stable and unstable features in other kinds of fluid flow. Whether the Red Spot has counterparts closer to home remains to be seen, but scientists are looking at several odd structures that develop, for weeks or for months, in the earth's atmosphere and oceans.

A STAR IS DYING

THE cataclysmic explosion of a nearby star in 1987 has dramatically strengthened major theories of how such explosions occur, and astrophysicists say the supernova has given them a new sense of confidence in their ability to interpret stellar events. In particular, scientists believe they were correct in predicting how and in what sequence an exploding star would combine simple atomic nuclei with protons and neutrons to make the heavy atoms essential to complex chemical processes, including the creation of life.

They have evidence that the explosion blasted dense layers of freshly created elemental matter into rings that are flying outward from a small remnant of the former star. The atoms of that remnant have been crushed into a compact, superdense globe of neutrons. The supernova explosion that occurred over the Southern Hemisphere some 160,000 light-years from earth was the closest such event in four centuries. "There's an enormous intellectual change taking place because of this object," W. David Arnett of the University of Chicago said. Dr. Arnett is a leading astrophysicist and is one of the scientists whose theoretical explanations have been confirmed by the event. "Before this thing happened," Dr. Arnett said, "we had a very different way of looking at our work. We'd developed good theories, but we didn't know they were correct, and we didn't act as if they were correct. Now that we're certain we're on the right track, we know what kind of instruments we should build to measure the next supernova in our galaxy. It probably won't be visible like this one, but if it's the same type, it will emit a characteristic pulse of neutrino particles that we can detect."

Supernova SN1987A, as the explosion was designated, flared into view on February 23, 1987, and a Canadian astronomer, Ian Shelton, working at Las Campanas Observatory in Chile, discovered it a few hours after its appearance. Dr. Robert E. Williams, director of the Cerro Tololo Inter-American Observatory in Chile, reported that the supernova brightened continuously until May 20, when it peaked at magnitude 2.97, the equivalent of a fairly bright star easily visible to the naked eye. And then it declined in brightness.

The supernova has scintillated with cosmic clues that astronomers have had little difficulty fitting into the theoretical models devised over the years to account for the behavior of exploding stars. Despite the glaring brightness of the supernova as viewed by such giant instruments as Cerro Tololo's telescope, whose light-focusing mirror is four meters in diameter, the explosion was not yielding its secrets easily. The glare the scientists saw was intense light emitted by the outermost shell of the blast—a kind of luminous globe that conceals what is happening within. The hot shells of gas, analogous to nested balloons, were expanding into space at thousands of miles per second, and as they did so, they grew cooler and more rarefied. But for the time being, these expanding gases, mostly in the form of electrically charged atoms called ions, were opaque; neither light nor any other form of electromagnetic radiation could leak out of the depths of the explosion to be observed on earth. "We're not really sure exactly when the supernova will become transparent to various wavelengths of radiation," Dr. Williams said shortly after the explosion, "and this is why it's essential to measure the light spectra with great care every night."

In the early months Dr. Williams's group had begun to detect features of the supernova spectrum (absorption lines) characteristic of the presence of barium and strontium, and this suggested that the observers were already looking deeper into the explosion than was possible before. "The beautiful results obtained at Cerro Tololo," Dr. Arnett commented, "not only suggest that the gas in the outermost hydrogen shell is becoming transparent. They confirm our prediction that barium and strontium would be created in this kind of process, and would turn up in the underlying helium shell."

SN1987A is a type of supernova, Dr. Arnett and other theorists believe, in which the progenitor star had been synthesizing heavy elements for a long time before the explosion occurred. In this hypothetical process, called the "S Process" (for slow process), the nuclear fusion that fuels the stellar furnace grafts protons and neutrons on atoms even heavier than iron, creating new elements as heavy even as uranium. The explosion itself creates other elements, and successive layers contain characteristic proportions of various elements.

The explosion starts with a very hot, short-lived star some twenty times the mass of the sun, which has consumed most of the hydrogen of which it was initially composed, fusing hydrogen nuclei together to create helium. As the core cools it can no longer support the crushing gravitational pressure of the star's huge mass, and the core collapses. The inward-rushing matter generates new heat, however, which initiates the fusion of helium. The process of cooling, collapse, reheating, and initiation of the new stages of fusion can continue through successively heavier elements up to iron, according to theory.

But iron is the end of the line for a star, because the fusion processes needed to create still heavier elements absorb energy rather than create it. Once a star has burned most of its fuel to iron, it quickly cools and suffers a core collapse so violent that a supernova explosion ensues. In the case of such supernovas as SN1987A, a small, ultradense neutron star is left at the center, a star so dense that atomic nuclei themselves are crushed together into a compacted mass of neutron particles. At the instant this happens, according to theory, an avalanche of neutrino particles rushes out of the superdense matter in the collapsed star, speeding outward in all directions at the speed of light. Neutrinos have no electrical charge and probably have no mass, and they therefore scarcely interact with matter; a single neutrino speeding through the entire earth would stand very little chance of colliding with an atom or particle.

But when great floods of neutrinos pass through a substance, the chances become greater that one of them will hit something and register its existence. When the supernova first appeared, it happened that several sensitive neutrino detectors were working on another project involving particle physics. Two of these detectors, one at Kamioka in Japan and another near Cleveland, Ohio, independently detected huge surges of neutrinos at the same moment. Subsequent analysis of the data by the group in Japan and by scientists at the University of California at Irvine, the University of Michigan, and Brookhaven National Laboratory on Long Island left no doubt that the neutrino burst had come from the supernova.

Astronomers everywhere exulted over that result. "The neutrinos reached earth in just the way theory predicted," Dr. Arnett said, "telling us that the core of the supernova had collapsed into a neutron star." At that point, Dr. Arnett and other theorists calculated the exact characteristics of the progenitor star that would account for such an explosion, and it wasn't long before observers found exactly what was needed. Intensive study of photographic plates made prior to the explosion revealed that a single star, discovered in 1969 by Nicholas Sanduleak of Case Western Reserve University, was at the right spot. Moreover, the star, designated Sk −69.202, was a hot supergiant some twenty times the mass of the sun—just what the theorists needed to produce a supernova with the characteristics of SN1987A.

Observations made by the International Ultraviolet Explorer satellite suggested that Sk −69.202 had survived the explosion and could therefore not have been the supernova's progenitor, but re-analysis of the data showed that the progenitor star had definitely disappeared. [MWB]

SMOG ON PLUTO

The latest observations of distant Pluto reveal what appear to be smoggy hazes in the thin atmosphere over the polar ice caps. The phenomenon reminds scientists of atmospheric patterns seen over the poles of Mars and in the Los Angeles basin. After a new analysis of the observations, Dr. Von R. Eshleman, a space scientist at Stanford University, said that, contrary to earlier reports, the Plutonian atmosphere probably does not contain a globe-circling layer of haze. Instead, the haze seemed to be confined to the polar regions.

In a report in *Icarus,* an international journal of solar system studies, Dr. Eshleman interpreted the data to indicate that a temperature inversion occurs at Pluto's poles, similar to what happens over the polar caps of Mars. The frozen carbon dioxide at the Martian poles is colder than the carbon dioxide in the atmosphere. Consequently, temperatures at higher altitudes are warmer than they are near the surface. On earth, when warm air is thus trapped above an area, this produces what is known as a temperature inversion. One result is static weather and an accumulation of smog. Dr. Eshleman said the Pluto observations were consistent with earlier estimates that the planet's atmosphere is thin and composed mainly of methane at a temperature of 370 degrees below zero Fahrenheit at the surface and minus 335 degrees at an altitude of twenty thousand feet.

CHICKEN LITTLE

A ROCK the size of a basketball plunged from the sky over Nakhla, Egypt, in 1911 and killed a dog. The dog thus became the highest form of terrestrial life known to have been struck down by a meteorite. As for the meteorite, its strange history is only now beginning to be understood. The Nakhla meteorite and a handful of similar specimens, a total of eight out of the thousands in the world's collections, now seem almost certain to have come from Mars. If so—if they are fragments of the Martian surface, providentially hurled earthward—they have unique value for those interested in the Martian equivalent of geology.

They are chief exhibits in an increasingly active exploration of cosmic dynamics: the collisions and orbits and flows that move matter from place to place. And the apparent existence of pieces of Mars is also helping to revitalize the business of studying meteorites for clues about the earth's neighbors. "The whole science of meteorites and what they mean has become much more exciting, because you might be getting all sorts of material left over from different stages in the solar system," says Charles A. Wood, a planetologist at the Johnson Space Center in Houston, one of the first proponents of the Martian theory. "To understand any place, we really have to have a sample."

The notion of Martian meteorites—initially quite outlandish—first arose in 1979, when data from the *Viking* landers showed a chemical resemblance between these odd samples and the Martian soil. Since then the chemical evidence has grown stronger and stronger, with researchers analyzing the traces of inert gases like argon and helium and putting together a mineralogical thumbprint. Furthermore, the samples in question are far younger than the usual run of meteorites—just 1.3 billion years, compared with more than 4 billion years. From a traditional point of view, that needed explaining.

"The standard wisdom was that all meteorites come from the asteroid belt and if these look weird then that's because we've made some mistakes in our measurements," Dr. Wood says. Rock from Mars, by contrast, could easily be young. Like homicide detectives lacking a motive, however, proponents of the Mars theory were confounded by a problem of dynamics: how did the rocks get here from there? Microscopic examination of the meteorites' crystal structure made the problem worse by showing that they had never suffered a particularly severe shock. Astronomers and physicists were left to explain what sort of cataclysm could have blasted large rocks off the surface of Mars, instantly accelerating them to an escape velocity of three miles per second, without unduly shocking them, let alone pulverizing them.

The dynamical problem now appears to have been solved. Two researchers at the University of Arizona's Lunar and Planetary Laboratory have shown that all eight meteorites could have been propelled into space by the collision of a twenty-mile-wide asteroid with Mars relatively recently, about 200 million years ago. The impact must have created a huge crater, at least sixty miles across. The researchers, Ann M. Vickery and H. Jay Melosh, calculated that the resulting shock wave—a tight band of extremely high pressure moving outward from the impact—would have left its mark indelibly on rocks deep underground.

At the surface, however, the shock wave would have behaved quite differently, they found. Their model shows a shallow zone where rocks, even boulders, can be kicked upward at high speed without ever feeling the full pressure of the wave. The calculations were published in the journal *Science*. Other researchers have proposed other models for getting the meteorites off the planet, including, most recently, a team at the California Institute of Technology who put forward a mechanism based on jets of hot vapor spewing forth after an impact. The Arizona scientists believe that their model works best when it comes to launching large boulders—and the Nakhla meteorite was apparently thirty feet across during its 200-million-year voyage in space, before fragmenting on the way to Earth.

In general, these are boom times for scientists puzzling out the dynamics of astronomical catastrophes. Bursts of vapor, sprays of rocks, and patterns of fragmentation have all come within reach of high-powered computer models. Many details remain troublesome, but, on the whole, the physics of such processes is coming into focus. Dr. Melosh refined his modeling techniques in simulating the birth of the moon, now believed to have occurred in a giant collision between earth and a smaller planetlike body. Dr. Vickery is studying the role of asteroid impacts in either adding gases to planetary atmospheres or blowing gases away; both possibilities have been proposed.

On larger scales, too, dynamical models are helping solve problems of how things form and how they move. The creation of structures as large as galaxies resembles patterns in turbulent fluid flows, according to some researchers. Techniques for understanding such flows and other complex motions are providing a strong complement to more usual techniques of analyzing the composition of astronomical objects in terms of different wavelengths or isotopes. Besides the eight meteorites thought to have come from Mars, scientists discovered in 1983 that several meteorites the size of golf balls had come from the moon. There was no doubt about the lunar meteorites because, unlike the Martian missions, the lunar missions brought back samples for direct comparison.

A part of the meteorite puzzle has been why Mars should have provided more samples than the moon. Mars, after all, is farther away and has much stronger gravity for rocks to break free of. On average, more debris should find its way from the moon. "Our model gets around that by postulating a large and quite rare event," Dr. Melosh says. "The inner solar system was sort of flooded with Martian debris, and it will stay flooded for several hundred million years." As he sees it, the collision was a singular incident. It was a catastrophe—a thing that buys an exemption, at least temporarily, from the law of averages. [JG]

HALLEY'S COMET

New observations of the chemical composition of Halley's comet suggest that the illustrious traveler through the solar system may be an interloper from deep space, astronomers have reported. The general assumption has been that Halley's, like other comets, came from the Oort cloud of icy material on the fringe of the solar system. The comet is composed mostly of water, ice, and frozen carbon monoxide. But using a new technique of chemical analysis, astronomers have found traces of elements in Halley's tail that suggest that the comet originated elsewhere and was captured by the sun's gravity as it passed nearby.

A team of scientists led by Susan Wycoff, an astronomer at Arizona State University, reported in the *Astrophysical Journal* that in the comet tail the ratio between two radioactive forms of the element carbon, carbon 12 and carbon 13, differed markedly from that found in other solar system objects, including terrestrial and lunar rocks, meteorites, and the atmospheres of the giant outer planets. "If our data are correct," Dr. Wycoff said, "this would be one of the most important discoveries about the comet." The observations were made with ground-based telescopes when Halley's comet flew by earth in 1986.

DID LIFE BEGIN IN
OUTER SPACE?

A MAJOR new round of probes to the inner and outer reaches of the solar system could yield important clues about how life began on earth, say scientists who study the mystery of life's origins. Space flights in the late 1980s and through the 1990s should provide the data that will allow scientists to study comets, planets, moons, and other cosmic destinations for evidence of the transition of chemicals from inanimate matter to living things. The far-flung search will be guided by a new theory that holds that life was the almost inevitable outcome of "chemical evolution" following the formation of the solar system. The theory has been driven by new evidence, most recently from the spacecraft that flew near Halley's comet, that the universe is awash with the chemical precursors of life.

"There is astonishing potential that clues to the origin of life will be found elsewhere in the solar system and in other stars and galaxies," says Lynn Griffiths, chief of the life sciences division of the National Aeronautics and Space Administration (NASA) in Washington. "Everywhere we look, we find biologically important processes and substances." While the main goal of the studies is to learn more about the process by which chemicals became organized into ever more complex forms, the latest findings have also revived hopes that signs of primitive life, possibly extinct, might be found on Mars.

The current theory suggests that, some 4 billion years ago, huge amounts of the elements essential to life and perhaps also such complex organic molecules as amino acids were showered onto earth and other planets by comets, meteorites, and interstellar dust. "Organic synthesis is going on everywhere, in the atmosphere of planets, on the surface of Pluto, and between the stars," says Dale Cruikshank, an astronomer at the NASA-Ames Research Center in Mountain View, California. "The universe is full of organic chemistry. However life originated on earth, this stuff has been raining on us from time immemorial."

Fossil evidence indicates that bacteria appeared on earth about 3.5 billion years ago, only 1 billion years after the solar system formed. The great challenge is to learn how, within that first billion years, simple organic chemicals evolved into more complex ones, then into proteins, genetic material, and living, reproducing cells. No record is believed to exist of chemical evolution in the earth's first billion years. But scientists think they can find clues elsewhere in the solar system, where similar evolution may have occurred in the past, or where such changes may now be brewing. Good places to look, the scientists say, include inside comets and asteroids, under frozen Martian lakebeds, deep inside the smoggy atmosphere of Saturn's moon Titan, and beneath the smooth frozen surface of Jupiter's moon Europa.

The advances in theory about life's origins are coming just as the country is poised for new interplanetary exploration. "The climate is good for solar system exploration," says Glenn Carle, chief of the solar system exploration branch at the Ames center. "It seems the queen has given us new ships." While research on the origins of life is not the main goal of the missions, the prospect of important new data is already generating excitement among the band of scientists around the country who are devoted to this problem.

American scientists watched from the sidelines while other nations flew spacecraft past Halley's comet in 1986, missions that found the comet to be carrying far more organic matter than expected. But planetary scientists were optimistic with the beginning of the second great phase of solar system exploration marked by the shuttle launching of *Magellan* and *Galileo*. In the 1990s most planetary spacecraft will be launched using expendable rockets, according to Dr. Geoffrey Briggs, director of NASA's solar system exploration division, thus avoiding the shuttle delays that can hamper precisely timed planetary missions.

An orbiting mission to Mars is planned for launching in 1992, Dr. Briggs says, and a new spacecraft is on the drawing boards, to be launched in 1995. Plans call for the craft to rendezvous with a comet in the year 2000, to send a probe into its core, and to fly alongside the comet for three years as it whips around the sun. An almost identical spacecraft with different instruments might be launched in 1996 for a four-year stay around Saturn and its moons. It would send a probe into the atmosphere of the moon Titan. NASA is also discussing possible ventures with the Soviet Union for further exploration of Mars. Soviet officials have said that studies relevant to life origins will be included in their Mars missions.

Some scientists are becoming increasingly optimistic that signs of extraterrestrial life will be discovered, says Mr. Carle, who organized a symposium on biological questions to be posed on future missions into deep space. Although the 1976 *Viking* mission to Mars found no evidence of life, he said, new ideas and observations about the chemical history of planets and stars are fueling the current optimism. One observation is that vast dust clouds floating between stars contain many of the precursor organic molecules that can, under the right conditions, be transformed into the building blocks of proteins, genetic material, and cell membranes. When the clouds collapse to form new stars and planets, it is theorized, some of these generally lightweight and fragile precursors are preserved in comets and in the atmospheres of outer planets.

The inner planets get so hot that the molecules are blown apart into their constituent atoms. Thereafter, the lighter elements that are critical to life as we know it—carbon, hydrogen, nitrogen, oxygen, phosphorus, and sulfur—are blown toward the outer planets. No one knows what the earth's early atmosphere was like, says Dr. Harold Klein, an expert in cosmological evolution at the University of Santa Clara. And they never will, he adds, because not long after the planets finished taking shape the sun, like many newly formed stars, emitted an intense burst of solar wind that literally knocked the atmospheres, if indeed there were any, off Mercury, Venus, the Earth, and Mars. "The Earth was bald," Dr. Klein says. "The paradox has been, where did our atmosphere come from?"

The increasingly seductive answer is that comets, meteorites, and dust contain all those lighter elements critical to life. In fact, comets seem to be 40 percent water, 40 percent fine sand, and 20 percent these vital chemicals, sometimes joined into complex organic molecules, like formaldehyde, which are also found in interplanetary and interstellar dust. When Halley's comet neared the sun and produced its enormous tail, scientists found that it lost an astonishing two hundred thousand pounds of material per second, presumably including plenty of the vital chemicals, some of which will eventually "rain" onto planets. Soon after the solar system formed, after the solar wind died down to normal velocities, scientists believe that the solar system was bombarded by comets and meteorites for more than a half-billion years.

Although most arriving comets would be blown apart on impact, and complex molecules would again be blasted apart, many scientists now theorize that the atomic building blocks for earth's air, water, and living creatures arrived via comets. It is also possible that more complex molecules, like amino acids and nucleic acid bases, might have occasionally survived the crash. The evidence for such a bombardment remains scattered throughout the solar system—in the craters of Mercury and the Moon, in the asteroid belt, in the large amounts of water and impact craters observed on moons around the giant outer planets, and in a steady rain of dust particles from space.

Once the bombardment of comets and meteorites stopped, scientists say, each planet followed a different path of chemical evolution. For reasons not yet fully understood, however, only earth followed a path that eventually led to primitive life followed by higher life forms. Again according to the theory, this chemical miracle happened amazingly fast if, as most scientists believe, the solar system is 4.5 billion years old. The oldest fossils on Earth are bacterial mats called stromatolites that are 3.5 billion years old.

This means that life arose and took form on earth within about half a billion years, Dr. Klein said, between the end of the heaviest bombardment and the appearance of stromatolites. It also means that some form of early primitive life may also have developed on Earth's closest neighbors. Early conditions on Venus, Earth, and Mars were probably very similar and very conducive to life, Dr. Klein says. The outer planets followed a different chemical path. Jupiter and Saturn are gas giants whose turbulent atmospheres, while filled with the vital chemicals of life, are almost certainly inhospitable to life. Some of their moons, however, retain atmospheres rich in organic material and water. Uranus and Neptune appear to have methane-rich stratospheres with frozen moons containing hydrocarbons and at least one moon with an atmosphere.

Given this model of how the solar system formed and has been flooded with organic molecules, space scientists want to take a second, closer look at all the bodies in the system. Here are some of the topics they hope to learn more about from each destination:

COMETS: What organic materials are present in comets? To what extent, and how, has this material been synthesized into complex, biologically interesting molecules? Is there any evidence that substantial amounts of material from comets have been delivered to the planets?

ASTEROIDS: Do they have large amounts of organic molecules? How much water do they contain? How many are burned-out comets?

MERCURY: How old are the craters, and what can they tell us about the heavy bombardment early in solar system history?

VENUS: Is there evidence that the planet had an early ocean? Because Venus is closer to the sun and hotter than Earth, scientists think the planet lost its water over about 100 million years and then experienced a runaway greenhouse effect. Early chemical processes leading to life would have been nipped in the bud.

MOON: How old are impact craters? What are meteorites on the surface like? Does the record of early bombardment exist on the moon? Did later impacts occur in any particular pattern or are they random?

MARS: Because it was once so much like early Earth and had abundant amounts of water, Mars might have produced primitive life. It is not known how Mars lost most of its atmosphere, although one theory holds that asteroids literally knocked it loose. In any case, fossils of early life forms, not unlike those that developed during those unrecorded half-billion years on Earth, could lie beneath frozen lake beds on Mars. According to Dr. Chris McKay at the Ames center, it is also remotely possible that life could exist on Mars if there is water in hidden niches. Life is very tenacious once it takes hold. Moreover, scientists have determined that turtles and cockroaches can survive in jars that simulate Martian conditions.

EUROPA: Does this Jovian moon have a liquid ocean beneath its smooth icy surface? Are there internal sources of energy that would promote life? Can cell membranes form in such cold temperatures?

TITAN: This moon of Saturn has a thick, smoggy atmosphere. What kind of prebiotic chemistry is taking place in its lower reaches? Do these processes give clues as to how precursor molecules led to life on Earth?

While the search for the origins of life is not the primary focus of planetary exploration, answers will be forthcoming as space scientists study basic physical and chemical processes in the solar system, says Dr. Lew Allen, director of the Jet Propulsion Laboratory in Pasadena, California. Because the universe is full of organic molecules that can lead to life, some scientists think that life must be the rule rather than the exception. But, Dr. Allen says, it remains to be settled whether earth is "a common thing or an extraordinary accident." [SB]

The Search
for Origins

THE BIRTH OF CONTINENTS

A REVOLUTION is taking place in attempts to understand how the continents were born and grew. Geologists once drew a simple picture of a single great land mass, assembled from earlier fragments, breaking into pieces 150 million years ago to form the continents of today. Now they say the process was much more complicated. Through much of the earth's history, they believe, continental fragments have been tossed back and forth, sometimes traveling several thousand miles. These pieces of the earth's crust, which come in varying shapes and sizes, ride atop the semimolten interior. They creep along at only a few inches a year, but that is fast enough to cover great distances when a journey lasts tens of millions of years.

The revised theory has evolved over the last decade and picked up strong support in some of the newer studies. The continents are patchwork affairs, the theory says, because they have repeatedly broken up, scattered their pieces, and then re-formed in different combinations. By one analysis, North and South America are composed of at least seventy fragments. In Asia, David B. Rowley of the University of Chicago reported, there are no fewer than sixteen continental blocks and other formations.

The North Slope of Alaska and its structural continuation in the Seward Peninsula of Alaska and the Chukotskiy Peninsula of Siberia are now believed to have been a single block that came from elsewhere. Its collision with the rest of Alaska produced the Brooks Range. Some geologists say the north slope of that mountain range was previously attached to the Canadian Arctic Archipelago, then swung seventy degrees counterclockwise to its current position around a "hinge" in the Mackenzie River delta. At a professional meeting in San Francisco, Warren J. Nockleberg of the United States Geological Survey identified nine slivers of landscape pushed together in the Yukon region by a collision 100 million to 142 million years ago.

He said remnants of an ocean basin and an arc of volcanic islands that were swept up by the collision have been found, and he reported on some of the most recent findings of a survey across a north-south line the length of Alaska.

A succession of additional fragments of landscape, or "terranes," drifted north to form southern Alaska and the rest of western North America. One, now the panhandle of Alaska, came from as far away as Australia, Jason B. Saleeby, a geologist at the California Institute of Technology, has reported.

The key to the new concept is recognition that a series of continental formation cycles occurred, rather than the one cycle proposed in 1962 by Tuzo Wilson of Canada. He said then that an earlier Atlantic Ocean had closed about 400 million years ago, bringing together Eurasia, Africa, and the Americas. They pulled apart 150 million years ago to form the present ocean, he said, completing a process now known as the Wilson cycle. But geologists now believe that there were many Wilson cycles and that they began much earlier than had been thought. For example, when scientists developed the theory of plate tectonics, which holds that continents and oceans are carried on huge plates that move about on the earth's surface, they deduced that India had drifted thousands of miles north from Antarctica and Australia before reaching Asia; the continuing collision has produced the highest mountains on earth. But now studies by scientists from America, Britain, China, and France indicate that this collision was only the most recent addition to the continent.

A French group reported that India's northward pressure continues to distort Asia, pushing Tibet east relative to the main body of the continent. The result is the Altyn Tagh fault, which runs for fourteen hundred miles along the southern edge of the Tarim block in western China. Satellite images suggest that since the most recent ice age, ten thousand years ago, the Tibetan side of the fault has slipped as much as twelve hundred feet east. Peter Molnar of the Massachusetts Institute of Technology, another specialist in Asian tectonics, described the assembly of Tibet in an issue of the *American Scientist*.

From 130 million to 300 million years ago, he said, at least three narrow strips of continent were successively added to southern Asia. They were presumably carried on the same plate that would smash India into the continent much later, about 40 million years ago. The rest of Asia was assembled during the past billion years, according to

Dr. Molnar, Dr. Rowley, Lev P. Zonenshain of the Institute of Ocean-
ology in Moscow, and others.

The blocks that now make up north and south China were once
separated by a wide ocean; they came together in the Late Triassic
period, 190 million years ago, Dr. Rowley said. They then collided
with Eurasia, according to reports at the San Francisco meeting; this
conclusion is based on magnetic orientations in ancient rocks of those
areas. Much earlier, or about 250 million years ago, the huge Tarim
basin, forming the heart of northwest China, collided with the region
to the north, said Shangyou Nie, Dr. Rowley, and their colleagues.
The Tarim collision helped build the ranges of the Tian Shan on its
northern flank, now the Chinese-Soviet border.

The older generation of Soviet geologists has been slow to accept
plate tectonics, but a large number of their younger colleagues told an
International Geological Congress in Washington how they believe
Eurasia came together, including successive arcs of volcanic islands
plastered against its eastern frontier as the floor of the Pacific bur-
rowed under it. At least ten such terranes were identified by the Soviet
group, some having come from as far as "several thousands of kilometers" away, one report said.

The geologists said the continuing northward pressure by India
has caused a succession of Asian blocks to slip northeast, east, and
finally south, which may have helped create the long north-south
valleys of Southeast Asia. Farther west, one such rift extends north
into Siberia and holds Baikal, the world's deepest lake. The collision
in Tibet has produced the world's highest plateau. Although rimmed
by mountains, it is remarkably flat, Dr. Molnar noted. "Off-road vehicles can travel hundreds of kilometers without encountering serious
obstacles," he wrote in the *American Scientist*. The bedrock of the plateau is rugged, but material eroded from its ring of mountains and
deposited on the plateau has filled the gaps and made it level, Dr.
Molnar believes.

Evidence of continental movements has also been observed in
North America. In fact, Dr. Wilson developed his cycle theory when
he realized that a stretch of land running from the Avalon Peninsula
in Newfoundland to Connecticut had originated on the far side of the
ancient Atlantic. Some geologists consider a similar slice of the Car-
olinas and Georgia to be part of this same Avalon terrane. Scientists
are now trying to decipher how inland portions of North America

were put together. Early in the debate on plate tectonics it was deduced that there had been a collision about a billion years ago that left the Grenville formation, a segment that runs from Newfoundland through the Adirondacks to Mexico.

Paul F. Hoffman of the Geological Survey of Canada said he believes he has seen this same formation as far away as Sweden, indicating it was at one time attached to North America when the continents were assembled a few hundred million years ago. But in an issue of the journal *Geology*, he suggested that seven "microcontinents" came together far earlier, about 1.8 billion to 2 billion years ago, to form the central core, or craton, of North America.

Debate on where the North Slope of Alaska came from has been hampered by the absence of detailed mapping of magnetism on the Arctic Ocean floor. In other oceans magnetic formations in the oceanic bedrock indicate the region's past direction of movement. But the ice cover of the Arctic has impeded such surveys. Nevertheless, Susan L. Halgedahl and Richard D. Jarrad of Columbia University's Lamont-Doherty Geological Observatory, studying rock extracted from three North Slope oil wells, found evidence that the North Slope had swung counterclockwise away from the Canadian Archipelago. This idea was proposed as long ago as 1955 by S. W. Carey of Tasmania. At the San Francisco congress, David B. Stone of the University of Alaska said magnetic evidence seemed to show a small amount of reversed motion after such rotation, perhaps in response to buffeting by terranes arriving to form the rest of Alaska.

Arthur Grantz of the United States Geological Survey in Menlo Park, California, reported further evidence for North Slope rotation around a hinge in the Mackenzie Delta. He and his colleagues proposed that the break from northern Canada and start of rotation occurred 150 million to 190 million years ago. Dr. Nockleberg said the terrane along the Yukon south of the Brooks Range is a fragment of a much larger terrane scattered for more than a thousand miles along the western margin of the mountains from Alaska to British Columbia. Its origin was very distant, he said. It contains pieces of zircon 2.1 billion to 2.3 billion years old, derived from its former parental continent.

Ample evidence for a collision by the North Slope block has been found along the south flank of the Brooks Range and the north edge of the Yukon Basin. Nevertheless, not all geologists have accepted the rotation hypothesis. They argue that there is no evidence for the proposed hinge near the mouth of the Mackenzie. Some have suggested that the North Slope broke away from the Lomonosov Ridge, which crosses the Arctic Ocean through the North Pole, or slid west from the Atlantic side of the continent. One proposal holds that the Lomonosov Ridge itself broke away from the continental shelf north of Europe. Dr. Grantz believes that the organic material under the North Slope that eventually formed one of the world's major oil reservoirs accumulated on the nearby ocean floor before the region broke loose, and that this rifting then created the tilted geologic structures in which the oil accumulated.

It was much warmer then. Fossil plants indicate that at the time of the North Slope collision, the region, now barren, was covered with forests. Fossils have been found along the Colville River, which flows into the Arctic Ocean. They indicate that 65 million to 70 million years ago, or shortly after the North Slope collision, dinosaurs were thriving there, even though the region may have been even nearer the North Pole and subject to very long winter nights. Fossils have been found in southern Australia showing that dinosaurs lived there when the region was farther south and also cold.

Despite such clues, there is probably no region on earth whose geologic history remains as uncertain as that of the Arctic Ocean and the lands around it. In 1970 Robert S. Dietz and J. C. Holden, in an early analysis of continental drift, suggested that there has always been an ocean there, Oceanus Borealis. In a new hypothesis, Dr. Zonenshain proposes that a continent, Arctida, occupied the Arctic Ocean, then dispersed to surrounding continents.   [ws]

Q&A

Q. *How do geologists determine the age of rocks?*

A. If rocks contain identifiable fossils, they can be dated from the age of the fossil. Otherwise, the radioactivity of elements in the rocks can give clues to their age, according to Carl Turekian, a professor of geology at Yale University. Radioactive elements break down into lighter elements at a constant rate, unaffected by heat, pressure, and the passage of time. By measuring the amount of the radioactive element in the rock and the amount of the element to which it has changed, scientists can tell how long decay has been under way and hence the age of the rock.

One of the oldest age determinations has been of zircon specimens, found to be 3.8 billion years old, said Dr. Turekian. Zircon's age can be calculated by measuring the extent to which a radioactive form of uranium has decayed into lead. This, he said, indicates the time that has elapsed since the rock was last molten.

The use of radioactive carbon for dating is largely limited to rocks containing fossils and other samples of once-living material and is not applicable to samples much more than sixty thousand years old.

THE GOOD EARTH

MICROSCOPICALLY, it comes in sheets, pipes, plates, and tendrils. It shimmers with electronic energy. It triggers intricate chemical reactions and sucks poisons from the environment. That much is known for certain. But scientists obsessed with clay, seemingly among the most ancient and ordinary of earthly materials, now go even further. This somber, clammy, doughy substance may be capable of storing information and replicating pieces of itself, some believe. And they speculate that those abilities may provide an answer to the mystery of how life began.

The surprising complexity of clay is beginning to come into focus with the help of new microscopes and particle beams for probing structure on the smallest scales. Some of clay's properties have long been known; it was the original catalyst in oil refining, for example, and small amounts can speed chemical processes by a factor of ten thousand or more. But as physicists, chemists, and geologists come closer to understanding how atoms organize themselves on surfaces, the puzzles of clay's behavior have formed a crucial frontier of materials science. "When you talk about clays in the natural world, you're talking about the most complex area of all geochemistry," says Hyman Hartman of the Massachusetts Institute of Technology (MIT). "Our understanding of clays is worse than our understanding of biology."

Clay is the product of thousands of years of weathering—the result of pounding, cracking, and crushing rocks, dissolving them in water, and crystallizing them again as particles. Water readily fills the tiny spaces, giving clay its familiar malleable feel. In a sense, clay is the wild, undomesticated precursor of the semiconductors that set off the computer revolution and the newly discovered superconductors that have begun transforming technologies of electricity and magnetism. Like those modern ceramic materials, clay is a crystal, with its molecules arranged in orderly arrays, and it has been found to have startling electronic properties. "If you take a lump of clay and hit it with a hammer it blows ultraviolet energy for a month," says Leila M. Coyne of San Jose State University in California. Dr. Coyne has shown that molecular irregularities—"defects"—in the crystalline lattice of clays give them the ability to store energy and then re-emit it. That is one of the clues tantalizing scientists who believe clay, rather than the primordial ocean, may hold the key to the origin of life. "If you think about what a life form is," Dr. Coyne says, "you have to be able to take energy from the environment and use it to drive chemistry. Energy storage, collection, and transfer is probably the most fundamental requirement of a living system."

Like most semiconductors, clay is silicon-based, containing by definition aluminum and oxygen atoms as well. Like the new semiconductors, clay's crystals form in layered sheets, fundamentally two-dimensional rather than three-dimensional. Although geologists have classified many different kinds of clay, some containing iron or magnesium, all are oxides of silicon and aluminum, and they share the basic layered molecular structure. On slightly larger scales, still visible only with powerful microscopes, the layers can build up in many shapes, resembling piles of jewels or weird, weedy gardens. Recent thinking about clay formation has looked less at how rocks are ground down and more at how such structures arise. The layering, which is more akin to a deck of microscopic playing cards than a bucket of sand, gives clay a phenomenally large surface area. A lump weighing one pound can have as much total surface as fifty football fields.

All that surface makes clay a powerful chemical engine, because it is on the surface of a substance that the most interesting molecular events take place. Apart from its use as a catalyst, clay's surface also makes it effective at neutralizing toxic chemicals, including dioxins and radioactive waste. "Heavy metals which are toxic and radioactive stay in the clay forever because they are attracted to the electrical charges on the clay sheets," says Pierre Laszlo of the University of Liège in Belgium. But it is the peculiar ability of clay to mix disorder with order that most intrigues scientists. Clay's checkerboard-like surfaces provide many sites at which one kind of ion can be replaced with another, subtly changing the behavior of the whole crystal. "Clays are not ideally crystallized," Dr. Laszlo says. "They have microdomains which are amorphous, and these amorphous domains—domains of disorder—are where the catalysis is. You can have dislocations because atoms are missing, or a fracture in platelets, and these local irregularities are where chemicals bind and chemical reactions occur."

So clay sits somewhere on a middle ground between the dull checkerboard regularity of crystals like salt or ice and the willful, changing forms of living organisms. That gives it a seductive appeal to those seeking to explain the origin of life. The central problem in origin-of-life science is always how nature made the leap: how it got from simple combinations of atoms in the early earth to highly complex organic molecules that not only process energy but also reproduce themselves. Any seemingly intermediate link attracts attention.

In the 1950s the "primordial soup" idea, still the dominant theory, received considerable help from a famous experiment that sent some simulated lightning bolts through a mixture of simple gases and managed to produce amino acids, the relatively complicated organic molecules that make up proteins. But since then the gap between amino acids and the self-reproducing machinery of DNA has remained vast. "There's been quite a lot of skepticism of the classical 'first have a soup and then the bits come together and so on,' " says Graham Cairns-Smith of the University of Glasgow, a leading advocate of the idea that life began with clays. The problem with closing the gap is that life as we know it is inherently "high tech," he said. "It's like a videorecorder—you just can't make a simple one." Clay, by contrast, is low tech. No one has been able to show that clay can reproduce itself, except in the trivial sense that any crystal reproduces its basic structure as it grows. Experimenters are looking for a more provocative kind of replication, the replication of "information."

In DNA, information resides in a sequence of molecules arranged like beads on a string that has the ability both to make copies of itself and to control the formation of proteins. As electrical engineers and computer scientists know, information can take many forms, from the waves in a radio transmission to the pits on a compact disk. In clay, that information could be stored as patterns of different ions on the crystal surface or as flaws in the crystal. "There is no information in the regular crystal structure," Dr. Cairns-Smith says. "A perfect crystal is a blank object. You have to create a pattern of changes, a disorder imposed on the crystal. Disorder is precisely the thing which can hold information." Experimenters hope to see whether bits of clay with particular patterns of ions can impose those patterns on other clay. Dr. Hartman at MIT is synthesizing clays in the laboratory and then "seeding" them with other clays. The theory is that patterns in one layer will perpetuate themselves as other layers pile up, and then a layer like the original will separate, as strands of DNA separate. "This is the beginning of proving that clays can replicate like DNA," he says. "The big breakthroughs that are going to be coming up now are from new microscopes, when they can be adapted to clays, that will pick up the structure atom by atom, not just average properties."

Dr. Cairns-Smith talks about a different sort of replication of information. He hypothesizes that a certain kind of defect in the crystal structure—even some pattern of misaligned molecules—can perpetuate itself in pieces of crystal that then cleave from the original. Either way, the ability to make copies of some kind of information is only part of a chain of properties needed for the creation of life. The information must somehow be able to change its environment, which is where the ability to catalyze reactions comes into play. In living organisms, the key catalysts are enzymes.

And somehow the creative evolutionary force of natural selection must come into play. Dr. Cairns-Smith suggests that even a feature like the size of a crystal could be grist for natural selection. If certain big crystals tended to make more big crystals, while small crystals made small crystals, he says, "one could imagine situations, at the bottom of a stream bed perhaps, where to be small is to be successful." Or perhaps a particular defect structure could affect a crystal's tendency to be absorbed in water and thus to be moved from place to place. Eventually, with certain "crystal genes" propagating themselves and evolving, a sort of "proto-organism" might form. The more sophisticated genetic machinery that now seems universal in living things might have piggy-backed itself onto such nearly living things, later taking over. Or so the theory goes.

Many find the speculation appealing; many others are far from convinced. "There's no experimental evidence," says Leslie E. Orgel of the Salk Institute for Biological Studies in San Diego—no evidence, that is, of a clay that can replicate itself and perform creative chemistry. "If someone could come up with a clay that had one of those properties, it would be interesting," Dr. Orgel says. "If they could show that it had both, that would be spectacular." [JG]

CHARLOTTE'S WEB

A fossil recently discovered near Gilboa, New York, indicates that spiders may have spun webs as long as 385 million years ago. An almost completely preserved spinneret found in Middle Devonian rocks is the earliest evidence yet of silk production by spiders, according to a report in the journal *Science*. Spinnerets are the silk-spinning organs in spiders' abdomens. Two fossils found earlier in Devonian rocks have been described as spiders but lack certain characteristics of the order Araneae, to which spiders belong, and the researchers suggested that those fossils were not spiders. The Devonian period was from 405 million to 345 million years ago.

In this report, William A. Shear of Hampden-Sydney College in Virginia and his colleagues said the fossil's spigots, through which a spider discharges silk, resemble those of a living spider suborder, Mesothelae, but the number and distribution resemble those of another suborder. The researchers speculated that the Devonian spider may have belonged to a sister group of the three living suborders of spiders.

They said it was unclear how the Devonian spider may have used its silk, speculating that it was used to line and enclose burrows in which it lived or as material to wrap its egg sacs. Although flying insects do not appear in the fossil record until much later, the researchers said it was possible that fossils of winged insects may eventually be found in Devonian rocks. That would suggest that the Devonian spider may have spun aerial webs to trap prey.

INSIDE A DINOSAUR'S EGG

EVERY gigantic dinosaur started out as a tiny embryo in an egg, but only recently have paleontologists, applying advanced medical imaging technology, discovered some of those embryos preserved inside fossilized eggs. The new findings are yielding important insights into the full life cycle of dinosaurs. Scientists in Montana found nineteen fossilized embryo skeletons from a previously unnamed type of dinosaur and the fragments of seven embryos of another species. They found the embryos through X-ray examinations of the six-inch-long eggs, revealing the intact skeletons within. With infinite care one embryo was partly separated from its mineralized surroundings—hatched, as it were, after 75 million years. In the fetal position, it resembled a little seahorse.

In Utah, scientists were almost sure they had detected a recognizable embryo inside a 150-million-year-old egg. Scientists believed the embryo, whose shape reminds them of a tadpole, was only a few days old, and so the egg was probably still inside the mother when she died. Other eggs containing preserved embryos have been excavated in Alberta by Canadian paleontologists. From two nearly identical eggs, each eight inches long, they have extracted bones and assembled a composite skeleton of an eighteen-inch dinosaur fetus. It was identified as Hypacrosaurus, a member of the hadrosaur, or duckbill, dinosaur family.

Suddenly, fossil embryos are the hottest properties in the laboratories of dinosaur paleontologists. They are exceedingly rare, and finding them intact inside eggs provides for the first time a way to study the early skeletal development of dinosaurs and what can be inferred from that about their growth patterns and physiology. The findings may help paleontologists tell if a small skeleton is an example of a mature animal of a small species or a juvenile of a large species, clarifying species identification.

So many discoveries are being made now, scientists said, because they have learned where they are most likely to find dinosaur nests and how to examine the interiors of ancient eggs with the medical X-ray technique of computerized axial tomography, or CAT scanning. "Mostly, there's been a change in exploring attitudes," says Philip Currie, the paleontologist who made the discoveries in Alberta.

"You go looking for big dinosaurs and that's what you find. You look for eggs and embryos and, if you're looking in the right areas, you're going to find them."

Digging in 1987 at Devil's Coulee in southern Alberta, Dr. Currie's team from the Tyrrell Museum of Paleontology in Drumheller, Alberta, was looking especially for fossil eggs. "We wound up with eggs and embryos along with hatchlings, juveniles, and adults," he says. The scientists are thus able to trace the growth stages of at least one dinosaur species. This is expected to help them determine the anatomical differences, for example, between young and mature dinosaurs. This should avoid mistaken identities and the proliferation of species designations.

Dr. Currie identified the embryo found in Alberta as Hypacrosaurus based on its skull, which already bore the bony crest of that species. These herbivores, which lived about 75 million years ago, grew to a length of twenty-five feet.

The richest lode of known dinosaur eggs is in western Montana, near Choteau. Since 1978, John R. Horner, curator of paleontology at the Museum of the Rockies at Bozeman, Montana, has been finding nests of eggs and hatchling skeletons there. In the journal *Nature*, Mr. Horner and David R. Weishampel, an anatomist at the Johns Hopkins University School of Medicine, reported the discovery of embryonic skeletons in a clutch of nineteen unhatched eggs. Based on a study of the embryos and adult bones near the nest, they concluded that this was a new genus and species of the hypsilophodontid family. They named it *Orodromeus makelai*. The first name means "mountain runner." Mr. Horner says that this reptile grew to only eight feet in length but that its long thigh and other anatomical features suggest it was one of the more fleet dinosaurs that ever lived. It was probably a herbivore or an omnivore, eating both plants and meat. The species name was Latinized for Robert Makela, an associate of Mr. Horner's who died in 1987.

Analysis of the skeletons has revealed some striking differences between *Orodromeus* and the embryos of another species, a duckbill named Maiasaura, found in the same area. The joints of limbs in the *Orodromeus* fetus were well formed and composed of calcified cartilage, suggesting advanced development. Mr. Horner and Dr. Weishampel concluded that the fast-developing *Orodromeus* young were probably able to leave the nest and fend for themselves almost immediately after hatching. In this respect, they were like nearly all other reptiles. The newly discovered fragments of maiasaur embryos, by contrast, seemed to confirm Mr. Horner's previous suggestion that these dinosaurs were late developers, spending many months in the nest like birds and being fed by their foraging parents. Despite their slow early growth, the maiasaurs reached thirty feet when fully grown.

A careful microscopic examination of some of the *Orodromeus* bones, Mr. Horner said, showed them to contain a dense network of fossilized blood vessels. This, he said, was an indication of rapid growth potential and the higher metabolic rates characteristic of warm-blooded mammals and birds. Some scientists believe dinosaur embryology may point to key evidence in the long-running debate over whether these animals were ancestors of modern birds and if some of them, unlike other reptiles, were warm-blooded creatures.

More conservative in his reaction, John H. Ostrom, a paleontologist at Yale University, doubts that embryo studies would yield such far-reaching results. "The identification of the resident of that egg is about the only really important information that can be gained." That identification would be useful because it would enable paleontologists in the future to make a quick identification, sometimes with nothing more than pieces of eggshell, of the species an egg or nest belonged to.

More eggs with dinosaur embryos are likely to reach the laboratories soon. Mr. Horner said he had already found several other specimens. Now that scientists know what to look for, David B. Norman, a research associate at the University Museum at Oxford, England, recommended that paleontologists return to the sites where dinosaur eggs were first found in the 1920s: the Gobi Desert of Mongolia and Aix-en-Provence in France. The hunting might be even better this time.   [JNW]

FAMILY AFFAIR

A team of Chinese and Canadian paleontologists has found a fossilized cluster of baby dinosaurs that apparently died together 75 million years ago in what is now Inner Mongolia.

Dr. Philip Currie of Alberta's Tyrrell Museum of Paleontology identified the sheep-size babies as ankylosaurs (armored dinosaurs) of the genus *Pinacosaurus*. The fact that at least five of the babies were killed together, he reported, is the first indication that ankylosaurs were gregarious and may have lived as families. The likelihood that some dinosaurs, maiasaur duck-bills, looked after their young after hatching was established in the 1970s by Jack Horner, a curator of the Museum of the Rockies in Bozeman, Montana.

Adult ankylosaurs of the *Pinacosaurus* (plank lizard) type were about eighteen feet long and had slimmer feet than other ankylosaurs. All ankylosaurs had squat, barrel-like bodies covered with thick protective plates. Their tails were tipped by massive, bony clubs that they may have used as weapons. Ankylosaurs were vegetarians, but paleontologists believe they were well equipped to defend themselves against the huge carnivorous dinosaurs of their time.

The Chinese-Canadian group of which Dr. Currie is a member was unable to establish what killed the baby dinosaurs, but he surmised that rapidly shifting sand might have overwhelmed them. Although Inner Mongolia was formerly believed to have been dotted by shallow lakes in the age of dinosaurs, the Chinese-Canadian expedition found evidence that the region had actually been a sparsely vegetated desert.

EVOLUTION AT WORK

WHEN Peter Sheldon chipped his first fossil trilobite from some rocks in Wales a decade ago, two American scientists were transforming standard evolutionary theory with a radical idea. Well-established species, they argued, tend to change little or not at all as the aeons pass; most real evolution takes place in rapid bursts at the rare moments when new species are born. Fifteen thousand trilobites later, Dr. Sheldon, now a research fellow at Trinity College, Dublin, is putting forward a sharp challenge to this conception of evolution. He contends that his research, an unusual, exhaustive study of a fossil family drawn from one place in the earth's crust, shows life evolving as Darwin imagined it, steadily and gradually.

Each ancestral trilobite, the remains of a small, crablike creature, blends smoothly into its evolutionary successor, Dr. Sheldon argues. He sees no gaps or jumps. Instead of long periods of equilibrium punctuated by dramatic changes, the "punctuated equilibrium" model, he finds a picture of continuity over 3 million years. "This is possibly the most thorough study that somebody's made which appears to answer the question," says John Maynard Smith, an influential British biologist who has long opposed the notion of punctuated equilibrium. "It's rather impressive." Dr. Maynard Smith wrote a strongly worded commentary to accompany the trilobite study in the journal *Nature*, rekindling what has been the most passionate debate in evolutionary theory over the last decade. Proponents of punctuated equilibrium, however, argue that Dr. Maynard Smith is exaggerating the importance of the trilobite research and misrepresenting their positions in the process. "He should know better," says Niles Eldredge of the American Museum of Natural History. "This is just bad scholarship on his part." "I really resent what he has said," says Steven M. Stanley, a Johns Hopkins University paleontologist. "It makes me seem like some kind of kook."

Dr. Eldredge and Stephen Jay Gould of Harvard University, the original authors of the punctuated equilibrium model, cite two major recent studies that offer support of their view. At best, Dr. Gould says, the Welsh trilobites represent a single, possibly misleading episode in the richly varied history of life. "It's a case," he says. "Even if it's totally correct, it's a case."

Steady flow or fits and starts—the division between these conceptions of evolution has dominated the debate over evolutionary theory. The punctuated equilibrium model has stimulated much research and drawn many adherents. Some of its central notions have taken firm hold. Even the most traditional Darwinians, for example, acknowledge that equilibrium has become an important part of the picture of evolution. Some species do little of evolutionary interest for millions of years at a time. Most theorists now agree that species can evolve at very different speeds under different circumstances. But the debate continues to rage, because it concerns far more than speed itself. At stake are the fundamental questions of evolution: When and why does a creature change from one form to another? Is most evolution the slow, unceasing accumulation of the small changes a geneticist sees in laboratory fruit flies, or does it occur in episodes, when a small population, perhaps isolated geographically, suddenly changes enough to give rise to a new species? Suddenly, in paleontological terms, can mean hundreds of thousands of years. A few creatures, theorists imagine, get cut off from the main body of a population. Their reduced numbers, perhaps combined with some changed circumstances in their environment, undermine their evolutionary stability, leading to rapid change. A new, distinct species is formed. Only later, when the population multiplies and spreads, does it leave behind enough evidence of its existence to impress fossil hunters of a later era. By then the pace of change has slowed—as theorists believe it has for the human species, millions of years after its birth somewhere on the African continent.

Proponents of punctuated equilibrium take pains to stress that such events rely mainly on the Darwinian principles of natural selection among individuals varying randomly from one another. Even so, to some biologists, punctuated equilibrium seems like a resort to some process apart from the usual rules—"mutations that appear to be magic," Dr. Maynard Smith says. "They have argued that their results mean that evolution as seen on the large scale is not just the summing up of small events but a series of quite special things that people like me"—population geneticists—"don't see. We don't want to be written out of the script."

As Dr. Sheldon collected his trilobites, chipping endlessly through soft Welsh rock with his hammer and chisel, he had no intention of intruding on this debate. He began, as field researchers do, by trying to fit his specimens into the proper taxonomic slots, dividing them up according to their Latin names. To his surprise, the task proved impossible. The lines dividing older species from their evolutionary successors quickly came to seem arbitrary, because he found so many intermediate forms, the "missing links" so often absent from the geological record. Darwin and his followers argued that the accumulation of fossils gave a fragmentary picture of evolution, preserving the occasional shell and skeleton but inevitably omitting long stretches of gradual change. Proponents of punctuated equilibrium contend that missing intermediate forms are often missing for a good reason: they belonged to events that were too rapid and isolated to compete in the fossil record with billions of years of established species, flourishing mostly unchanged.

Trilobites were animals with three-lobed shells, roughly the size of a silver dollar, thriving a half billion years ago on the ocean floor. Every so often they would molt, shedding a shell they had outgrown, so not every fossil represents a dead trilobite. Dr. Sheldon, dating his specimens according to the layers of rock from which he retrieved them, has assembled a history of eight different lineages in his particular location. Overall, he finds a clear trend. The youngest members of each line had several more ribs than their ancestors had three million years before. But he also finds an occasionally unsteady, directionless kind of change, with surprising reversals. "I haven't any particular macroevolutionary ax to grind," he says, "but I suspect that many gradualistic patterns have been obscured by the descriptive process. The way they are named as discrete species gives the impression of abrupt appearance. I'm sure that in some organisms changes take place abruptly, but I suspect that many more organisms show gradual change than we currently perceive." Trilobites, as it happens, are a specialty of Dr. Eldredge's, and he reads a very different message in the Welsh results: "This is much ado about nothing. Actually, I'm cheered by this." To him, the addition of a few ribs over a span of three million years amounts to little change, little enough to be considered stasis, which does not necessarily require "lock-rigid invariance." He notes that the differences between trilobites of the different lineages were far greater than between the early and late members of each line. Such tiny changes, in one geographic locale, cannot account for the great trends in the broad tapestry of trilobite evolution, he says. Dr. Eldredge agrees that the standard taxonomic scheme is faulty, arbitrarily creating species where none really existed. He says that the Welsh specimens seem to add up to just eight species—one for each line—with the oldest members, the youngest, and the intermediate forms all belonging to the same group.

That assessment irritates Dr. Maynard Smith: "Eldredge really cannot say that. It really won't wash. This is the kind of change that separates existing species." Even he acknowledges, however, that punctuated equilibrium correctly describes the tendency of many species to remain static over long periods, a tendency that few recognized a generation ago. Two American studies, both gathering detail from thousands of fossils, have lent strong weight to the punctuational model. One, by Alan H. Cheetham of the Smithsonian Institution in Washington, D.C., focused on bryozoa, small ocean animals that encrust rocks. He measured many different characteristics and found that most change occurred in the branching events that gave rise to new species. Within species, there was little evolution. The other study, by Dr. Stanley and Xiangning Yang at Johns Hopkins, looked at nineteen lineages of clams and came to the same conclusion. "Clearly there is some gradual evolution in the evolution of life," Dr. Stanley says. "We shouldn't take extreme positions. What we really want to know is where does most evolutionary change occur. If it turns out that most change is abrupt, then we have a very interesting pattern that people haven't appreciated." [JG]

THE WANDERERS

NEW fossil discoveries and genetic evidence have fueled a resounding debate among anthropologists over the timing and circumstances of the last major event in human physical evolution, the emergence of the anatomically modern *Homo sapiens*. One view, which is gaining adherents, holds that modern humans evolved in one place—almost invariably identified as Africa—and then migrated elsewhere, gradually developing slight racial differences in response to various regional conditions. In the other model, stoutly defended by some prominent scientists, modern humans are seen as arising virtually simultaneously and independently in different places in Africa, Europe, and Asia.

No one is disputing the substantial evidence that the earliest human ancestors evolved in Africa, with the ape-men 2 million to 3 million years ago, and that some of their more adventurous descendants, *Homo erectus*, spread through the warmer regions of the Old World. At issue, instead, is when, where, and how their descendants, archaic *Homo sapiens*, made the fateful transition to thoroughly modern humans. There is no question, either, that all modern races are members of the same species with the same fundamental genetic heritage. But fossil findings in an Israeli cave, showing that modern-looking *Homo sapiens* lived in the Middle East as long ago as 92,000 years, are being cited as support for the out-of-Africa theory. Proponents of this theory say that these cave people were most likely the descendants of original modern *Homo sapiens* who had migrated from Africa.

According to this interpretation, the findings also threaten to displace the Neanderthals, a type of archaic *Homo sapiens,* from a central place on the human family tree. The Neanderthals, whose fossil remains were the first humanlike ones to be unearthed, lived across Europe, in the Middle East, and as far east as Uzbekistan in central Asia from about 125,000 to 30,000 years ago, when they mysteriously disappeared. They were the cavemen of popular lore, stooped and brutish, but some scientists in recent decades had come to think of them as much less primitive beings and likely close relatives, perhaps even direct ancestors, of modern humans.

But fossils from another Israeli cave show that Neanderthals inhabited the area as recently as 60,000 years ago, raising doubt whether they could have been the ancestors of modern *Homo sapiens* who were present 30,000 years earlier. Proponents of the out-of-Africa hypothesis contend that this discovery supports their view that Neanderthals were a distinct and parallel species that came to a dead end and, therefore, played no role in modern human evolution. Christopher B. Stringer and Peter Andrews, paleontologists at the British Museum of Natural History in London, writing in the journal *Science,* fueled the controversy by asserting that the collective evidence now "favors a recent African origin of *Homo sapiens.*"

Nothing of the kind, counters Milford Wolpoff, a professor of anthropology at the University of Michigan, who is an unbending advocate of the multi-regional model. The African model, he argues, overlooks marked paleon tological evidence of anatomical features of Neanderthals and other regional archaic *Homo sapiens* that persist in early modern humans, indicating considerable interbreeding and hence suggesting that the two branches could not be too dissimilar. Apparent similarities between the cultural aspects of the Neanderthals and the anatomically modern humans also testify to a close relationship, Dr. Wolpoff says. For example, the two groups used similar stone tools and practiced ritual burials of their dead.

In an assessment of the conflicting theories, Fred H. Smith, a professor of anthropology at the University of Tennessee at Knoxville, agreed that "reasonable" transitional fossils and archaeological samples like tools exist outside Africa. For this and other reasons, he and his colleagues conclude that multi-regional evolution "is the best explanation for modern human origins."

But Dr. Smith cautions that the data so far are not sufficiently unequivocal to warrant dogmatic assertions by either side in the debate. "People are going mad," says Ofer Bar-Yosef, a professor of Old World Paleolithic archaeology at Harvard University, commenting on the growing ferment. The debate came to a boil in 1988 with a report that modern-looking *Homo sapiens* lived in the Middle East as long ago as 92,000 years, 50,000 years earlier than had been estimated. The date was established by a technique called thermoluminescence, which is mainly used to date pottery. Flint found in the Qafzeh cave near Nazareth in Israel was heated to release the energy of electrons trapped inside since the stone chips were last burned long ago, presumably when they fell from the hands of toolmakers into a campfire. Analysis of the light emissions determined when the burning last occurred, and since the skulls of modern humans were found in the same sediments, it was concluded that these people had occupied the cave at the same time. The dating was conducted by a team of French and Israeli scientists headed by Hélène Valladas, a physicist at the Institute for Low-Level Radiation in Gif sur Yvette, France. Most scientists have accepted the reliability of the 92,000-year date.

This was good news for the out-of-Africa forces. If modern humans evolved in Africa, the oldest fossils with modern features would be expected to be found in or near Africa. Some such fossils have been excavated in South Africa at the Border Cave and Klasies River Mouth sites and have been dated at about 100,000 years. But many anthropologists dispute the validity of these findings and say it is difficult to put a date on the earliest modern humans in Africa any more precise than somewhere between 40,000 and 100,000 years. Even Dr. Stringer, a leader of the out-of-Africa school, concedes that it remains unclear why it then apparently took 50,000 years for the Qafzeh people to spread into Europe and eastern Asia. And Dr. Smith noted that there is no known compelling reason—population pressures or marked changes in technology or food supply—for African populations to have left their homeland about 92,000 years ago.

For that matter, Dr. Wolpoff asks, why assume that the Qafzeh people originated in Africa? "If that date's correct, it means that Eden may not be in Africa," he said. "It might also be right back where the Bible says." According to the model proposed by Dr. Wolpoff and others, the early *Homo erectus* hominids spread out of Africa more than a million years ago. Then regional groups living in relative isolation slowly evolved into several archaic versions of *Homo sapiens*, one of these being Neanderthals. There must have been some "gene flow" from mixing of these regional populations, anthropologists say, or else they might have evolved into distinct species.

The opposing theory holds that the modern species developed in Africa and moved into the rest of the world. People like the Neanderthals were driven into extinction. Out-of-Africa theorists have gained valuable support from molecular biologists who are adept at tracing genetic lineages. Their research showed that every person living today can trace his or her maternal ancestry to one woman who lived in Africa 100,000 to 200,000 years ago. Allan C. Wilson and Mark Stoneking of the University of California at Berkeley and Rebecca L. Cann of the University of Hawaii deduced the "African Eve" by studying a key set of genes, mitochondrial DNA, passed down only from mothers to their children. They examined these inherited genes from native Africans and people in the United States whose ancestors were African, Asian, European, or aboriginal Australian.

Finding a greater number of mutations in the genes from people of African descent, the scientists concluded that Africans have been diversifying longer and, therefore, represented the earliest modern branch of the family tree. In a computer analysis of the mutation rates, they determined when the last common ancestor of all people had lived in Africa. Dr. Stringer and Dr. Andrews, in arguing their African-origin theory, set great store by these findings. But other researchers have produced results indicating that the common ancestor might have been Asian. "There is still much disagreement among the experts about what the genetic data can and do tell us about modern human origins," Dr. Smith and his colleagues observed.

A second discovery in Israel, at Kebara Cave at Mount Carmel, is critical to the debate. There scientists found skeletons identified as Neanderthal and dated at 60,000 years. Some anthropologists point out that the timing makes it difficult to explain how the Neanderthals could have evolved into modern humans. The fact that Neanderthals and modern humans probably coexisted there and elsewhere in Eurasia for thousands of years casts doubt on any ancestor-descendant relationship and makes it more likely that the two were separate species.

Examination of the Kebara skeletons by Yoel Rak, a professor of anatomy and human evolution at Tel Aviv University, revealed striking differences in the structure and orientation of the sockets into which the thigh bones fit. They face sideways more than in modern *Homo sapiens*, which Dr. Rak and associates attributed to differences in

"locomotion and posture-related biomechanics." These and other anatomical features, he said, convinced him that Neanderthals were a
separate species.

Whatever their biological relationship, the Kebara Neanderthals
and the Qafzeh modern humans were leading lives of remarkable
similarity. "We found the same archaeological indicators—stone tools,
hearths, burials, use of ochre in body painting—in both sites," Dr.
Bar-Yosef says. To Dr. Bar-Yosef, among other out-of-Africa advocates, the findings showed that different species could develop some
of the same technologies and cultural attributes, which would mean
that these are not necessarily reliable indicators of close biological
affinity. To Dr. Wolpoff the cultural similarities strongly suggest that
the Neanderthals and early modern humans were no more than different races of the same species and interbred to produce the line of
fully modern *Homo sapiens*.

Perhaps the picture is much more complex than imagined in either theory, as Dr. Bar-Yosef suggested. "Modern *Homo sapiens* may
have emerged more than once," he says. "Many populations in prehistoric times were small and struggled and failed and just died out.
We have a tendency, because we want to tell a story, to make up a story
that sounds like a continuum. Some of the modern-looking humans
we find 92,000 years ago may have been unsuccessful. Maybe there
have been many more branches along this evolutionary tree. We need
to know more about this and about the chronology of these evolutionary events. Then we can fight over interpretations." [JNW]

BUT DID THEY FLOSS?

The simple toothpick may have been one of the first "tools" of
human design. As anthropologists imagine it, early human ancestors sat by the fire after a hard day at the hunt, chewing on
roasted mammoth and picking their teeth with sticks cut to
sharp points. At other times they just picked their teeth idly,
while contemplating what a daub of paint might do for drab
cave walls.

Evidence for Stone Age toothpicks is indirect but compelling, anthropologists say. Fossil teeth, the most durable relics of
early life, seem to tell the tale. Analysis of grooves on ancient

teeth has led to a consensus that these are the marks of heavy toothpick use by human ancestors. The journal *Nature* reported that the earliest known example of the grooved-teeth phenomenon was found in 1.8-million-year-old fossils of *Homo habilis,* an ancestral species, excavated at Omo in Ethiopia.

The grooves were especially common in the teeth of Neanderthals and other archaic *Homo sapiens* of Europe and Asia between 130,000 and 35,000 years ago. Researchers considered whether the grooves could have been the result of tooth decay, dietary grit, or stripping and processing fibers in making domestic goods. "None of these, however, really fits the evidence," *Nature* reported.

The similarity of the prehistoric grooves to toothpick-caused abrasions in historical and modern populations of American Indians and Australian Aborigines argued for the toothpick interpretation of the data. In an article in *Current Anthropology,* Christy G. Turner II, professor of anthropology at Arizona State University, concluded, "As far as can be empirically documented, the oldest human habit is picking one's teeth."

WHEN WOMEN RULED

MARIJA Gimbutas is an eminent archaeologist who says she believes the world once lived in peace. It was during the Stone Age, she says, when goddesses were worshiped and societies were centered on women. Then, about six thousand years ago, this Old European culture, in which the two sexes lived in harmony with one another and with nature, was shattered by patriarchal invaders who installed their warlike gods in place of the life-generating Great Goddess. It is a thesis that has made the sixty-eight-year-old professor of archaeology at the University of California at Los Angeles a heroine among many feminist social critics and religious thinkers and a controversial figure, to say the least, among her colleagues.

References to Dr. Gimbutas's theories are sprinkled liberally throughout a growing literature about goddess-based religion. For some time feminist writers have been seeking nonpatriarchal mythologies and rituals in Jungian psychology, reconstructed notions of witchcraft, or even in pure creations of the imagination. But Dr. Gimbutas gives them something more: the seeming stamp of science and the reassurance of history. Her work was a major scholarly source for Riane Eisler's *The Chalice and the Blade,* a sweeping analysis of cultural evolution that has become a minor classic in the women's movement. In *The Once and Future Goddess,* a book on goddess symbols and images, Elinor W. Gadon calls Dr. Gimbutas's research germinal and fundamentally important.

"Marija Gimbutas is the one world-class scholar showing that what feminists wished were true is in fact true," says John Loudon, a senior editor at what was then Harper & Row who worked on the archaeologist's latest book. Dr. Gimbutas is indeed a prolific scholar, the author of twenty books (including a monumental study of Bronze Age Indo-European cultures) and more than two hundred articles. She has directed five excavations in Europe, reads more than twenty languages, and brings to her work an extraordinary knowledge of European folklore and mythology.

But the skepticism about this thesis by many leading archaeologists and anthropologists is unmistakable, though it is almost always comes with expressions of respect for Dr. Gimbutas's other contributions and with concern for her struggles with lymphatic cancer. Yet the growing acceptance of her theories among nonexperts has led some of these scholars to feel that they should make their own criticism more widely known. In the end, they say, Dr. Gimbutas's work raises sensitive questions not only about prehistoric civilization but also about the relations between speculation and scholarship and between scholarship and social movements.

Dr. Gimbutas argues that between 7000 B.C. and 3500 B.C. the people of Europe lived in sedentary agricultural societies that worshiped the Great Goddess, delighted in nature, shunned war, built comfortable settlements rather than forts, and crafted superb ceramics rather than weapons. The social system was matrilineal. Women headed clans or served as queen-priestesses. Men labored as hunters and builders. But neither men nor women dominated the other sex. Death was not absent from this world. Death-wielding goddesses are found among the female deities that predominated in this period, but their worship, Dr. Gimbutas theorizes, was always closely linked to themes of life and regeneration.

Between roughly 4000 B.C. and 3500 B.C. this peaceful and harmonious "Old Europe" was shattered by waves of Indo-European invaders on horseback, the theory holds. These marauders from the Russian steppes transformed Europe. Their warrior gods dethroned the nurturing Goddess. Her various manifestations were incorporated into the male-dominated pantheons as wives, daughters, or consorts, sometimes eroticized like Aphrodite or militarized like Athena. Patriarchy and hierarchy replaced sexual and social egalitarianism. The Goddess religion and its symbols went underground, putting on the masks of subordinate but still powerful female Greek and Roman deities, of the Virgin Mary, and of mysterious figures in folklore and fairy tales. But a substratum of Old Europe survived in harvest customs, peasant beliefs about springs, rocks, trees, and animals, medieval magic, and the practices that Christian authorities persecuted as witchcraft.

This secret stream is essential to decoding much of Western culture, Dr. Gimbutas maintains, but it is also a source of ancient wisdom that modern civilization must tap to counter its own alienation from nature. It is a dramatic story of paradise lost and now rediscovered. Originally set forth in a 1974 book, *The Goddesses and Gods of Old Europe*, it inspired, besides feminist thinkers, a number of women artists who were captivated by the remarkable images of ancient female figurines in Dr. Gimbutas's book and later publications.

Her ideas have been welcomed by eminent figures like the late mythologist Joseph Campbell, who wrote a foreword to Dr. Gimbutas's latest volume before he died in 1987, and the anthropologist Ashley Montagu, who hailed that book as "a benchmark in the history of civilization." But many other investigators of prehistoric Europe have not shared the enthusiasm. Bernard Wailes, a professor of anthropology at the University of Pennsylvania, says that most of Dr. Gimbutas's peers consider her "immensely knowledgeable but not very good in critical analysis."

"She amasses all the data and then leaps from it to conclusions without any intervening argument," Mr. Wailes says. "Most of us tend to say, oh my God, here goes Marija again."

Ruth Tringham is a professor of anthropology at the University of California at Berkeley who is an authority on the same period and geographical area of prehistoric Europe as Dr. Gimbutas. Choosing pages at random from *The Language of the Goddess,* she repeatedly voices dismay over assertions that demand, she says, serious qualification. "No other archaeologist I know would express this certainty," Ms. Tringham says. Linda Ellis, an archaeologist at California State University at San Francisco, who took courses from Dr. Gimbutas as an undergraduate at UCLA and has worked at some of the same archaeological sites in southeastern Europe, says, "We're looking at small agricultural villages of no more than between ten and twenty-five homes. We can tell a lot about their economy, what they traded, what they made, what they ate, how they built homes, with whom they had contacts. The problem becomes how far you can push that, how much you can say about how a person thought." Ms. Ellis makes it clear that she thinks Dr. Gimbutas has gone too far.

David Anthony, an assistant professor of anthropology at Hartwick College in Oneonta, New York, whose areas of research also coincide closely with Dr. Gimbutas's, says that, contrary to her claims, the cultures of Old Europe built fortified sites that indicate the presence of warfare. There is also evidence of weapons (including some used as symbols of status) and of human sacrifice, hierarchy, and social inequality, Mr. Anthony adds. The collapse of Old Europe is not fully understood, he says. It may have been due to a change in climate, the exhaustion of the soil, internal social dissent, or population pressures. Mr. Anthony agrees that the arrival of migrants and the emergence of less sedentary, stock-grazing societies around 3000 B.C. was a principal component of later European developments. But there was no real evidence that the migrants caused the collapse rather than followed it, he says. He also sees no evidence that women played the central role in either the social structure or the religion of Old Europe. These were "important and impressive societies," but rather than Dr. Gimbutas's "Walt Disney version" they were "extremely foreign to anything we're familiar with."

Dr. Gimbutas said in an interview that the nineteenth-century Swiss scholar Johann Jakob Bachofen and his followers Robert Briffault in France and James Frazier in England were among her forerunners. In ancient laws and legends and in the beliefs of isolated tribes or communities, they found vestiges of an earlier matriarchal stage of human civilization centered on worship of a mother goddess. Dr. Gimbutas tempers her forerunners' notion of strict matriarchy, but her similarities to these predecessors divide her from many colleagues.

Anthropologists are now uncomfortable with what, a century ago, were popular schemes of civilization passing from one well-defined stage to another. Archaeologists reject Dr. Gimbutas's reliance on the folklore and mythology of historical times to interpret objects from many millennia earlier. But what others see as Dr. Gimbutas's weaknesses, she sees as weaknesses of her discipline. Archaeologists are not interested in religion, she says, and they lack the knowledge and language skills to use comparative mythology and folklore to interpret their findings.

Her own interpretations involve extended chains of associations. Using stripes or hatchmarks or other motifs as clues, she identifies figurines, often from societies thousands of years apart, with specific animal deities, which in turn are assigned specific religious functions. All of this she subsumes into the "cohesive and persistent ideological system" of the Great Goddess. It is a method that Mr. Campbell, in an introduction to her book, compares to the one used in deciphering the Rosetta Stone, but her critics find it fanciful or arbitrary. "In a way she's a very brave woman, very brave to step over the boundary and to take a guess," says Ms. Ellis. But Ms. Ellis strongly rejects Dr. Gimbutas's detailed assertions.

Dr. Gimbutas calls the enthusiastic reception of her work by artists and feminists "an incredible gift" coming late in her life. But "I was not a feminist and I had never any thought I would be helping feminists." Still, *The Language of the Goddess* rings with a fervent belief that knowledge about a Goddess-worshiping past can guide the world toward a sexually egalitarian, nonviolent, and "earth-centered" future. This immediate, some would say ideological, agenda is one thing that disturbs some archaeologists and anthropologists. They recall how easily their fields have been bent to ideological purposes, including the use Nazi scientists made of Indo-European prehistory to bolster claims of Aryan superiority.

In endorsing *The Language of the Goddess,* Gerda Lerner, a professor of history at the University of Wisconsin who has written on the origins of patriarchy, said that although Dr. Gimbutas's theory "can never be proven," by simply providing an imaginative alternative to male-centered explanations, the theory can "challenge, inspire, and fascinate." Dr. Gimbutas's critics agree that speculation and intuitive leaps are essential prods to scholarly progress. "But it is the responsibility of the leapers to be aware that they are leaping and to make other people aware of it also," Ms. Tringham says. [PS]

THE SHINING PAST

ARCHAEOLOGISTS working in Peru have unearthed stunning evidence that monumental architecture, complex societies, and planned developments first appeared and flowered in the New World between five thousand and thirty-five hundred years ago—roughly the same period when the great pyramids were built in Egypt and the Sumerian city-states reached their zenith in Mesopotamia. Stepped pyramids and huge, U-shaped temples more than ten stories high. Bright multicolored friezes with jaguar and spider motifs. Broad plazas flanked by residential areas. All these have been uncovered, or partly so, in the first few excavations, made in the last decade, by a number of archaeologists working among literally scores of sites in more than fifty narrow river valleys that plunge from the Andes to the Pacific along the length of Peru.

The earliest sites are a thousand years older than their counterparts in Central America, traditionally regarded as the cradle of civilization in the Western Hemisphere, and they are consequently forcing reassessment of pre-Columbian accomplishments in the Americas. Their antiquity also narrows the gap between the emergence of civilization in the New World and the Old. "This idea of the Old World being ahead of the New World has to be put on hold," says Richard Burger, a Yale University archaeologist who is one of several working to uncover and interpret the remains of the ancient Peruvian civilization.

The "whopping great monuments" of the Andes "invite favorable comparison" with what was going on in the Middle East, says Michael Moseley, an archaeologist at the University of Florida who has long worked in the region. The culture of the early Andeans never reached the heights of Mesopotamia and Egypt. Neither they nor their descendants, including the Incas three thousand years later, developed writing or used the wheel. Nor does even the latest evidence show the Andean culture to be nearly as ancient as that of the Fertile Crescent, whose origins date back at least eight thousand years. But some archaeologists say that in their ability to construct planned communities and grand structures, organize labor, create art, and devise a political system, the Andeans of four millennia ago might have shown up well beside their Old World contemporaries.

However that may be, the evidence of human settlement back then is "much more spectacular on the West Coast of Peru than it is anywhere in Mexico," says Kent Flannery, an archaeologist at the University of Michigan who has long worked in Central America. "I think a lot of people who have worked in Mexico and haven't been to Peru will be surprised at how early and spectacular the stuff is down there."

The biggest Andean monuments predate the rise of the Maya in Central America by nearly two thousand years and the Aztecs by nearly three thousand. Those civilizations, like that of the Incas, came relatively late. The emerging picture of this earliest American civilization is that of a people tied initially to the sea, living off its bounty, but then moving abruptly—no one knows why—into the Andean highlands to build a flourishing economy based on irrigated agriculture that prospered in spite of the harsh, cold, and arid climate at altitudes around ten thousand feet. Based on the remains of buildings, art, and contents of refuse piles, archaeologists say these people dressed in cotton tunics, breechcloths, shawls, and capes, and that they lived well on sweet potatoes, peanuts, beans, manioc, fish, shellfish, and an occasional deer or sea lion, with guinea pigs a delicacy—"sort of like caviar," Dr. Burger says.

They were no strangers to bloody terror, and may, like the later Aztecs farther north, have practiced human sacrifice. Archaeologists have done only a fraction of the work in the Andes that they have done in the Middle East, and the discoveries of the 1980s in Peru are so fresh that scientists are only beginning to understand the nature of the society that was built there.

Some believe the most advanced settlements were urban city-states governed by a powerful elite that controlled economic assets, principally its food stocks. Others believe the settlements fell short of being truly urban. They also say there is no evidence so far of the force, oppression, and exploitation that are commonly thought to have been a part of Old World city-states. The early Andeans, some say, may have relied instead on economic cooperation, with religion as the binding force, to organize their society.

In any case, there is wide agreement that the early Andeans took a developmental path different from that of the Old World and even from that of Central America. Not least, the Andean civilization appears·to have been more diffuse and decentralized because the isolated valleys forced it to develop that way. All of this is bringing about a third radical revision of thought about the origins of Andean civilization. First, the Incas took credit for having invented it. Then, early in this century, the Inca claim was disproved by the German archaeologist Max Uhle, who found a long succession of pre-Inca cultures stretching back some fifteen hundred years. Uhle theorized that the advanced cultures of Central America had touched off the evolution of South American civilization.

This was challenged beginning in the 1920s by the Peruvian archaeologist Julio C. Tello, who contended that Andean civilization was homegrown and that its progenitor was the Chavín culture, named for Chavín de Huantar, a monumental ceremonial center that flourished in the highlands of central Peru some twenty-five hundred years ago. There, Tello contended, Andean civilization crystallized and spread to less developed regions in the highlands and on the coast. Archaeologists accepted Tello's explanation for half a century. But now, on the basis of the last decade's discoveries, they are concluding that Chavín de Huantar was not the cultural fountainhead, but rather the culmination of developments that began some two thousand years earlier in other areas.

The revision does not stop there. Donald Lathrap, an archaeologist at the University of Illinois at Urbana, believes on the basis of archaeological finds of items dating back eight thousand years that both the Mesoamerican and Andean civilizations have common roots in Neolithic chiefdoms that existed in the central Amazon seven thousand to eight thousand years ago. "The cradle was there," says Dr. Lathrap, and the culture "spread out both to Mexico and Peru."

The Andean temples and secular structures of five thousand to three thousand years ago that are now being uncovered have burst on the archaeological scene with what Dr. Moseley calls "thunderingly dramatic" effect and are the most compelling force in the reappraisal. They are even more dramatic because the dry cold of the Andes has preserved them exquisitely. Adobe friezes and sculptures that might have been destroyed in another climate are preserved almost intact, with their vivid reds, blues, and blacks still showing. Seeds, pollen, and animal skeletons are not fossils, but real.

It is something of a mystery to archaeologists why any major civilization would develop in the Andean valleys and on the Peruvian coast. The region's altitude and aridity make it "grossly hostile," says Dr. Moseley, who adds: "That anyone ever lived there is a bit of a surprise." But they clearly did, as evidenced by the remains of scores of ancient communities, from the Jequetepeque Valley of Peru in the north to the Lurin Valley in the south, that have been investigated in the last decade. Typically, each community has at its center a monumental public structure, usually built in a U shape.

The earliest of these, documented by radiocarbon dating as having been built some time from 5,000 to 4,100 years ago, were established along the coast on sites as large as 140 acres. At most sites, the temples were surrounded by residential areas and refuse deposits. A representative temple of this period, at Aspero on the coast, was about three stories high and measured 98 by 131 feet at its base. Around 4,000 to 3,700 years ago, activity abruptly shifted inland and irrigated agriculture replaced fishing as the main economic resource.

The temples themselves appeared to have both a religious and a secular function. One of the few to be radiocarbon dated is at a site 3,100 to 3,400 years old called Cardal, south of Lima, excavated by a team headed by Dr. Burger. A huge three-story stairway leads up to the entrance to the four-story temple. At the top, flanking the entrance to the temple's inner chamber, are two enormous mouths, each with three-foot-long fangs and interlocking teeth, painted red and yellow. When the Andeans moved inland, they increased the size of their monumental buildings. The largest to be discovered so far is a colossal stepped pyramid, more than ten stories high, at a site called Sechin Alto in the Casma River Valley north of Lima. On the basis of preliminary radiocarbon dating, it is thought to be 3,400 to 3,700 years old.

Sechin Alto may have been in some sort of loose regional association with another Casma Valley complex, the most spectacular site to be excavated and dated so far. Called Pampa de las Llamas-Moxeke and dating back 3,500 to 3,800 years, it has been excavated by Dr. Sheila and Dr. Thomas Pozorski, a husband-and-wife archaeological team at Pan American University in Edinburgh, Texas. The site contains a U-shaped temple ten stories high. Up a hill to the north is a three-story secular public building. Between them is a huge plaza, about two-thirds of a mile long, flanked by smaller buildings and dwellings.

Based on their excavations, the Pozorskis believe that the large secular building, which archaeologists call Huaca A, or Mound A, was a food storehouse that held the community's accumulated wealth. The arrangement of the rooms was such that access would have been easy to control, and the Pozorskis say that some sort of a bureaucratic elite probably exercised power by controlling these economic assets.

"If our interpretation is true about this warehouse," says Sheila Pozorski, "it was a very powerful economic weapon over the people." This, along with the complexity of the site, its size, its precise planning, and the high degree of labor mobilization and organization required to build the monuments, leads the Pozorskis to conclude that Pampa de las Llamas-Moxeke was an early city-state.

But Dr. Burger and others believe that the Peruvian centers followed what Dr. Burger calls "a special evolutionary path." He says there is no evidence in Peru of any kind of social stratification along economic lines. The right to use irrigation canals was probably tied to "your responsibilities in construction of the monumental architecture, and in ceremonies. The glue would be religion."   [WKS]

BAMBOO POWER

Archaeologists have long puzzled over the means Egyptian pyramid builders used to raise and move immense blocks of stone without the help of wheels, pulleys, or domestic animals. One study suggests that the basic stone-raising tool along the Nile five thousand years ago was the simple bamboo pole. Writing in the British journal *Nature*, John Cunningham of Skidmore College reported that slender, flexible poles were often represented in Egyptian art. Many modern scholars have proposed that the Egyptians used large, rigid levers to raise stones, but "it is revealing that levers such as these were never portrayed as being used to move and transport things in Egyptian art," he said.

Instead, he wrote, long flexible poles would have permitted builders to exert a lifting force that could not have been easily attained with ordinary levers. When a simple lever is used for lifting it bears the entire weight of the object, while flexible poles lifting in concert share the load.

In practice, Mr. Cunningham wrote, the pyramid builder would have arranged a row of poles in parallel underneath a stone block with the pole ends extending beyond the edge of the block. The builder would then have propped up each pole end, forcing the pole to bend. Bending a single pole upward would not in itself have exerted enough force to raise the block. But the collective action of raising the ends of many poles would have joined the lifting forces exerted by each one, thereby providing the necessary mechanical advantage.

Mr. Cunningham reported that he had demonstrated the principle, using twelve slender oak poles to lift a block weighing twenty-six hundred pounds. The Inca builders of Peru may also have used flexible poles to raise stones weighing up to three hundred tons that they used in their fortresses, he said.

THE BODY IN THE BOG

ABOUT 2,200 years ago, an aristocratic young Celt was brutally slain and thrown into a bog near what is now Manchester, England. His body, almost perfectly preserved, has become the focus of an intense scientific effort. Several archaeologists have already inferred from the study of the body that ancient Celtic theocracies governed by a priestly caste of Druids probably dominated much more of Europe than had been previously assumed. The condition of the body and the manner of the man's death, moreover, suggest that he may have much in common with similar bodies found in peat bogs in Denmark during the 1950s. If the Denmark bodies are also those of Druids, the implication, one of the chief researchers suggests, is that Celts rather than Germans ruled Scandinavia in the second and third centuries B.C. The Germans, in this view, may have been merely a subsidiary branch of the dominant Celtic society.

The recent burst of research and scholarship started at an ancient bog near the Manchester airport, where sphagnum peat moss has been growing for thousands of years. The peat, once commonly used as fuel, is now harvested mainly as a medium for cultivating plants. On August 1, 1984, a day many archaeologists now regard as a milestone, a commercial peat cutter named Andy Mould was about to throw a load of peat into a shredding machine when some of the crumbling moss fell away, revealing a human foot. Archaeologists and other experts subsequently recovered the torso, head, and arms of one of the most perfectly preserved ancient bodies ever found. The leathery flesh was stained deep brown and had been distorted by long burial in the wet, iron-rich peat, but facial features, skin texture, physique, and even stomach contents were intact.

To study the remarkable find, the British Museum and other institutions quickly organized a research task force that included forensic pathologists, anthropologists, chemists, historians, ethnologists, and other experts. A flood of discoveries followed, and it soon became evident that "Lindow Man," named for Lindow Moss, the locality where the corpse was found, had been no ordinary Celt. Several pieces of evidence, including the serene expression of the man's face, reconstructed from analysis of muscles and skin, suggested that he had gone willingly to his horrible death. His executioners had cut his throat, crushed his windpipe with a thong, bludgeoned his head, and held his face under water.

In 1986 the British Museum published a compendium of the initial findings in a book, *The Body in the Bog*. In another book, *The Bog Man and the Archaeology of People*, Dr. Don Brothwell, a zooarchaeologist who participated in the research, presented extensive evidence that Lindow Man must have been a member of the Celtic ruling caste. His healthy physique suggested good nutrition, and the absence of calluses on hands and feet indicated that Lindow Man, who was under thirty years old when he died, had not been a common laborer. No wounds or marks were found on the body other than the injuries that caused the man's death, and experts therefore concluded he had not been a soldier. But the identity of Lindow Man and the meaning of his death remained elusive.

Scientific reports about the corpse intrigued two British investigators who had not participated in the early research. Dr. Don Robins, a chemist who frequently analyzes material from mummies and other archaeological objects, and Dr. Anne Ross, an archaeologist specializing in Celtic history, both became interested in Lindow Man's last meal. Their collaboration in examining the meal has led to the startling conclusion that Lindow Man was a Druid priest who drew the losing ticket in an ancient lottery. He apparently accepted his martyrdom and willingly gave up his life to propitiate three of the gods worshiped by Celts before the Romans invaded Britain.

The detective work carried out by Dr. Ross and Dr. Robins, summarized in a paper in the journal *New Scientist*, focused on a partly digested cake that pathologists found in Lindow Man's stomach and small intestine. In an interview, Dr. Ross recalled, "The BBC had a television program about Lindow Man, in which Dr. Ian Stead of the British Museum said he was puzzled by the fact that some of the cake in the man's stomach had been badly scorched. That started me thinking." Through her knowledge of Celtic lore and the ancient languages of the British Isles, Dr. Ross realized that the scorched cake might be burned bannock, a coarsely ground barley griddle cake used in Druidic rituals. Following tradition, one section of the thin, flat cake was allowed to scorch. Priests of the order would break the finished cake into pieces, place it in a leather bag, and pass the bag around. Each participant would take out a piece, and the man who drew the burned piece would be sacrificed to the gods. "Even today," she says, "the burned bannock ritual survives, in nonfatal form, in parts of Derbyshire, Staffordshire and Cheshire, near where Lindow Man was found."

But proof that Lindow Man had actually consumed the sacrificial burned bannock was lacking until Dr. Robins joined the investigation. Dr. Robins, who is affiliated with Queen Mary College of the University of London, has frequently used a technique called electron spin resonance in his archaeological work. The process relies on the fact that heating and subsequent cooling causes freed electrons to become trapped in the structure of objects, he said. Their abundance can be measured with the use of a strong magnetic field. The number of trapped electrons is related to the duration and degree of heating the object underwent.

"The technique enables us to tell how hot an ancient object became and for how long it was kept hot," he said. Dr. Robins determined that Lindow Man's last meal had been cooked for no more than eight minutes, just the time it takes to prepare burned bannock in the traditional way. The pathologists had meanwhile concluded that Lindow Man had eaten the cake about thirty minutes before his death, a time consistent with what is known of the sacrificial ritual. Dr. Ross says the multiple means used to kill the victim were consistent with religious precepts of the Druids. The god Tarainis was traditionally honored by beheading or bludgeoning the sacrificial victim, while the god Esus called for a cut throat, and Teuttates was propitiated by drowning. Dr. Ross believes that Lindow Man must have been considered so valuable by his priestly order that he was offered in sacrifice to all three of these gods.

The fact that some of the bodies found in the Danish bog were strangled in a manner similar to that used to kill Lindow Man suggests to archaeologists that Druid authority extended to Scandinavia as well as the British Isles, France, and most of central Europe. "Some bog bodies have been found that were clearly victims of simple murder rather than religious rites," Dr. Robins says. "For example, a tax collector murdered in the fourteenth century turned up in Danish peat, and since he was fully dressed, we know the murder had nothing to do with religion." By contrast, Lindow Man and other bog bodies that were probable victims of Druidic rites apparently were naked when they were slain. Lindow Man did, however, wear a curious armband made of fox fur.

Ancient bodies have frequently been found in peat deposits in England, Ireland, and Scandinavian countries, but only in recent years have scientific techniques been available for exploiting the information they contain. Once unearthed, these bodies begin to decay and disintegrate in a matter of hours and few have been preserved. (Chemical techniques are being used to preserve the remains of Lindow Man, however.) "The peat not only waterlogs the bodies but it keeps out oxygen," Dr. Robins says. "The water is saturated with iron and sulfur that replaces components of the flesh, and when all the conditions are right, the peat may preserve a body for thousands of years. The bones dissolve in the acidic peat water and eventually disappear, but the protein in the flesh is converted into stable material."

Archaeologists lament the fact that ancient Celtic civilization left no written history. Much of what is known about the history of early Celts, or Gauls, as they were known to the Romans, was recorded by their Roman enemies, notably Julius Caesar. The Romans sought to destroy Druidism because of the threat it posed to their power, and at least some Roman descriptions of the Celts must be discounted as propaganda. Scholars therefore welcome indirect evidence of the kind that Lindow Man has brought to light.

"Despite their lack of literature, the early Celts could certainly write," Dr. Ross says. "In fact, their written language may even have preceded the runic writing used by the Norse. But the Celts left records only of commercial transactions. They chose not to record anything having to do with religion, social rites, or history. These subjects were evidently taboo, and the tradition was passed along orally by sages, poets, and priests." Dr. Ross hopes that many more artifacts and bodies will be found in European peat deposits to help illuminate the history of the continent's myth-shrouded antiquity. [MWB]

THE LANGUAGE OF JESUS

AMERICAN scholars have begun work on the world's first comprehensive dictionary of Aramaic, the ancient language presumably spoken by Jesus. The Aramaic dictionary, which is expected to take twenty years to compile even with the help of computers, is the latest of four monumental projects undertaken in the United States to compile dictionaries of ancient Middle Eastern languages. Together these dictionaries are expected to give archaeologists, historians, and linguists the best tools yet devised for deciphering the cultures of the people who spoke them.

Scholars from Johns Hopkins University in Baltimore, Catholic University in Washington, and Hebrew Union College in Cincinnati have joined forces for the Aramaic dictionary project, which will define about forty thousand words, all of those known from inscriptions and texts. Dr. Delbert R. Hillers of the Department of Near Eastern Studies at Johns Hopkins University, director of the work, says the project could easily outlive some of the scholars who had begun it. But the dictionary, he says, will provide vital new insights into the empires of ancient Assyria and Persia, whose administrators and diplomats spoke Aramaic. Why the Assyrians and Persians chose Aramaic instead of the Akkadian and Persian languages of their own people is one of the puzzles modern scholars hope eventually to penetrate.

The Aramaic dictionary's compilers have decided to restrict their focus to the period before A.D. 1400, even though Aramaic has survived to modern times. The language declined in importance after 1400, although it is still spoken in parts of Syria and Iraq as well as a few Christian communities in other parts of the world, including the congregation of St. Mark's Syrian Orthodox Church of Hackensack, New Jersey.

By including all the meanings of each Aramaic word as they changed over time, the group expects to illuminate the evolution of a great language and to improve the translations of biblical and secular texts. Noting that the other major dictionary projects for ancient Middle Eastern languages are in varying stages of completion, Dr. Hillers describes Aramaic as "the last unclimbed peak of ancient Middle Eastern languages."

"The great University of Chicago Assyrian and Hittite dictionaries are nearing completion, and we already have excellent Hebrew dictionaries and a great Egyptian dictionary," he says. "Work has begun on a Sumerian dictionary. The Aramaic dictionary will complete the picture." Although earlier dictionaries have been compiled for certain intervals during the evolution of Aramaic, none covers the language through the span of its major use, about 900 B.C. to A.D. 1400. "It is as if compilers had created separate English dictionaries for the languages of Shakespeare, Hemingway, and the great English-language writers of other periods, without ever producing a dictionary compiling the language as a whole throughout its history," Dr. Hiller said in an interview.

The golden era of Aramaic was between the sixth and fourth centuries B.C., when it was the official language of the Persian empire, which extended from Egypt to India. Even after the Greek conquest of Persia, Aramaic remained in common use throughout the realm, borrowing many words from Greek and other languages. The modern "square" Hebrew alphabet, Dr. Hiller says, is actually the Aramaic alphabet as it was adopted by Hebrew-speaking peoples. A close linguistic relative of Hebrew, Aramaic gradually replaced Hebrew as a religious language, and parts of the Bible, including the story of Belshazzar's feast in the book of Daniel, were written in Aramaic.

In some cases, the meaning of biblical Aramaic phrases remain obscure. The ominous and supernatural "handwriting on the wall" that appears in the passage concerning Belshazzar's feast reads: "Mene, mene, tekel, upharsin," which is usually translated as "Thou art weighed in the balance and found wanting." But the Aramaic words could also mean, "Numbered, numbered, weighed and divisions," or the names of weights, "A mina, a mina, a shekel and half-shekels."

The Aramaic dictionary may help clear up such ambiguities and provide improved translations of such archaeological material as the Dead Sea Scrolls, papyrus scrolls written by scribes of a Jewish sect at about the time of Jesus.

Paradoxically, the new scholarship in Aramaic is unlikely to improve understanding of early Christian texts. Although Jesus is presumed to have spoken Aramaic, the gospels of the Christian Bible were recorded in Greek translations, and nearly all of the original Aramaic works were lost.

Two other great dictionary projects are nearing completion after many years' work. The oldest of these, the University of Chicago Assyrian dictionary project, began in 1921. Dr. Erica Reiner, who has served as editor of the work for the last thirty-six years, said the dictionary of Assyrian, also known as Akkadian, is now virtually finished, with about fifteen thousand words. From the project's inception, new tablets written in cuneiform, or with a wedge-shaped wooden stylus, have continued to be uncovered by archaeologists, and their words have been incorporated in the dictionary.

Another vast project that is nearing completion is the University of Chicago's Hittite dictionary, which has been in preparation since the mid-1930s. The Hittite language was spoken and recorded in cuneiform writing on clay tablets between 2000 B.C. and 1200 B.C. The Hittite empire, with Egypt and Mesopotamia, was one of the greatest Middle Eastern powers during the second millennium B.C.

Work on the third of the comprehensive dictionaries, the University of Pennsylvania's Sumerian dictionary, has barely begun. Scholars started the project in 1976 and published the first section, the letter B, in 1984. They expect to publish a second and longer section covering the letter A this year. With luck, the entire eighteen-volume Sumerian dictionary will be completed in twenty-five years, said Dr. Ake W. Sjoberg, director of the project.

The popularity of the Sumerian dictionary, even in fragmentary form, has surprised its editors. The 750 copies printed of the letter B, priced at $40 each, were sold almost immediately, and the University of Pennsylvania group has had to publish successive editions to keep up with demand. Brisk sales and backing by the National Endowment for the Humanities have helped pay for continuing research. Sumerian is the oldest known written language. The University of Pennsylvania owns roughly 60 percent of all known Sumerian literary texts, including the earliest known account of the biblical flood. Some of these texts date from as early as 3200 B.C.

The language is also an enigma, unrelated to any other known. Luckily for scholars, Sumer became bilingual in Sumerian and Akkadian before the death of the civilization, which was centered in Mesopotamia, and many texts are written in both languages. Most Sumerian tablets are fragmentary, and a large part of current research is devoted to assembling completed texts from fragments that may be half a world apart. "Many of the texts we study are the legal and commercial records kept by scribes and accountants," Dr. Sjoberg explains. "In common with the cash-register slips supermarkets hand out, they have to be placed in context to be understood. This is one of the ways in which computers are especially useful to our work. They help us sort out dates, names and other common features that can link together the separated parts of single documents."

Human memory also plays an important role. Dr. Bendt Alster, one of the scholars on the team, can often remember the ragged outlines of tablet fragments he has seen in photographs or the collections of other museums. He sometimes realizes that a fragment he is examining would fit together with some fragment he knows to be in the Louvre, the British Museum, or one of the great museums of Istanbul or Baghdad. He says this enables him to assemble texts as if they were jigsaw puzzles, clearing up mysteries that had defied scholars for years.

A large proportion of the words and phrases in the new dictionaries of ancient languages deal with earthy subjects. Under the letter *B* in the Sumerian dictionary, more than a full page is devoted to the word *bi*, which can mean excrement, anus, or any of a variety of related words. Many examples of the word's use are included, such as "gu sa-sur bi nu-ha-za sila-a KU," meaning "an ox with diarrhea leaves a long trail of dung." Although the four languages covered by current dictionary projects differ from each other, overlapping influences and times interrelate them. "We are all building on the work of each other," Dr. Sjoberg says, "and together we are uncovering the roots of Western civilization." [MWB]

DIE HARD

Discoveries on Cyprus have provided a "snapshot" of everyday life in the late Roman Empire. Archaeologists from the University of Arizona have found the skeletons of two men who must have been killed instantly by the earthquake in the year 365 that destroyed the Roman city of Kourion in one of the greatest disasters of antiquity. Seven other skeletons and a wealth of artifacts were found there in earlier excavations.

One of the skeletons was of a man in his early twenties who was apparently alone at a workbench when the roof caved in on him. Coins and ceramic pots lay about. In the next room of what was probably a row of live-in shops was the skeleton of another man. A bronze pot was sitting on an open oven. A finely fashioned bronze oil lamp was lying at the doorway.

"At this rate, we estimate that at least five hundred people must have been killed," said David Soren, a University of Arizona archaeologist. "In every place where people might have lived, we're finding skeletons."

CULTURE SHOCK

IN the traditional tales of the Maori of New Zealand, their ancestors arrived in 1350 in seven magnificent canoes after a heroic migration from the distant islands of Polynesia. They believed in a supreme being known as Io. And so, ever after, the legend of the Great Fleet and the cult of Io have been centerpieces of Maori culture. This story of Maori origins is now the subject of heated controversy, a scholarly echo of colonialism with a twist.

Anthropologists have determined that the tradition was more an invention of European anthropologists than an authentic heritage handed on by Maoris from the past. But today's Maoris accept the tradition as historical fact—never mind its dubious provenance—and angrily resist any revisionist assaults on their revered culture by foreign anthropologists. The Maoris argue that anthropologists may have created and imposed this culture but it is theirs now, they are proud of it, and so let them believe what they want to.

The episode and others like it raise some troubling questions for anthropology. Is there such a thing as unchanging tradition in societies? Is tradition largely fixed and static or is it forever in flux? With what certainty can anthropologists identify the principal characteristics of a culture? These questions strike at the very heart of anthropology. The study of culture—how a distinct set of myths, rituals, and living practices defines a people—has been fundamental to most anthropological research. For decades anthropologists strived in their fieldwork and writings to make their cultural studies a more rigorous science. But the profession is now rent with self-criticism based on a growing skepticism about assumptions of the very reality of the concept of culture.

"It's a watershed moment," says Dr. George E. Marcus, an anthropologist at Rice University who is a leader of the skeptics. "We're not talking about the breakup of anthropology, but the reconception of its central concepts." Dr. Clifford Geertz, an anthropologist at the Institute for Advanced Study in Princeton, New Jersey, says: "There's much more of a concern over how anthropologists know what they say they know about a culture. There's a lessening of confidence in being able to model anthropology, or any human science, on the natural sciences."

More conventional anthropologists, while accepting some of the new criticism as valid, fear that it overlooks solid accomplishments and will have a paralyzing effect on the profession. Cultures may not be the timeless phenomena they were once thought to be, but scholars insist that there is a range of knowledge about societies that can be documented as fact and that certain ideologies underlie the way a society thinks and acts. "There's a danger we will do too much of this navel watching," warns Dr. Stanley J. Tambiah, an anthropologist at Harvard University. "I'm not willing to say that there is no reality out there for us to observe and represent. We have put together a certain fund of knowledge that cannot be dismissed."

A new assessment of the Maori experience at the hands of colonial powers and their anthropologists points up the problems associated with the controversy. In a study published in the journal *American Anthropologist,* Dr. Allan Hanson, a professor of anthropology at the University of Kansas, said the accepted Maori tradition was a "cultural invention" of the Europeans that "raises fundamental questions about the nature of cultural reality and whether the information that anthropologists produce can possibly qualify as knowledge about that reality."

Early this century, anthropologists in New Zealand constructed the Maori tradition out of fragments of art, oral legend, and some manuscripts of questionable authenticity. A British scholar, S. Percy Smith, was instrumental in developing a single historical account about the migration in seven canoes and how most Maori tribes trace their origins to one or another of the canoes. The Europeans, Dr. Hanson and other modern scholars argue, had several reasons for fostering this invented tradition. In part, they were influenced by the now-discredited anthropological theory of diffusionism and long-distance migration. Many scholars of the day sought to trace the various primitive peoples being encountered back to a few cradles of civilization. Theory had a way of coloring observations. Long before the Great Fleet invention, the first missionary to visit New Zealand, in 1819, suggested that the Maoris had "sprung from some dispersed Jews." Later, it was theorized, based on supposed linguistic affinities, that they were of Aryan stock.

Such notions arose, consciously or subconsciously, because of scholarly preconceptions and an eagerness to assimilate the indigenous Maoris into the European culture being planted in New Zealand. Accordingly, Dr. Hanson says, Europeans "were pleased to discover in the Maori race the capacity for sophisticated philosophy, as demonstrated by the Io cult, and a history of heroic discoveries and migrations that included the Great Fleet." This seemed to ennoble Maoris in European eyes to the point, he adds, "where it became possible to entertain the possibility of a link with themselves."

There was a grain of truth in the tradition. The Maoris did have stories of their ancestors arriving in canoes in the indeterminate past. In 1950, however, Te Rangi Hiroa, an anthropologist who is half Maori, observed that the god Io's creative activities—bringing forth light from primordial darkness, dividing the waters, and forming the earth—had too much in common with Genesis to be convincing as a purely Maori tradition. Both European and Maori scholars in recent years have discredited the tradition's authenticity. But the Maoris had their own reasons for clinging fervently to "their" tradition.

They accepted the invented heritage not as a step toward assimilation into the European culture but to bolster a sense of their own ethnic distinctiveness and value. "This sense has grown dramatically in strength and stridency of expression in recent years," Dr. Hanson reported. Other anthropologists acknowledge Dr. Hanson's expertise in Maori studies and accept his findings. Indeed, they cite many other examples of invented culture, particularly by colonial scholars and explorers who thought they understood how certain tribal patterns should be. The tribal people, powerless, then had to play by those rules.

The land-tenure system in the Fiji Islands, for example, is the product more of anthropological interpretations than of actual practices in the dim past; yet Fijians would not hear of changing it, anthropologists say. "We're finding that happened all over the place," says Dr. Richard Handler, an anthropologist at the University of Virginia.

It even turns out that the Scottish kilt hardly sprang from primeval heather; it was an invention by the English, of all people, in the eighteenth century, several anthropologists say.

Dr. James Clifton, an anthropologist at the University of Wisconsin at Green Bay, has recently completed a book called *The Invented Indian,* which makes the case that the image of the American Indian is largely a creation of white Americans, beginning with the romantic tale of Squanto teaching the Pilgrims to plant corn and thus inspiring the first Thanksgiving day. And Dr. Handler is studying the implications of historical restorations at places like Williamsburg, Virginia, where, he said, "people are reinventing what they consider to be their culture."

A recognition that culture is not a timeless, changeless reality has led in the last decade to a critical ferment among anthropologists. Dr. Marcus calls it a "thorough critique of the discipline—not an intellectual movement with a new paradigm at its center, but a critique calling for a redesign, a reconceptualization of the way anthropology studies culture." Early anthropology, Dr. Handler notes, came out of the romantic tradition in which "alienated Westerners or moderns were searching for the pristine and original cultures, they were attracted by the romance of the primitive and unspoiled, the Noble Savage." After World War II, anthropologists took a more scientific approach, searching for general principles of culture that might correspond to the laws of other sciences. But still the assumption was that there existed some primeval tradition for a society that was the changeless essence of the people.

In today's revisionist criticism of anthropology, called postmodernism, everything is in a state of flux. Tradition is not the same from one time to the next. It is always being invented and re-invented, from within a society or by outside forces. A few anthropologists suspect that postmodernism is no more than a buzzword and that the revisionist criticism could be only an intellectual fad. "It is a minority, but a very vocal and very talented minority," Dr. Tambiah says of the critics. The controversy is not strictly a generational matter, though more of the critics are young anthropologists. Dr. Handler says the new thinking has gained a wide following in the last five years and "is almost mainstream."

Three recent books are considered influential presentations of the new thinking. They are *Anthropology as Cultural Critique* by Dr. Marcus and Dr. Michael Fischer; *Writing Culture: The Poetics and Politics of Ethnography*, edited by Dr. James Clifford and Dr. Marcus; and *The Predicament of Culture* by Dr. Clifford. Anthropologists hailed Dr. Hanson's new interpretation of Maori tradition as a clear expression of the new critique. Dr. Marcus says it represented the "positive element of the critique that does not just leave everything in ruins."

As Dr. Hanson wrote: "The fact that culture is an invention, and anthropology one of the inventing agents, should not engender suspicion or despair that anthropological accounts do not qualify as knowledge about cultural reality. It follows that the analytical task is not to strip away the invented portions of culture as inauthentic, but to understand the process by which they acquire authenticity."

Dr. Hanson argued that the Great Fleet and cult of Io, though invented by Europeans, represented authentic Maori tradition because, the source notwithstanding, it is believed by the society. For anthropologists, Dr. Marcus said in commenting on the Maori study, it is important to appreciate that culture is not something that weathers the vicissitudes of history but is something created by history. [JNW]

THE FIRST AFRICAN
AMERICANS

DECORATED clay tobacco pipes, excavated in Virginia and Maryland and long thought to be Indian artifacts, have now been identified as the earliest preserved examples of craftsmanship by black people in North America. The discovery is considered a rich addition to black history and important evidence in understanding the close working and living relationship of blacks and whites on the English plantations of seventeenth-century America before slavery became fully and rigidly established. In those days, according to historical interpretations emerging from recent archaeological research, the few blacks in the colonies worked as indentured servants—bound by contract and not necessarily for life—and usually lived in the plantation house with whites. Not until the last years of the century, with the institutionalization of slavery based on race, were they moved to slave quarters and used primarily as field workers.

"The pipes reflect close contact and cooperative craftsmanship between Africans and English on seventeenth-century plantations," says Dr. Matthew Emerson, an anthropologist at the University of California at Berkeley, who identified the pipe decorations as African. "They are fashioned in European form, but decorated in a West African art style." The finding is thus seen as an illuminating contribution of archaeology to the social and economic history of a period when written documents give little or no account of the lives of ordinary people, particularly blacks.

Dr. Emerson identified the decorations by comparing the style and workmanship of hundreds of the pipes with that of contemporary West African pottery. The resemblance was striking and not likely to be a coincidence. One typical example is the *kwardata* motif, a diamond shape on a banded background. Another is a six-pointed star, with three tiny circles around the tip of each point. Dr. Emerson says he found these to be distinctive images widely used on pottery in Nigeria in the seventeenth and eighteenth centuries, but not in European or American Indian work of the time. The *kwardata* motif was still being applied to pots made in the early twentieth century by the Ga'anda people of north-central Nigeria.

Moreover, Dr. Emerson says, the pipe makers apparently used the same techniques as the African potters. In both cases the designs were inscribed in the clay with toothed instruments that left dotted lines, and they were highlighted with white clay rubbed into the lines. The only thing European about the clay pipes was their shape. They were presumably made on molds imported from Europe or on molds produced in the colonies from European designs. The stems were usually six and a half inches long and the bowls about two inches tall. Archaeological evidence, based primarily on other materials found with the pipes, indicates that they were made between 1650 and the end of the century.

Many of the pipes were discovered over the last eighty years at scattered sites near the Chesapeake Bay in Virginia and Maryland, and others were found recently by Dr. Emerson while he was working on his doctoral dissertation. These were uncovered as part of excavations at the Flowerdew Hundred plantation, on the south bank of the James River, midway between Richmond and Williamsburg. It is one of the earliest plantation sites in Virginia. Dr. James Deetz, professor of anthropology at Berkeley and leader of archaeological research at the Flowerdew Hundred site, believes the pipes were previously misidentified because they were primitive and everything primitive from that period was assumed to be Indian. "It was an uncritical assumption," he says, especially considering that few Indians remained in that area in the last half of the seventeenth century.

Although historical records document the arrival of the first blacks in British North America in 1619 at Jamestown, the pipes were apparently the first material evidence of black life on the continent. Some blacks were also present with the Spanish in the early years of St. Augustine, founded in Florida in 1565. They are described in documents as carpenters, ironsmiths, and builders of fortifications. Recent excavations at St. Augustine, directed by Dr. Kathleen Deagan, an archaeologist at the Florida State Museum, have uncovered the site of a fort built and occupied by blacks, but have yet to find any artifacts whose designs are distinctively African.

The pipe findings reinforce other research into the transformation of Virginia plantation life with the introduction of slavery laws in 1665. Of the first twenty-five blacks to arrive at Jamestown, brought by a Dutch man-of-war, fifteen went to work as indentured servants of Sir George Yardley, Virginia's first governor, who owned the thousand-acre Flowerdew Hundred plantation. Historical records indicate that they lived in the white household and were sometimes able to negotiate their freedom. In time, some blacks even had whites as their own indentured servants. Some scholars argue that for a time white Virginians seemed to be ready to accept blacks as full members or potential members of the community.

After a study of archaeological findings, Dr. Dell Upton, an architectural historian at Berkeley, concluded that separate slave quarters did not typically appear on the plantations until the last two decades of the seventeenth century. This, he says, was the most dramatic physical manifestation of the transformation of plantation life brought about by slavery. Dr. Emerson says the African-style pipes disappeared soon afterward. The blacks, he surmises, no longer had access to the pipe-making molds at the house and were occupied almost full time in field work. Also, with slavery came the greater prosperity made possible by large-scale tobacco harvests, and white planters began importing more goods, including European-made pipes. In the new social order, the planters became more paternalistic and provided the slaves with many of their needs, also including imported pipes.

Likewise, Dr. Deetz says, the change in plantation life can be seen in the appearance, toward the end of the 1600s, of locally produced, unglazed pottery in a variety of European shapes. Here, too, archaeologists originally assumed that these pottery remains were Indian artifacts. Writing in the journal *Science*, Dr. Deetz said most scholars now agree that the pottery, which archaeologists call Colono ware, was made and used by black slaves. When they lived in the plantation house, they had had no need for their own housewares. Smoking pipes, unglazed pottery, and the buried foundations of early slave quarters—these are the humble pieces of evidence from archaeology that are being used to flesh out life at a crucial turning point in colonial America. [JNW]

Understanding Human Behavior

HOW YOUR SEX AFFECTS
YOUR BRAIN

RESEARCHERS who study the brain have discovered that it differs anatomically in men and women in ways that may underlie differences in mental abilities. The findings, although based on small-scale studies and still very preliminary, are potentially of great significance. If there are subtle differences in anatomical structure between men's and women's brains, it would help explain why women recover more quickly and more often from certain kinds of brain damage than do men, and perhaps help guide treatment. The findings could also aid scientists in understanding why more boys than girls have problems like dyslexia, and why women on average have superior verbal abilities to men. Researchers have not yet found anything to explain the tendency of men to do better on tasks involving spatial relationships.

The new findings are emerging from the growing field of the neuropsychology of sex differences. Research on sex differences in the brain has been a controversial topic, almost taboo for a time. Some feminists fear that any differences in brain structure found might be used against women by those who would cite the differences to explain "deficiencies" that are actually due to social bias. And some researchers feel it is important to affirm that many differences in mental abilities are simply due to environmental influences, such as girls being discouraged from taking math seriously.

The new research is producing a complex picture of the brain in which differences in anatomical structure seem to lead to advantages in performance on certain mental tasks. The researchers emphasize, however, that it is not at all clear that education or experience do not override what differences in brain structure contribute to the normal variation in abilities. Moreover, they note that the brains of men and women are far more similar than different. Still, in the most significant new findings, researchers are reporting that parts of the corpus callosum, the fibers that connect the left and right hemispheres of the brain, are larger in women than in men. The finding is surprising because, overall, male brains—including the corpus callosum as a whole—are larger than those of females, presumably because men tend to be bigger on average than women.

Because the corpus callosum ties together so many parts of the brain, a difference there suggests far more widespread disparities between men and women in the anatomical structure of other parts of the brain. "This anatomical difference is probably just the tip of the iceberg," says Sandra Witelson, a neuropsychologist at McMaster University medical school in Hamilton, Ontario, who did the study. "It probably reflects differences in many parts of the brain which we have not yet even gotten a glimpse of. The anatomy of men's and women's brains may be far more different than we suspect."

The part of the brain that Dr. Witelson discovered is larger in women is in the isthmus, a narrow part of the callosum toward the back. Dr. Witelson's findings on the isthmus are based on studies of fifty brains, fifteen male and thirty-five female. The brains examined were of patients who had been given routine neuropsychological tests before they died.

"Witelson's findings are potentially quite important, but it's not clear what they mean," says Bruce McEwen, a neuroscientist at Rockefeller University. "In the brain, bigger doesn't always mean better."

In 1982 a different area of the corpus callosum, the splenium, was reported by researchers to be larger in women than in men. But that study was based on only fourteen brains, five of which were female. Since then, some researchers, including Dr. Witelson, have failed to find the reported difference, while others have. Since such differences in brain structure can be subtle and vary greatly from person to person, it can take the close examination of hundreds of brains before neuroanatomists are convinced. But other neuroscientists say the findings are convincing enough to encourage them to do tests of their own.

Both the splenium and the isthmus are located toward the rear of the corpus callosum. This part of the corpus callosum ties together the cortical areas on each side of the brain that control some aspects of speech, such as the comprehension of spoken language, and the perception of spatial relationships. "The isthmus connects the verbal and spatial centers on the right and left hemispheres, sending information both ways—it's a two-way highway," Dr. Witelson says. The larger isthmus in women is thought to be related to women's superiority on some tests of verbal intelligence. It is unclear what, if anything, the isthmus might have to do with the advantage of men on tests of spatial relations.

The small differences in abilities between the sexes have long puzzled researchers. On examinations like the Scholastic Aptitude Test, which measures overall verbal and mathematical abilities, sex differences in scores have been declining. But for certain specific abilities, the sex differences are still notable, researchers say. While these differences are still the subject of intense controversy, most researchers agree that women generally show advantages over men in certain verbal abilities. For instance, on average, girls begin to speak earlier than boys, and women are more fluent with words than men, making fewer mistakes in grammar and pronunciation. On the other hand, men, on average, tend to be better than women on certain spatial tasks, such as drawing maps of places they have been and rotating imagined geometric images in their mind's eye—a skill useful in mathematics, engineering, and architecture. Of course, the advantages for each sex are only on average. There are individual men who do as well as the best women on verbal tests, and women who do as well as the best men on spatial tasks.

One of the first studies that directly link the relatively larger parts of women's corpus callosums to superior verbal abilities was reported by Melissa Hines, a neuropsychologist at the University of California at Los Angeles (UCLA) Medical School. Dr. Hines and her associates used magnetic resonance imaging, a method that uses electrical fields generated by the brain, to measure the brain anatomy of twenty-nine women. They found that the larger the splenium in the women, the better they were on tests of verbal fluency. There was no relationship, however, between the size of their splenium and their scores on tests of spatial abilities, suggesting that differences in those abilities are related to anatomical structures in some other part of the brain or have nothing to do with anatomy. "The size of the splenium," Dr. Hines says, "may provide an anatomical basis for increased communication between the hemispheres, and perhaps as a consequence, increased language abilities."

Researchers now speculate that the larger portions of the corpus callosum in women may allow for stronger connections between the parts of women's brains that are involved in speech than is true for men. "Although we are not sure what a bigger overall isthmus means in terms of microscopic brain structure, it does suggest greater inter-hemispheric communication in women," Dr. Witelson says. "But if it does have something to do with the cognitive differences between the sexes, it will certainly turn out to be a complex story."

Part of that complexity has to do with explaining why, despite the bigger isthmus, women tend to do less well than men in spatial abilities, even though the isthmus connects the brain's spatial centers, too. "Bigger isn't necessarily better, but it certainly means that it's different," Dr. Witelson says.

A variety of other differences in the brain have been detected by the researchers in their recent studies. For instance, Dr. Witelson found in her study that left-handed men had a bigger isthmus than did right-handed men. For women, though, there was no relationship between hand preference and isthmus size. "How our brains do the same thing, namely use the right hand, may differ between the sexes," Dr. Witelson says. She also found that the overall size of the corpus callosum, particularly the front part, decreases between forty and seventy years of age in men, but remains the same in women.

Several converging lines of evidence from other studies suggest that the brain centers for language are more centralized in men than in women. One study involved cerebral blood flow, which was measured while men and women listened to words that earphones directed to one ear or the other. The research, conducted by Cecile Naylor, a neuropsychologist at Bowman Gray School of Medicine in Winston-Salem, North Carolina, showed that the speech centers in women's brains were connected to more areas both within and between each hemisphere. This puts men at a relative disadvantage in recovering from certain kinds of brain damage, such as strokes, when they cause lesions in the speech centers on the left side of the brain. Women with similar lesions, by contrast, are better able to recover speech abilities, perhaps because stronger connections between the hemispheres allow them to compensate more readily for damage on the left side of the brain by relying on similar speech centers on the right. Roger Gorski, a neuroscientist at UCLA, reported finding that parts of the hypothalamus are significantly bigger in male rats than in female ones, even though the size of the overall brain is the same in both sexes. And Dr. McEwen, working with colleagues at Rockefeller University, has found a sex difference in the structure of neurons in part of the hippocampus that relays messages from areas of the cortex.

Dr. McEwen, working with rats' brains, found that females have more branches on their dendrites, which receive chemical messages from other neurons, than do males. Males, on the other hand, have more spines on their dendrites, which also receive messages from other neurons. These differences in structure may mean differing patterns of electrical activity during brain function. "We were surprised to find any difference at all, and, frankly, don't understand the implications for differences in brain function," Dr. McEwen says. "But we'd expect to find the same differences in humans; across the board, findings in rodents have had corollaries in the human brain." [DG]

LOOK-ALIKE COUPLES

Science is lending support to the old belief that married couples eventually begin to look alike. Couples who originally bore no particular resemblance to each other when first married had, after twenty-five years of marriage, come to resemble each other, although the resemblance may be subtle, according to a research report.

Moreover, the more marital happiness a couple reported, the greater their increase in facial resemblance. The increase in facial similarity results from decades of shared emotions, according to Robert Zajonc, a psychologist at the University of Michigan, who did the research.

In the study, people were presented a random array of photographs of faces, with the backgrounds blacked out, and were instructed to match the men with the women who most closely resembled them. Two dozen of the photographs were of couples when first married; another two dozen were of the same couples twenty-five years later, most taken around the time of their silver wedding anniversary.

All the couples in the photographs were white, lived in Michigan or Wisconsin, and were between fifty and sixty years old at the time of the second picture. The young couples showed only a chance similarity to each other, the study found, while the judges found a definite resemblance between the couples who had been married a quarter-century. While the resemblances were not dramatic (some seemed to involve subtle shifts in facial wrinkles and other facial contours, for instance), they were marked enough that the judges were able to match husbands and wives far more often when the couples were older than when they were younger. And the resemblances were greater in some couples than in others, the study found.

Dr. Zajonc (pronounced ZI-onz), in explaining the findings, holds that factors such as similar diets—and thus deposits of fatty tissues—may contribute to the resemblance but are not crucial: when the photographs were evaluated for facial fat, the older couples were found to have less similarity than the younger ones. Instead, he proposes, people often unconsciously mimic the facial expressions of their spouses in a silent empathy and that, over the years, sharing the same expressions shapes the face similarly.

THE WAR BETWEEN THE SEXES

IN the war between the sexes, virtually all combatants consider themselves experts on the causes of conflict. But now a systematic research project has defined, more precisely than ever before, the points of conflict that arise between men and women in a wide range of relationships. The things that anger men about women, and women about men, are just about the same whether the couple are only dating, are newlyweds, or are unhappily married. Although a vast body of literature cites heated arguments over money, child-rearing, or relatives as frequent factors in disintegrating marriages, those conflicts seldom emerge in the new studies.

Instead, the research often found more subtle differences, like women's feelings of being neglected and men's irritations over women's being too self-absorbed. There were also more pointed complaints about men's condescension and women's moodiness. Some forms of behavior bothered both sexes about equally. Both men and women were deeply upset by unfaithfulness and physical or verbal abuse. But the most interesting findings were several marked differences between men and women in the behaviors that most disturbed them. Sex, not surprisingly, was a major problem, but men and women had diametrically opposed views of what the problem was. Men complained strongly that women too often turned down their sexual overtures. In contrast, the most consistent complaint among women was that men were too aggressive sexually. This conflict may be rooted deep in the impact of human evolution on reproductive strategies, according to one theory, or it may simply reflect current power struggles or psychological needs.

Understanding the sources of trouble between the sexes, psychologists say, could do much to help couples soften the impact of persistent problems in their relationships and help therapists in counseling couples having difficulties. "Little empirical work has been done on precisely what men and women do that leads to conflict," says David M. Buss, a psychologist at the University of Michigan who conducted the studies. The results were published in the *Journal of Personality and Social Psychology*.

Dr. Buss conducted four different studies with nearly 600 men and women. In the first, he simply asked men and women in dating relationships about the things their partners did that made them upset, hurt, or angry. The survey yielded 147 distinct sources of conflict, ranging from being disheveled or insulting to flirting with others or forcing sex on a partner.

In the second study, Dr. Buss asked men and women who were dating or who were newlyweds how often they had been irked by their partner's doing any of those things. From these results, Dr. Buss determined that the complaints fell into fifteen specific groups. He then had another group of men and women rate just how bothersome those traits were. Men said they were most troubled by women who were unfaithful, abusive, self-centered, condescending, sexually withholding, neglectful, or moody. Many men were bothered, for example, if their partner was self-absorbed with her appearance, spending too much money on clothes and being overly concerned with how her face and hair looked.

Women complained most about men who were sexually aggressive, unfaithful, abusive, condescending, emotionally constricted, and by those who insulted the woman's appearance, neglected them, or openly admired other women. Many women were also bothered by inconsiderate men. For instance, they complained about a man who teases his partner about how long it takes to get dressed, who does not help clean up the home, or who leaves the toilet seat up.

Other research has produced supporting findings. "We've seen similar points of conflicts in marital fights," says John Gottman, a psychologist at the University of Washington whose research involves observations of married couples while they fight. "Many of these complaints seem to be due to basic differences in outlook between the sexes." He cites men's complaints that women are too moody or that they dwell too much on their feelings. "That is the flip side of one of women's biggest complaints about men, that they're too emotionally constricted, too quick to offer an action solution to an emotional problem," he says.

"Generally for women, the natural way to deal with emotions is to explore them, to stay with them," Dr. Gottman says. "Men, though, are stoic in discussing their emotions; they don't talk about their feelings as readily as women. So conflict over handling emotions is almost inevitable, especially in marriages that are going bad."

For couples in the first year of marriage, Dr. Buss found, the sexual issues were far less of a problem than for most other couples. Instead, women tended to complain that their new husbands were inconsiderate and disheveled. "You'd expect that sex would be the least troubling issue for a couple during the honeymoon year," Dr. Buss says. "Even so, newlywed men were still bothered somewhat about their wives' sexual withholding, but the wives didn't complain much about their husbands being sexually aggressive."

Nevertheless, the overall finding that men tend to see women as being sexually withholding, while women see men as too demanding, also fits with other findings. Researchers at the University of New York at Stony Brook found in a survey of close to a hundred married couples that the husbands on average wanted to have sex more often than did the wives. In the last of Dr. Buss's series of tests, married men and women were asked about their main sources of marital and sexual dissatisfaction. A new set of complaints emerged, along with the previous ones cited by dating couples and newlyweds. The more dissatisfied with the relationship, the longer the list of complaints. For example, the more troubled the couple, the more likely the husband was angered by his wife's being too possessive, neglecting him, and openly admiring other men.

The dissatisfied wives, by contrast, added to their list of complaints that their husbands were possessive, moody, and were openly attracted to other women.

Sex was especially problematic for unhappily married men and women. "The sexual complaints are standard in unhappy marriages," Dr. Gottman says. "But it tends to crop up even in otherwise happy marriages. Generally, women have more prerequisites for sex than men do. They have more expectations about what makes lovemaking OK. They want emotional closeness, warmth, conversation, a sense of empathy. Sex has a different meaning for women than for men. Women see sex as following from emotional intimacy, while men see sex itself as a road to intimacy. So it follows that men should complain more that women are withholding, or women say men are too aggressive."

Dr. Buss sees his results as affirming the importance of evolution in shaping human behavior. "The evolutionary model that I use holds that conflicts occur when one sex does something that interferes with the other's strategy for reproduction," he says. His view is based on the theory put forth by Robert Trivers, a social scientist at the University of California at Santa Cruz, who proposed that women are more discriminating than men about their sexual partners because biologically women have to invest more time and energy in reproduction than do men.

Men on the other hand stand to gain in terms of reproductive success by having sexual relations with as many women as possible, the theory holds. "To some degree the sexes are inevitably at odds, given the differences in their strategies," Dr. Buss says. Even so, he does not see evolution as explaining all his findings: "The sources of conflict between men and women are much more diverse than I predicted."

Some of that diversity may be caused by sex roles. "Men and women are socialized differently as children," says Nancy Cantor, a social psychologist at the University of Michigan. "Men, for example, are not expected to be as open with their emotions as women, while women expected to be less aggressive than men. So you'd expect a list very much like he found."

Dr. Gottman offers another explanation. "The categories sound like much of what we see in couples' fights. But they miss what underlies all that: whether people feel loved and respected. Those are the two most important dimensions in marital happiness." [DG]

THE SCIENCE OF COURTSHIP

The flippant lines that some men use to impress women may actually ruin their chances of landing a date, according to research by Dr. Michael Cunningham, a psychologist at the University of Louisville in Kentucky. But women can say just about anything and men will keep talking to them, his studies found.

In two of the studies, Dr. Cunningham recruited three male and two female undergraduates to go to bars and randomly approach unaccompanied people of the opposite sex and similar age. The undergraduates were told what opening line to use. One variety of opener was "cute/flippant," such as "Bet I can outdrink you." Another category was the direct approach, "I'm a little embarrassed, but I'd really like to meet you." The third type of opening line was innocuous, such as "What do you think of the band?" or a simple "Hi."

The men found that the type of line they used greatly affected their chances of starting a conversation. Only about 20 percent of those who used cute/flippant approaches got a positive response, such as a smile or a few words. In the other encounters, the women turned away or asked the men to leave. In contrast, the innocuous and direct approaches had 50 to 80 percent success rates.

But the women got a positive response at least 80 percent of the time, no matter what kind of line they used. In fact, when the women coupled a "Hi" with a light touch on the man's arm, every man they approached reacted positively.

In a follow-up laboratory study, Dr. Cunningham found that women looked for sociability and intelligence; they did not consider flippant come-ons a sign of intelligence. Men tended to play down intelligence as a criterion for initial conversation, focusing instead on the woman's sexual attractiveness.

THE NEED TO LIE

WHILE recognizing that lying is a universal lubricant of social life, psychiatrists are seeking to determine when it becomes destructive and just which kinds of mental problems it can typify. An article in the *American Journal of Psychiatry* in 1988 was an attempt to call attention to the general neglect of lying as a topic for psychiatric research.

Psychologists who are studying how and why children learn to lie are finding that certain lies play positive roles in a child's emotional development. A child's first successful lie, for instance, is seen by some researchers as a positive milestone in mental growth. While the recent research sees lying that does damage as a matter for concern, it is pragmatic in taking the occurrence of lying in social life for granted. One study, for instance, found that, on average, adults lie—or admit to doing so—thirteen times a week. "Lying is as much a part of normal growth and development as telling the truth," says Arnold Goldberg, a professor of psychiatry at Rush Medical College in Chicago. "The ability to lie is a human achievement, one of those abilities that tend to set them apart from all other species."

Psychiatrists see lying as pathological when it is so persistent as to be destructive to the liar's life or to those to whom he lies. The most blatant lying is found in the condition called "pseudologia fantastica," in which a person concocts a stream of fictitious tales about his past, many with a small kernel of truth, all self-aggrandizing. "One patient blithely told me that he spoke his first complete sentence at three months, at three years gave sermons to crowds at his church work, and had a job at a news magazine where he made $8 million a week," says Bryan King, a psychiatrist at the University of California at Los Angeles School of Medicine. Dr. King wrote the article on pathological lying, published in the *American Journal of Psychiatry*, with Charles Ford, a psychiatrist at the University of Arkansas Medical School, and Marc Hollender, a psychiatrist at Vanderbilt University School of Medicine. "Pathological liars seem utterly sincere about their lies, but if confronted with facts to the contrary, will often just as sincerely reverse their story," Dr. King says. "Their stories have a believable consistency, but they just do not seem able to monitor whether they are telling the truth or not."

Research suggests that this most extreme form of lying is associated with a specific neurological pattern: a minor memory deficit combined with impairment in the frontal lobes, which critically evaluate information. In such cases the person suffers from the inability to assess the accuracy of what he says and so can tell lies as though they were true.

Not all pathological lying stems from such neurological difficulties. Psychiatrists are also grappling with lies that typify certain emotional disorders and are told by people who know they are lying. Dr. King's article describes five varieties of lies, each of which comes more naturally to those who suffer from one or another of five common personality problems. While such lies could be told by anyone, they are far more likely in those with the following personality problems because each kind of lie springs from the pressing psychological needs at the core of the disorders:

- Manipulative lies are the hallmark of the sociopath, or "antisocial personality," who is driven by utterly selfish motives. Such people are not necessarily criminals; they may gravitate toward the fringes of trades like sales, where their bent toward lying may serve them well. Since sociopaths feel no remorse or empathy for their victims, they are capable of the most coldhearted of lies.

- Melodramatic lies that make him or her the center of attention are natural to the hysteric, or "histrionic personality." Such people are searching desperately for love. They are also more taken with emotional truths than with the facts of a situation. "Casual lies are to the hysteric what license is to the poet," according to Dr. King.

- Grandiose lies typify the narcissist, whose deep need to win the constant approval of others impels him to present himself in the most favorable light. Narcissists are prone to exaggerate their abilities or accomplishments in order to seem more impressive. Because narcissists feel entitled to special treatment—for instance, believing that ordinary rules do not apply to them—they can be reckless in their lies.

• Evasive lies are typical of the borderline personality, whose wildly vacillating moods and impulsive actions constantly get him into trouble. Many of the borderline personality's lies are told to avoid blame or shift responsibility for his problems to others.

• Guilty secrets account for many lies of the compulsive person, a type who generally is scrupulously honest. Compulsives pride themselves on following the rules and on attention to facts and details. But they also suffer from a fear of being shamed and hence lie to prevent other people from finding out about things they feel would meet with disapproval. Their lies are often mild, concerning things most others would find no cause for lying about; one man, for instance, lied to his wife to keep her from finding out about his being in therapy.

This research and the other recent studies of the topic were done in America and England; the psychologists do not know to what extent the findings apply to other cultures.

Along with the new focus on lying in psychiatric problems, there is intense research on the role of lying in normal development. Researchers feel they must first understand what is normal about lying before they can know what leads to pathological lying. Oddly, the research has resulted in an appreciation of the positive role lying plays in psychological growth, with some child experts seeing great significance in a child's first lie.

In this view, which is part of psychoanalytic "self" theory, the child's first lie, if successful, marks the initial experience that his parents are not all-knowing. And that realization, which usually occurs in the second year of life, is crucial to the child's development of the sense that he is a separate person with a will of his own, "that he can get away with things," Dr. Goldberg says. But that lie also is the beginning of the end of the idealization of one's parents that all infants feel. "The first time you see a limit to your parents' powers is a developmental step forward, toward a more realistic view of others," he says.

The ability to lie, in the view of other researchers, is a natural by-product of a child's psychological growth. "The crucial human skills are among those that equip children to lie: independence, intellectual talents, the abilities to plan ahead and take the other person's perspective, and the capacity to control your emotions," according to Paul Ekman, a psychologist at the University of California at San Francisco who wrote the book *Why Kids Lie.*

The years from two to four seem to mark a crucial period for children in mastering the art of the lie, according to studies by Michael Lewis, a psychologist at Rutgers Medical School: "In one study we've just completed with three-year-olds, we set up an attractive toy behind the child's back and tell him not to look at it while we leave the room. About 10 percent don't peek while we're gone. Of the rest, a third will admit they peeked, a third will lie and say they did not peek, and a third will refuse to say. Those who won't answer seem to represent a transition group, who are in the process of learning to lie, but don't do it well yet. They are visibly the most nervous. Those who say they did not look—who lie—looked the most relaxed. They've learned to lie well. There seems to be a certain relief in knowing how to lie effectively."

By and large, children lie for the same reasons adults do: to avoid punishment, get something they want, or make excuses for themselves. However, preteens usually have not learned to tell the white lies of adults, which work as social lubricants or to soothe another's feelings, researchers say. One of the more common kinds of lies for preteens is the boast, inventing or embellishing on one's deeds, which is meant to win the approval and admiration of one's peers. Grandiosity is frequent in children of this age, who may boast that they are able to do things like ski or speak a foreign language, when it is simply not true. "Children at that age are fine-tuning their superego, or conscience," says Dr. Goldberg. "The first evidence of pathological lying shows up during these years, in children who have a faulty superego and think they can get away with anything."

Sometime between the ages of ten and fourteen, most children become as capable as adults in their lies, according to Dr. Ekman: "If a child did not develop the abilities that allow him to lie, he would remain immature. The question is, will they lie, and if so, why?"

Adolescence marks another point in development where lying takes on a special psychological significance. "Adolescence is a time of a renewed search for ideals, when the child's ideals undergo a major transformation," says Dr. Goldberg. "The adolescent is seeking a model, a perfect person to emulate. It's much like the moment in infancy before they realized their parents' imperfections." This reassessment of values can lead the teen-ager to "a sense that he can do whatever he wants," Dr. Goldberg adds. "They start to test limits all over again, to see what they can get away with—and lying to parents can be a large part of that."

Although there are obvious problems in finding out exactly how common lying is, studies based on reports by parents and teachers suggest that about one in six of children lie more than occasionally. But more than that—up to one in four adults—will admit to having lied fairly frequently as children. In only about 3 percent of children, though, is lying so constant that it is a serious problem, according to Magda Stouthamer-Loeber, a psychologist at the Western Psychiatric Institute and Clinic in Pittsburgh.

Understandably, children who get into trouble frequently are also those who tend to lie the most. Thus children who are brought to clinics for mental health problems (mainly problems in conduct such as aggressiveness) are two and a half times as likely as other children to be chronic liars, according to a review of findings in the *Clinical Psychology Review* by Dr. Stouthamer-Loeber.

Studies have found that children who are chronic liars tend to get into more serious trouble as they grow older. For example, a British study found that 34 percent of boys who were rated by their parents or teachers as lying had criminal offenses fifteen years later. And in an American study of 466 men from the Cambridge, Massachusetts, area, those labeled as "liars" while they were in elementary school were significantly more likely than other boys to have had a conviction for a crime such as stealing by the time they reached their twenties. Still, researchers are uncertain just how much lying is a cause, and how much a symptom, of the problem. "We don't know if lying is a stepping stone that leads to maladjustment, a warning sign of later trouble, or just one feature of a larger problem," says Dr. Ekman. [DG]

THE TENACITY OF PREJUDICE

IN SEEKING to understand the tenacity of prejudice, researchers are turning away from an earlier focus on such extreme racism as that exhibited by members of the Ku Klux Klan to examine the pernicious stereotypes among people who do not consider themselves prejudiced. A troubling aspect of the problem, researchers find, is that many stereotypes seem to be helpful in organizing perceptions of the world. Recent studies of this cognitive aspect are proving useful in explaining the tenacity of prejudice as a distortion of that process. One finding is that people tend to seek and remember situations that reinforce stereotypes, while avoiding those that do not. Another troubling conclusion of the research is that simply putting people of different races together does not necessarily eliminate prejudice. For example, Walter Stephan, a psychologist at the University of Delaware, found in a review of eighteen studies of the effects of school desegregation that interracial hostilities rose more often than they decreased at desegregated schools.

Overt, admitted bigotry is on the decline, studies indicate. Yet they reveal that a more subtle form of prejudice, in which people disavow racist attitudes but nevertheless act with prejudice in some situations, is not declining. Such people justify prejudiced actions or attitudes with what they believe are rational, nonracist explanations. To those who have felt the sting of racial discrimination, the phenomenon is well known. An employer, for instance, may reject a black job applicant, ostensibly not because of his race but because the employer says he believes the person's education and experience are not quite right. Yet a white applicant with the same qualifications is hired.

Part of the difficulty in eradicating prejudice, even in those who intellectually view it as wrong, stems from its deep emotional roots. "The emotions of prejudice are formed early in childhood, while the beliefs that are used to justify it come later," says Thomas Pettigrew, a psychologist at the University of California at Santa Cruz and a noted scholar in the field. "Later in life you may want to change your prejudice, but it is far easier to change your intellectual beliefs than your deep feelings. Many southerners have confessed to me, for instance, that even though in their minds they no longer feel prejudice against blacks, they still feel squeamish when they shake hands with a black. The feelings are left over from what they learned in their families as children."

Psychoanalytic theories, too, point to the importance of childhood experience. "We distinguish between the familiar and the strange early in infancy," says Mortimer Ostow, a psychoanalyst and professor of pastoral psychiatry at the Jewish Theological Seminary in New York. "Then in childhood, when we join groups, we learn to draw boundaries between us and them. By adolescence the group identity becomes even more important, and out-groups become the place to deposit our own faults." The classic psychoanalytic literature on prejudice notes that a person's own sense of insecurity is often reflected in the need to find an out-group to despise, with the person's most loathed personality characteristics pushed onto someone else—thus, the "filthy" Jews or blacks, or Italians or whites. New work is adding to these theories.

Dr. Ostow and other psychoanalysts have studied people in treatment who explored their own anti-Semitic prejudices. "The inner dynamics are surprising," says Dr. Ostow. "We find that there almost always was a time in the past when the prejudiced person was attracted to the other group. The prejudice is a later repudiation of that earlier attraction." Often the attraction occurs in childhood or adolescence, according to Dr. Ostow. The child becomes fascinated by strangers, particularly by people in a group other than that of his own family. At the same time, though, the child may experience this as a betrayal of his family. The child then pulls back from the fascination, often after a rebuff or disappointment, or when he feels guilt at betraying his family. When the attraction happens later in life, the turning point is often rejection by a lover. "The prejudice that forms symbolizes a loyalty to home and its values," says Dr. Ostow. "But it is built on a deep ambivalence."

Much of the recent work on prejudice has moved from a psychoanalytic view to a cognitive one, in which most prejudice is seen as the by-product of the normal processes whereby people perceive and categorize one another. To a large extent the new work builds on that of Gordon Allport, the late Harvard University psychologist and author of the 1954 classic *The Nature of Prejudice*. Dr. Allport proposed that the roots of prejudice included the tendency to label people according to their membership in a group. Such labels, Dr. Allport observed, take on a "primary potency," whereby the individual is seen in terms of the group stereotype—Chinese, say—rather than in terms of his specific character. Some current work elaborates on this aspect of Dr. Allport's work, as well as on his proposal that specific kinds of contact between people of different groups could reduce their negative stereotypes of each other. Dr. Allport proposed that beneficial contacts would, for instance, be cooperative ones between people who are of equal status, that allow people to get to know each other personally. Other kinds of contact—between people who are not equal in status or power, such as between an upper-class homemaker and a minority-group domestic employee, for example—would tend to reinforce prejudices.

Many psychologists now believe that school busing and some other desegregation efforts have often failed to foster the beneficial sorts of contacts. Where they have occurred, however, prejudice has lessened. Too often, research has shown, children of different ethnic groups in newly desegregated schools fail to mix socially, instead forming hostile cliques. This, together with the frequent perception that one group is lower in status than the other, can intensify racial stereotypes. On the other hand, in cases where children are more likely to work together as equals to attain a common goal, as on sports teams or in bands, stereotypes do tend to break down. And new experiments are being tried to foster such beneficial interactions in the classroom. The new explorations of the cognitive role of stereotypes find them to be a distortion of a process that helps people order their perceptions. The mind looks for ways to simplify the chaos around it. Lumping people into categories is one. "We all need to categorize in order to make our way through the world," says Myron Rothbart, a psychologist at the University of Oregon. "And that is where the problem begins: we see the category and not the person."

The tenacity of people's stereotypes, both innocent and destructive, is a result of the pervasive role of categorization in mental life. And the stereotypes tend to be self-confirming. "It is hard to change people's preconceptions once they are established," says Dr. Stephan, who is one of those doing the new research. "Even if you present people with evidence that disconfirms their stereotypes—an emotionally open and warm Englishman, say, who breaks your image of the cold, reserved English—they will find ways to deny the evidence. They can say, 'He's unusual,' or 'It's just that he's been drinking.' "

In a study of a recently desegregated school, Janet Schofield, a psychologist at the University of Pittsburgh, found that many of the black students thought the whites considered themselves superior. When white students offered help to black students, the blacks often spurned the offers, seeing them as a confirmation of the attitudes they attributed to the whites. And research by David Hamilton, a psychologist at the University of California at Santa Barbara, shows that people tend to seek and remember information that confirms their stereotypes. So, a black who sees whites as haughty and unfriendly may notice more and remember better the whites who have acted that way than those who were warm and friendly. And if, for example, white people avoid black people, there is little opportunity for receiving information that might upset their stereotypes.

Even people who profess not to be prejudiced often exhibit subtle forms of bias, according to research by the psychologists Samuel Gaertner of the University of Delaware and John Dovidio of Colgate University. Many national surveys have shown, for example, that the racial attitudes of whites have become markedly more tolerant over the last forty years. But other research suggests that "although the old-fashioned, 'redneck' form of bigotry is less prevalent, prejudice continues to exist in more subtle, more indirect and less overtly negative forms," Dr. Gaertner and Dr. Dovidio assert in *Prejudice, Discrimination and Racism*, published by Academic Press. "People who believe they are unprejudiced will act with bias in some situations, but give some other, rational reason to justify the prejudiced act," Dr. Gaertner says.

According to research by John McConahay, a psychologist at Duke University, this more subtle form of prejudice is marked by ambivalence and exhibits itself most often in ambiguous situations where racism does not seem to be at issue. In one experiment Dr. McConahay found that whites who scored highest on a test of this subtle racism tended to reject more black applicants than white ones for a hypothetical job, though the applicants' qualifications were identical.

Some experts say social or historical facts play a role in justifying prejudice. Thus in the Southwest, negative stereotypes of Hispanic people fit traits often ascribed to migrant laborers, a role many Hispanic people held for decades. Years after such roles end, Dr. Stephan has found, the specific stereotypes still prevail. "America is full of realities from 350 years of discrimination against blacks that make blacks, in the minds of some, seem to be at fault when actually they are victims," says Dr. Pettigrew. "There are, for instance, very few black professors at Harvard, where I taught for many years. Why? The prejudiced person attributes that fact to something about blacks, rather than to something about Harvard or about the means by which tenure decisions are made." [DG]

HOW AND WHY PEOPLE DREAM

REM sleep, a paradoxical and mysterious state characterized by rapid eye movements, is leading scientists to new insights about the chemistry of the brain. While the brain registers intense activity during REM sleep, the body enters a temporary state of benign paralysis. It is the stage of sleep in which dreaming occurs, but most dreams elude memory's grasp.

In the new research on REM sleep, scientists are exploring its chemistry, the electrical state of the brain cells, and the location and physical characteristics of cells and nerve circuits involved. One objective is to answer a question that has perplexed researchers for generations: why do people dream? Scientists also want to know whether dreaming is necessary, as well as to understand the mechanics and chemistry of dreaming.

"It's a state in which the hard wiring of the brain doesn't change, but the state of the organism changes more dramatically than in any other situation," says Dr. Robert W. McCarley of Harvard Medical School in Boston and the Veterans Administration Hospital in Brockton, Massachusetts. "By studying it we can learn what are the mechanisms by which the brain alters its own excitability." The studies may help doctors better understand the brain in health and disease. Dr. McCarley and some other experts view the dreaming phase of sleep as a possible model for the study of bizarre thought processes that occur in mental illness. Indeed, a hallmark of serious depression is that REM sleep often appears earlier in the sleep cycle. Drug treatment that helps ease severe depression often has the effect of delaying that phase of sleep.

One of the circumstances that bring on the dreaming stage is an abundance of acetylcholine in the brain. This chemical is one of the brain's main neurotransmitters—substances that nerve cells use to signal to their neighbors. While brain cells that use acetylcholine are active, others that use different neurotransmitters are subdued. Dr. J. Allan Hobson of the Harvard Medical School believes the most important recent development is the ability to produce REM sleep experimentally with drugs.

The dreaming state can be brought on abruptly in animals when a substance that acts like acetylcholine is injected directly into certain nerve cells of the pons, a primitive part of the brain. Several studies of this kind, done at Harvard under the leadership of Dr. Hobson and Dr. McCarley, showed that drugs similar in action to acetylcholine could produce REM sleep and that other treatment that interfered with the natural breakdown of that nerve signal substance would have the same effect. In some of the experiments, done on cats by Dr. Helen A. Baghdoyan, now at Pennsylvania State University in Hershey, the responses were dramatic. REM sleep began within two or three minutes after injection, instead of the half hour it would take naturally. The phase of sleep lasted for two to four hours, rather than the usual six minutes.

Dr. J. Christian Gillin's group at the University of California at San Diego has evoked REM sleep in volunteers by giving them drugs that act like acetylcholine in the brain. Related studies have demonstrated that the drug-induced REM sleep produces dreams that are qualitatively no different from those that occur in natural sleep.

In research on REM sleep in animals, Dr. McCarley and colleagues have demonstrated that certain groups of brain cells in the pons become increasingly excitable as the first REM period approaches. That excitability is a chemical and electrical state that brings the cells closer to "firing," or sending signals to other brain cells. When the same cells of the reticular formation of the pons fire during the waking state, they cause physical movements, particularly the semi-automatic motions that contribute to walking, running, and many other physical acts. In the dreaming phase of sleep, eye movements continue but the nerve connections that govern limb movements are shut off.

Almost all of human dreaming occurs during REM sleep, and dreaming has held intense fascination in every civilization and every epoch. The actual stage of sleep and its key characteristics were discovered in the early 1950s. Today this phase of sleep is intriguing to scientists because it may hold clues to the crucial functions of sleep as well as to understanding of the brain.

One theory concerning REM sleep holds that this combination of signaling within the brain without resultant physical movements is actually the raw material of dreams. The sleeping brain forms images and scenes in trying to make sense of its own internal signaling. This theory was proposed ten years ago by Dr. McCarley and Dr. Hobson. In important respects it offered an explanation of dreaming totally different from Freud's theory that the dream was a disguised and censored expression of forbidden ideas. "We hypothesize that this perceived movement becomes part of the dream plot," Dr. McCarley and Dr. Edward Hoffman of Harvard said in a report that buttressed the theory by analyzing 104 dreams described by people who were awakened at the end of REM sleep episodes. A majority of the dreams involved running, walking, and similar behaviors that would be expected from a brain concocting stories to match internal signals. "The theory challenges the psychoanalytic idea that the many meaningless aspects of dream mentation are the result of an active effort to disguise the meaning of unconscious wishes," Dr. Hobson asserts in an article on dreaming. But he does not suggest that the content of dreams is random. An individual's dreams may contain "unique stylistic psychological features and concerns and thus be worthy of scrutiny by the individual to review life strategies."

Dr. McCarley notes that the bizarre features of dreams—including impossible feats such as flying under one's own power and sudden shifts in scene—may be worth study as a temporary and harmless, but possibly enlightening model of mental illness. The bizarre quality of dreams is probably itself rooted in the neurophysiology of REM sleep, in Dr. Hobson's view. He has argued that the bizarreness of a dream is probably a consequence of changes in properties of the brain and may have no particular psychological significance. Specialists say that REM sleep itself must have useful biological functions because it has been preserved throughout mammalian evolution and because it is difficult, perhaps impossible, to eliminate it. When it is delayed by waking the dreamer every time the rapid eye movements begin, REM returns more insistently in succeeding sleep episodes. Animals have died in experiments in which REM sleep was prevented for long periods.

Dr. Gillin and others have eliminated REM sleep for long periods by using drugs of the class of monoamine oxidase inhibitors for treatment of depression. Some patients have evidently gone without REM sleep for as long as a year and a half. The dream phase rebounds when the drug is discontinued. The depression did not reappear with the renewed dreaming.

In a normal person each episode of REM sleep is preceded by four earlier stages of sleep, spanning about ninety minutes, in which brain activity and physical activity become progressively more subdued. REM sleep itself is a paradox in that the brain abruptly becomes much more active, while the rest of the body (except the eyes) remains inert. In many respects the electrical activity of the dreaming brain is like that of the waking brain. But the chemistry is entirely different. In REM sleep the nerve cells that use acetylcholine are active, while those that depend on other neurotransmitters, like norepinephrine and serotonin, are quiet. REM sleep is experienced by virtually all mammals. The human infant spends sixteen hours a day asleep, and half of this time is REM sleep. This stage of sleep actually begins in the fetus at the twenty-third week of gestation.

Does the fetus dream? If so, of what? No one knows. What is the vital function of REM sleep that has apparently preserved it throughout mammalian evolution? No one knows that, either. All of human experience suggests that the nondreaming phases of sleep represent periods of rest and restoration, but this does not seem to fit the characteristics of REM sleep, in which the brain is highly active and uses substantial amounts of energy.

Yet it is a particularly deep stage of sleep. The sleeper's muscles do not respond to signals from the brain, and the brain responds hardly at all to the outside world. But the dreamer's brain waves are more like those of light sleep. The eyes dart back and forth and portions of the brain seem to be intensely active, in some respects as active as in wakefulness. Dr. Francis Crick of the Salk Institute and Dr. Graeme Mitchison of Cambridge University have proposed a theory that the dreaming phase of sleep helps the brain flush out the day's accumulation of unneeded and unwanted information.

Another concept assigns REM sleep a more constructive importance. Dr. Hobson summarized this in an article on current theories of sleep published in the new two-volume *Encyclopedia of Neuroscience*, edited by George Adelman.

"In the developing and in the adult animal REM sleep could guarantee maintenance of circuits critical for survival whether or not they were called upon for use during the wake state," he writes. The complex interconnections among nerve cells in the brain seem to be formed through use of the cells, and they need to be used to be maintained. In the fetus the kind of nerve cell activity that occurs in REM sleep could help establish nerve circuits in the brain that will be necessary after birth. Furthermore, the fact that use of some neurotransmitters like norepinephrine and serotonin is subdued during REM sleep may be relevant. These neurotransmitters are in continuous use during the waking hours. REM sleep might help replenish the brain's supplies.

Dr. McCarley believes the new research on the chemistry and nerve circuitry of dreaming sleep has exciting implications. "For physiology, the excitement is in learning how the brain shapes behavior," he says. For psychiatry, meanwhile, the research offers clues to the manner in which different behaviors, normal and abnormal, are controlled by the brain. [HMS]

STAY ALERT:
TAKE A NAP

The human body was meant to have a midafternoon nap, according to a new consensus among sleep researchers who are studying the biological rhythms of sleep and alertness. The judicious use of naps, sleep researchers now say, could be the key to maintaining alertness in people like truck drivers and hospital interns, whose urgent need for alertness must often battle with building drowsiness. Studies are also finding that an afternoon nap can significantly increase mental alertness and improve mood, particularly in the large number of people who sleep too little at night.

This interest in naps has grown almost accidentally, as researchers sought to track the cycles of sleepiness and wakefulness throughout the twenty-four hours of the day. To their surprise a wide range of studies using methods ranging from brain wave recordings to sleep diaries came to the same conclusion: there is a strong biological readiness to fall asleep during the midafternoon, even in people who have had a full night's sleep.

Although many people believe that midafternoon drowsiness is caused by eating a heavy lunch, the new research shows that is not the case. The midafternoon dip in alertness and intellectual ability occurs whether or not people eat lunch, according to Roger Broughton, a professor of neurology at the University of Ottawa. He says it depends purely on the time of day rather than on eating.

"It seems nature definitely intended that adults should nap in the middle of the day, perhaps to get out of the midday sun," says William Dement, director of the Sleep Disorders Clinic and Research Center at Stanford University.

THE MAKING OF MEMORY

SCIENTIFIC inquiry into personal memory is revealing the forces that create, distort, and sometimes erase the images that constitute each person's autobiography. Sorting through the fiction and fact with which each of us paints the canvas of our lives, the new research is assaying with new precision which aspects of a person's memory are likely to be most accurate and which aspects skewed or even erroneous. It examines the periods of life best remembered, those most often lost, and the factors that shape or contaminate memory. Through such findings researchers are coming to understand more about the strands from which personal past is woven.

"Most people would be quite surprised at how malleable their memory is—even those memories they feel most certain about," says David C. Rubin, a psychologist at Duke University. Dr. Rubin's research has shown that people remember some parts of their lives far more easily than others. Remarkably, the pattern of past memories tends to be the same for everyone. For instance, from middle age on, most people have more reminiscences from their youth and early adult years than for the most recent years of their lives.

Evidence on how the present paints the past is emerging in other research revealing people's propensity to forget parts of their life that no longer fit with their current images of themselves. This became clear in a study of the early home life of 310 men and women who as children had been so troubled they were treated in a child guidance clinic. Researchers who tracked the children down some thirty years later discovered that those who had adjusted well in adulthood had fewer memories of the painful events of childhood than did those who were currently suffering from emotional problems.

For the well-adjusted adults, the forgotten facts of childhood included family dependence on welfare and being in the care of foster parents or in a home for delinquents. "They have become conventional people after a troubled and disadvantaged childhood; they like to look back on life as though it were always that way," says Lee Robbins, a sociologist in the department of psychiatry at Washington University in St. Louis, who published the study in the *American Journal of Orthopsychiatry*.

The bias in memory can work both ways, Dr. Robbins points out. A present predicament can sensitize a person to parts of his past. For instance, one study showed that people with arthritis were more likely than their nonarthritic siblings to remember that a parent had also suffered from the disease. Similarly, research by Gordon Bowers at Stanford University and others has shown that depressed people, for example, remember sad events from their past more easily than happy people do, while happy people recall more pleasant moments.

When it comes to the emotional facts of childhood, people from the same family can remember almost opposite circumstances. In another study Dr. Robbins compared the childhood memories of patients being treated for alcoholism or depression with those of their siblings who were no more than four years older or younger. At the time of the study the subjects were thirty to fifty years old. Dr. Robbins found that the pairs of siblings agreed 71 percent of the time on such factual matters as whether the family had moved or whether the parents yelled at each other during arguments. But when the siblings were asked to say how often these things had occurred, the level of agreement fell dramatically, to 47 percent. And when the memories were of matters that required a value judgment or inference—such as whether a parent was hard on the children, whether a mother hid her anger, or whether the father's drinking embarrassed the family—the level of agreement plummeted to as low as 29 percent. The findings have strong implications for psychotherapy, where the patient's recall of his past is often central. "A clinician relies heavily on the ability of people to tell about their past," says Dr. Robbins. "But their answers are likely to be highly colored by their current view of themselves. And the vaguer the question—for instance, how happy was your childhood?—the more open to inaccuracy the answer will be." And because memory is so malleable, there may be a danger that a psychotherapist will cause a bias in his patient's memories by searching for particular kinds of events, according to Donald Spence, a psychoanalyst at Rutgers Medical School. "The patient's memory is fragmented already," he says. "If the therapist suggests something simply by asking about it, he may color the patient's recollection so that he weaves it into his memory."

Dr. Spence cites a study of mothers whose children had been seen at a well-baby clinic fifteen years earlier. The mothers were asked to tell when the child had reached developmental milestones such as talking and walking. Their answers tended to be earlier or later than was actually the case, depending on what they had learned, from child experts like Benjamin Spock, was better. While most alterations in memory have no deeper meaning, some are seen as having clinical significance. Freud saw biases and holes in memory as a clue to a person's deepest conflicts. Psychoanalysis tries to get past a patient's "screen memories"—false memories that hide painful truths—and retrieve a better understanding of those truths. "Although there is much forgetting that is simply forgetting, a rule of thumb is that the more psychodynamically important a memory is, the more prone it is to warping or forgetting altogether," says Theodore Shapiro, a psychoanalyst and professor of psychiatry at Cornell Medical College.

The relationship between disturbing events and their repression is suggested by a study of fifty-three women who had been victims of incest as children. The earlier, longer, and more violent the incest, the more it had been forgotten, according to a report by Judith Lewis Herman, a psychiatrist at the Women's Mental Health Collective in Somerville, Massachusetts. The extent of the incest was known because the women studied had been able to find independent corroboration of their experiences.

Apart from the impact of emotions and conflicts on memory, the lifelong contours of personal memory itself are uneven, with some periods of life standing out while others recede, according to research by Dr. Rubin of Duke reported in *Autobiographical Memory*, a collection of articles he edited that was published by Cambridge University Press. The spontaneous memories of those in their later years, Dr. Rubin reports, fall off steadily over the most recent two to three decades, but increase for the two to three decades before that and then recede again through childhood. Seventy-year-olds thus tend to remember more from their thirties than from their fifties. "It seems to be that reminiscence flows more freely about the period in life that comes to define you: the time of your first date, marriage, job, child," according to Dr. Rubin, who discovered the effect in an analysis of data from five studies. "It's not that life is duller from forty to fifty-five than from twenty to thirty-five, but that the patterns are more stable, and so less memorable."

The one period of life that virtually no one remembers is the years before the age of about four. While Freud attributed this childhood amnesia to the repression of infantile sexuality, some of the memory researchers today believe it occurs because mental abilities used to cue memory (language, for instance) have not yet matured.

The nature of a person's earliest memory is coming in for special scrutiny. Alfred Adler, one of Freud's disciples, proposed that the earliest memory a person has reveals the person's overall psychological stance in life. These early memories are seen by many psychoanalysts as retrospective inventions or convenient selections that express some psychological truth about one's life. Patients' earliest memories sometimes shift over the course of psychotherapy as psychological conflicts become resolved, according to clinical studies.

The relationships reflected in those memories, in this view, repeat themselves in a wide range of situations in the person's life. Evidence that the earliest memories play such crucial roles comes from a study by Jacob Orlofsky, a psychologist at the University of Missouri at St. Louis, published in the *Journal of Personality and Social Psychology*. College students were tested on the degree to which they had reached an adult identity, and those levels were compared against their earliest memories.

Those students who had reached fullest psychological maturity—having achieved commitments after a period of doubt and searching —tended to have early memories reflecting such themes as striving and mastery. This was also true of students who were still searching on the way to full maturity. But those students who had adopted their parents' values without independently seeking an identity had early memories that reflected such dependence. Their memories revolved around themes of a need for nurturance and safety, and of complying with authority to maintain closeness.

Not surprisingly, people's memories are better for the out-of-the-ordinary, special moments of their lives than for the mundane, according to new data on which events survive best in memory. In the study by William Brewer, a psychologist at the University of Illinois, students carried beepers and wrote down what they were doing and thinking about at random moments when the beepers went off. Most of the events were soon forgotten. The students' best recall for what they had been doing was when they were reminded of what they had been thinking; the students forgot about 20 percent of the events just one week later and could not recall half of them two months later. The students' worst recall was when they were simply told the time and date—a week later, they could recall only a third of the events and two months later just 15 percent. "The forgetting rate was much faster than had been expected," says Dr. Brewer.

The best-remembered moments were those that were exciting, unusual, or novel, like the first date with a particular person. "The events of everyday life are unremittingly dull, and people have little memory for their specifics," Dr. Brewer says. "These are the events that vanish in memory." [DG]

MUSIC AND REMEMBERING

Factors as subtle as music people hear or odors they smell can shade the emotional tone of their memories, research is showing. "The events of the day—a bus splattering mud on you, being yelled at by your boss, or seeing a movie you enjoy— create a cognitive bias that makes some memories more likely to come to mind while others become less probable," says Howard Ehrlichman, a psychologist at the Graduate Center of the City University of New York. "Things that make you feel good prime more happy memories, while things that get you down prime sad ones."

In an experiment done with Jack N. Halpern, a colleague at City University, Dr. Ehrlichman tested the power of pleasant and unpleasant odors to influence memory. They had volunteers reminisce while they smelled either almond extract, which most people find pleasant, or pyridine, a bitter-smelling chemical. The volunteers were told to bring to mind memories of specific incidents and describe them. Volunteers later rated how happy or unhappy they had been at the time.

Those who recalled their memories while smelling the pleasant odor recalled close to 15 percent more happy memories than unhappy ones. The unpleasant odor caused a bias in the opposite direction: those smelling it recalled about 10 percent more unhappy memories than pleasant ones. "The effect is certain to be stronger outside the laboratory, in natural settings—the smell of the sea, the aroma of a meadow, apple pie in Grandma's kitchen," says Dr. Ehrlichman, explaining that this is because people are more at ease in natural settings.

IQ ISN'T EVERYTHING

IN an effort to make up for some of the glaring limitations of IQ tests, researchers have begun to develop new ways to measure the kinds of emotional factors and psychological attitudes that lead to success in everyday life. While IQ tests remain excellent predictors of how well one will do in school, they have little or nothing to do with who will earn the most money or prestige or have the most satisfying social life or relationships. The new tests are intended to assess the more practical intelligence that underlies these accomplishments.

The new approach goes beyond purely mental skills to assess emotional factors and psychological attitudes that can either interfere with or facilitate the use of those skills. It has fostered new theories of what it means to be smart. The old theories focused on academic skills such as verbal or mathematical quickness. But the new theories describe a spectrum of practical talents such as the ability to pick up the unspoken rules that govern success in a corporate or professional career or the habits of mind that foster productivity. "IQ and success in living have little to do with each other," says Seymour Epstein, a psychologist at the University of Massachusetts. "Being intellectually gifted does not predict you will earn the most money or achieve the most recognition, even among college professors."

One factor emerging as crucial for life success is what might be called emotional intelligence. "How well people manage their emotions determines how effectively they can use their intellectual ability," Dr. Epstein says. "For example, if someone is facile at solving problems in the quiet of her office, but falls apart in a group, then she will be ineffective in a great many situations."

Dr. Epstein has developed a test that measures "constructive thinking," the ability to respond effectively to life. The test measures how well a person manages his emotions and challenging situations, as well as habitual responses to problems such as setbacks and failures. It differs from earlier alternatives to the IQ scale that attempted to measure such factors as creativity. Most of the constructive attitudes the test measures have the ring of common sense. People who think constructively, for instance, tend not to take things personally and not to fret about what others think of them. Rather than complaining about a situation, they take action.

Dr. Epstein has found that many academically bright people have self-destructive habits of mind, such as holding back from new challenges because they fear the worst possible outcome for themselves. Among these nonconstructive ways of thinking were: holding to private superstitions (such as believing that talking about a potential success would keep it from happening); a naive, unrealistic optimism (for instance, that people can do absolutely anything if they have enough willpower); and a generally negative, pessimistic outlook. "Typical of destructive ways of thinking is one student I recall who played a beautiful solo with a band," Dr. Epstein said. "He had been dreading it, convinced he would do terribly. When people praised him, he discounted it, saying that even if he did well that once, he'd certainly do terribly the next time."

How well people score on the test of constructive thinking, Dr. Epstein has found, predicts a great range of life success, from salaries and promotions, to happiness with friendships, families, and romantic relationships, to physical and emotional health. Among people bright enough to attend college he found that IQ was related to none of these sorts of success; it was simply irrelevant. "In a sense, there are two minds," Dr. Epstein says. "One, the experiential mind, has to do with how you react to the world emotionally; it makes instantaneous decisions and calls the shots day to day. It has nothing to do with IQ. The other, the rational mind, has to do with how we explain what we do, and how well we understand a novel or know math. It has little to do with success in living." Certain childhood experiences seem to shape constructive thinking, for better or for worse. Those who scored higher on the test of constructive thinking reported having parents who did not overprotect them, but rather trained them in independence. The sense of having been loved or rejected by one's parents, however, did not relate to scores on the test. "Constructive thinking depends to a large extent on having parents who teach you to be strong in the world, to learn to handle things on your own," Dr. Epstein says. "Love is not enough; it takes training in doing things yourself."

Still, many practical talents that lead to success in life are rarely taught explicitly. Rather, those who excel seem able to absorb this knowledge tacitly. In one recent study, psychologists at Yale developed a test that measures the knack of selling. The psychologists see the art of persuasion as essential to success in much of life.

"The ability to sell is a kind of persuasion everyone needs," says Robert Sternberg, the psychologist at Yale who is doing much of the new work. "You sell yourself when you meet someone, you sell your ideas or point of view, you sell when you negotiate a deal. Sales is a skill that demands a specific kind of practical intelligence." In research with Richard Wagner and Carol Rashotte at Florida State University, Dr. Sternberg found that the talent for sales included such things as knowing that when someone stalls in making a decision, the best approach is not to press him but rather to ask him why he's not prepared to make a decision at that moment. Another persuasion tactic used by those with sales talent was not to argue with the person one is selling to, but rather to acknowledge the validity of his position and then make one's own point.

While such rules of thumb may sometimes be taught as sales strategies, more often successful salespeople seem to grasp them intuitively. When the test was given to people who sell insurance, high scores were correlated with the number of years they had been in sales, the number of sales they had made, and the number of awards they had received. In another study with Dr. Wagner, included in *Practical Intelligence*, published by Cambridge University Press, Dr. Sternberg studied the kinds of tacit knowledge typical of successful business managers. The test assessed three kinds of practical intelligence. One was how well a person managed himself—dealt with procrastination, for instance. Another was the ability to manage others, such as knowing how to assign and tailor tasks to take maximum advantage of another person's abilities. The third was knowing how to manage one's career—how to enhance one's reputation, for example.

A typical question asks what should be the basis for selecting new projects to tackle; often there are more than a dozen choices, including that the project should be "fun," that it enable one to demonstrate one's talents, or that it require working directly with more senior executives. The people who are most successful tend to choose the same top priorities. Those executives who did best on the test tended to have more years of management experience and to have higher salaries than those who did less well, Dr. Wagner and Dr. Sternberg found.

One of the first techniques for assessing practical intelligence was developed by David McClelland, a psychologist at Boston University, and George O. Klemp, Jr., a consulting psychologist at Charles River Consulting in Boston, Massachusetts. By careful comparisons of outstanding performers in a given field with mediocre ones, they were able to uncover many of the specific competencies that set the two groups apart. In a study of managers, for instance, the problem-solving skills of the best managers included the tendency to push for concrete information when faced with ambiguity; another was the ability to seek information from as wide a range of sources as possible. They also displayed a curious knack for finding unusual analogies to explain the essence of a situation. The best managers were adept, too, at influencing people. They consistently anticipated the impact of their actions on others in the company and did not hesitate to confront people directly when there were problems. They also were adept at building a sense of collaboration by, for instance, involving subordinates in making decisions that would affect them, particularly controversial decisions, according to results obtained by Drs. Klemp and McClelland.

The new line of research was triggered by an influential critique of IQ tests Dr. McClelland published in 1973 when he was at Harvard, and by a series of equally skeptical articles written about the same time by Ulric Neisser, a cognitive psychologist then at Cornell. They were among the first prominent psychologists to argue that academic intelligence had little or nothing to do with success in life. Until then, it was widely assumed that a central core of intelligence, which was measured by IQ tests, could be applied by bright people to find success in almost any field.

But as Dr. Neisser pointed out, the questions on IQ tests are nothing like the challenges one meets in life. IQ questions tend to be formulated by other people, to offer all the information one needs to answer them, and to have nothing to do with people's own experience or interests. They are also well defined, Dr. Sternberg observes, with only one correct solution, and usually just one way to arrive at the right answer. But usually none of that is true of the problems people face in their daily lives, such as how to find a mate or a better apartment, handle personal finances, or get ahead in one's career. Dr. Sternberg has proposed a theory of intelligence that includes such traits as how well a person plans strategies for problem solving or handles novel situations. And a theory put forth by Howard Gardner of Harvard describes seven kinds of intelligence, including the body control displayed by athletes and dancers, musical talent, and interpersonal skills such as being able to read another's feelings, as well as more academic abilities like mathematical and logical reasoning.

Much of the new work examines attitudes that allow people to make the best use of whatever mental skills they may have. One such outlook is what psychologists call "self-efficacy," the belief that one has mastery over the events of one's life and can meet a given challenge. "People's beliefs about their abilities have a profound effect on those abilities," says Albert Bandura, a psychologist at Stanford University who has done the major research on self-efficacy. "Ability is not a fixed property; there is huge variability in how you perform. People who have a sense of self-efficacy bounce back from failure; they approach things in terms of how to handle them rather than worrying about what can go wrong."

In the study of exceptional managers by Drs. McClelland and Klemp, for instance, the best ones displayed a strong self-confidence, seeing themselves as the most capable person for their job and as being stimulated by crisis. Along similar lines, Dr. Martin Seligman, a psychologist at the University of Pennsylvania, has shown that people who are more optimistic do better than pessimists in a wide variety of endeavors, from selling insurance to achievement in school. Self-efficacy varies from one part of a person's life to another. A self-confident manager, for instance, may feel ineffective as a father. Dr. Bandura and other researchers have found that self-efficacy acts as a powerful force in people's choices of what they will try in life and what they avoid. Many women, they have found, have a low level of self-efficacy with regard to computers or math and hence tend to shy away from careers that depend heavily on those skills.

Some of the psychologists believe that although practical intelligence seems to come naturally to certain people, other people can be trained to be smarter in this way to some extent. Dr. Sternberg and Dr. McClelland, for example, have worked on developing training techniques to enhance different aspects of practical intelligence.   [DG]

THE BOY WHO WALKED IN CIRCLES AND OTHER TALES

WHEN James was three years old, he developed a strange habit: whenever he crossed the street, he would walk in circles around the manhole covers. In kindergarten, he would sit for hours drawing circles on pieces of paper. At the age of eight, he would stand up and sit down seventeen times before finally sitting in a chair, and would go back and forth through a door seventeen times before finally exiting.

James suffered from obsessive-compulsive disorder. Scientists have assumed that the problem is extremely rare. But a 1988 national survey of households found that the disorder is twenty-five to sixty times more common than had been thought, afflicting perhaps one in forty Americans at some point in their lives.

The disorder is vastly underreported, researchers say, because many who suffer from it keep their symptoms secret. While their symptoms are seldom as severe or obvious as those of James, millions of people find their lives dominated by desperate attempts to fight off strange impulses and bizarre thoughts.

The most widely accepted previous estimates put the prevalence of obsessive-compulsive disorder at about one person in every two thousand. The 1988 study estimates that from 1.9 to 3.3 percent of Americans will have the disorder during their lives.

"The obsessive-compulsive is at the extreme end of a continuum that begins with the person who, for instance, is scrupulously neat about his desk," says Blanche Freund, a psychologist at the Program for the Clinical Study of Anxiety Disorders at the Medical College of Pennsylvania, in Philadelphia. The program is one of several across the country that specialize in treating the disorder. "Being especially neat is not a problem; a compulsive style like that can be a distinct asset in some ways," Dr. Freund says. "More serious is the compulsive personality, someone who gets completely lost fussing over details, and who, for example, needs everyone to be equally neat in order to make him feel comfortable. And at the extreme are people who spend most of their day obsessing or performing compulsive rituals."

Obsessions and compulsions are symptoms of the same disorder. Obsessions are thoughts that completely preoccupy a person, while compulsions are actions that people feel they must perform. Although obsessive-compulsive disorder has long been regarded by psychotherapists as one of the most difficult problems to treat, new techniques using medication or behavior therapy have proven highly effective in recent years. "Between drugs and behavior therapy, about 70 to 80 percent of those with the most extreme symptoms can have significant improvement or be cured," says Dr. Judith L. Rapoport, chief of child psychiatry at the National Institute of Mental Health. James is a pseudonym Dr. Rapoport gave for one of the children who have been successfully treated for the disorder at her facility.

The most promising new treatments for obsessive-compulsive disorder are three drugs developed as antidepressants: clomipramine, fluoxetine, and fluvoxamine. In the most effective behavioral techniques the patient is methodically exposed to things he dreads while being forbidden to indulge in his compulsions.

The world of the obsessive-compulsive is often nightmarish and bizarre. One form it may take is the compulsion to pluck one's hair out strand by strand. Young women often pluck their hair until they become bald. Some obsessive-compulsive people will take hours to leave their homes because they check and recheck every appliance in the house to be sure that it is off. Others feel compelled to perform private rituals, such as the man who spent hours circling the same route in his car, fearing he had struck a pedestrian. "If you ask someone with extreme obsessions how much of the day their thoughts are free of the obsession, they will say 'almost never,' " Dr. Freund says.

Although people with obsessions realize that their thoughts are strange or unwanted, they feel unable to stop them. Similarly, people with compulsions say that though they often do not want to perform the act, anxiety builds until they give in. Often, though, the disorder is less debilitating, confining itself to compulsive rituals done in secret or those that interfere little with a normal life. Another of Dr. Rapoport's patients, a sixteen-year-old girl, spent hours every Sunday removing everything from her room, then washed the walls and floors. In other respects, her life was not unusual: she did well in school and led a normal social life. Researchers believe that such circumscribed symptoms are one reason that estimates for the prevalence of obsessive-compulsive disorder have been low: symptoms are often concealed from family or friends. The new, higher estimates come from part of a national survey of 18,572 men and women conducted door-to-door in New Haven, Baltimore, St. Louis, the Piedmont region of North Carolina, and Los Angeles.

One question to detect people with obsessive-compulsive disorder concerned common obsessions. These included persistent, unpleasant thoughts, such as that they might harm someone they love, that their hands were dirty no matter how much they are washed, or that relatives may have been hurt or killed. In trying to detect compulsions the interviewers asked if people felt they had to "do something over and over again," even if they considered it foolish or tried to resist it.

Respondents were also asked whether they felt they had to do something in a particular order, like dressing, and whether they felt they had to repeat the entire sequence if something was done improperly. Some people suffer mainly from obsession, others from compulsions, and still others from a combination of the two. While the vast majority of those people who come for treatment suffer from both problems, only 9 percent of those surveyed had both obsessions and compulsions.

To qualify as a symptom for the survey, the obsessions or compulsions must have been present for at least three weeks, must have persisted despite all attempts to get rid of them, and must have interfered to some extent with aspects of the person's life. In the survey, 468 people were identified as qualifying for the diagnosis. Although the national average was 2.5 percent, the percentage of sufferers varied from 1.9 to 3.3 among the five areas. The results were reported by Marvin Karno, a psychiatrist at the University of California at Los Angeles (UCLA), and colleagues at other institutions.

It is not necessarily abnormal for children to develop an obsession or compulsion, such as counting cars that pass by. But when these become such strong forces in their lives that they interfere with normal play or school, they are considered symptoms that should be treated.

Evidence from brain studies of people with severe obsessive-compulsive disorder, done by teams led by Dr. Rapoport and by Lewis Baxter, a psychiatrist at UCLA, suggest the involvement of a brain circuit that runs between the frontal lobes and the basal ganglia, a series of structures at the base of the brain.

Those brain areas are involved in coordinating what the senses register with how the body responds to it. Dr. Rapoport believes they may have a role, for instance, in patients who doubt the evidence of their senses. "These patients will turn a knob to check that a door is locked, then start to doubt it as they are going down the steps, and go back again and again to recheck," Dr. Rapoport says.

The medications that have been particularly effective for people who suffer from obsessions seem to act on the basal ganglia, Dr. Rapoport says. Clomipramine, fluvoxamine, and fluoxetine all block the uptake in the brain of serotonin, a brain chemical for which the basal ganglia have particularly dense receptors. While the drugs seem to be effective in reducing symptoms in 70 percent of patients, their side effects—for example, men may have difficulty with ejaculation—lead some patients to stop them.

On the other hand, behavior therapy techniques that work best with obsessive-compulsive patients are so demanding that about a quarter of those who seek help decline to go through the process, according to a report by Gail Steketee, a clinical social worker, and Edna Foa, a psychologist at the Eastern Pennsylvania Psychiatric Institute in Philadelphia.

"We recommend that patients try behavior therapy first, and then if that does not work for them, try medications," Dr. Rapoport says. "Some people respond best to drugs, some to behavior therapy. If people are motivated enough to stick with the treatments, the odds are excellent that they can be helped." For patients who go through a three-week program of behavior therapy, about 80 percent experience a significant decrease in symptoms, their report said. Of these, only about 20 percent have a serious return of symptoms.

One technique used in behavior therapy consists of exposing patients to the object of their fears. A patient who feared getting cancer was taken to a cancer ward. In another technique, patients who feel compelled to engage in specific rituals are forbidden to do so. A man who dreaded touching cigarette ashes and "low-class" people felt compelled to scrub his hands for three minutes as often as thirty times a day. But during the treatment he was kept from washing his hands and was exposed to the things he dreaded, ashes and "low-class" people. At the end of treatment he washed his hands just six times a day and then only for twenty seconds.

While those who undergo treatment typically retain vestiges of their obsessions or compulsions, treatment is considered successful if symptoms no longer interfere with the person's life. "They might still check the door an extra time or two to be sure it's locked," Dr. Freund says. "But they don't do it for hours on end, like they used to." [DG]

DON'T THINK ABOUT WHITE BEARS

What happens when a person is ordered not to think about a white bear? Predictably, psychologists find, it becomes difficult to think about anything else. A research team at Trinity University and the University of Texas (both located in San Antonio) examined people's ability—or inability—to suppress thought. It is an issue that drew much early attention from psychoanalysts, who focused on the role of thought suppression as a defense mechanism and on its link to neuroses. But the Texas team said few contemporary researchers have studied it.

The thought suppression experiments are important because the phenomenon has serious implications, according to the authors: Daniel M. Wegner, Samuel R. Carter III, and Teri L. White of Trinity, and David J. Schneider of the University of Texas. For example, studies have shown that jurors can be influenced by testimony they have been instructed to ignore and people in general can be influenced by news reports they are told are untrue.

The authors said it appeared attempts to suppress thoughts of something can lead to obsession about it. But they said people could reduce preoccupation with a thought by picking another subject as a "distractor" and concentrating on that instead. The researchers tested the white bear proposition on students who volunteered for a stream-of-consciousness experiment. The students spoke into tape recorders all the thoughts that entered their minds during a five-minute period.

Just before starting one such session, some students were instructed to avoid thinking about white bears, but to ring a bell if the thought did enter their minds despite the prohibition. They rang the bell early and often.

Later these students were instructed to think about white bears during another five-minute stream-of-consciousness recording. They reported the bear image more often than another group, which had been instructed to think about white bears but had not had the anti-bear conditioning. The effort to suppress thoughts of bears was not only unsuccessful initially; once freed to think about bears, subjects thought about them more often than if they had never tried suppression. The psychologists noted that there was an impressive literary precedent for their choice of subjects to think about: as a child, Dostoyevsky challenged his brother not to think of a white bear, and the other child, the authors said, was "perplexed for a long time."

NARCISSISTS IN TROUBLE;
OR, MIRROR, MIRROR ON
THE WALL

NARCISSISTS will be pleased to know that they hold the fascination of psychotherapists now more than ever. In fact, therapists are recognizing narcissism in more and more of their patients. The traits include an inflated sense of self-importance and an insatiable need to be the center of attention. While no one knows if the ranks of narcissists are truly growing or if the therapists are simply recognizing them more often, narcissism is increasingly being diagnosed as the underlying problem in patients who complain of other problems, such as an inability to sustain relationships or severe depression after minor failures on the job.

Along with the rising interest in treating those whose narcissism undermines their mental health is a growing appreciation of how "healthy" narcissism plays a major and useful role in the lives of successful people. Psychoanalysts, for example, see signs of narcissism in the drive to receive adulation that powers the careers of sports and entertainment figures. But narcissism may also help explain those petty tyrants who run companies or offices as though they were totalitarian states.

While narcissism has been known since ancient times and was described by Freud in his writings, diagnosis and treatment of the condition are now surging, particularly among therapists with a psychoanalytic orientation, who often seek further training to help them treat narcissists.

Why the intense interest? One reason is that narcissists are particularly hard to treat. They find it difficult to form the warm bond with a therapist that naturally evolves with most other patients. Instead, they often become cold or even enraged when a therapist fails to play along with their inflated sense of themselves. "A narcissistic patient is likely at some point to attack or devalue the therapist," Dr. Gerald Adler, a Harvard psychiatrist, says. "It's hard to have to sit with such people in your office."

But narcissism is not limited to the most extreme cases, who make their way to the therapist's office. Many psychoanalysts hold that a healthy adjustment and successful life is based to some degree on narcissism. Healthy narcissists feel good about themselves without needing constant reassurance about their worth. They may be a bit exhibitionistic, but do not need to play down the accomplishments of others to put themselves in a good light. And although they may like adulation, they do not crave it. "Normal narcissism is vital for satisfaction and survival; it is the capacity to identify what you need and want," says James Masterson, a psychiatrist at Cornell University Medical College. His book *The Search for the Real Self* describes the treatment of narcissism.

Pathological narcissists, on the other hand, need continual reassurance about their value; without such reassurance they feel worthless. Although they have a grandiose sense of themselves, they crave adulation because they are so unsure of themselves that they do not know they have done well or are worthwhile without hearing it from someone else. "The deeply narcissistic person feels incomplete, and uses other people to feel whole," says Dr. Adler. "Normally, people feel complete on their own."

With the study of narcissism the issue of self-esteem has moved to the center of psychoanalytic concern. "Self-esteem depends on how well-developed your sense of self is," says Paul Ornstein, a psychiatrist at the University of Cincinnati. "We're all exceedingly protective to the extent we feel vulnerable."

"Narcissistic vulnerabilities," as psychoanalysts refer to them, make people particularly sensitive to how other people regard them. "You see it in marriage, in friendships, at work," Dr. Ornstein says. "If your boss fails to smile when you greet him it may create a withdrawn, anxious feeling. If so, your self-esteem has been hurt. A sturdy self absorbs that; it has a bank account of self-esteem, so it's not unbalanced. But if you're vulnerable, then these seemingly small slights are like a large trauma."

On the surface, extreme narcissists are often brash and self-assured, surrounded by an aura of success. Indeed, they are often successful in their careers and relationships. But beneath that success, feelings of inadequacy create the constant need to keep inflating their sense of themselves. If they do not get the praise they need, they can lapse into depression and rage. Admiration, rather than the pure pleasure of doing things well, is what propels narcissists to their success, psychoanalysts say. Thus many workaholics put in their long hours out of the narcissist's need to be applauded. And, of course, the same need makes many narcissists gravitate to careers such as acting, modeling, or politics, where the applause is explicit.

Many difficulties in intimate relations are due to narcissism, according to David A. Berkowitz, a psychiatrist at Tufts University: "Narcissism makes someone vulnerable to the least failure to be loved or accepted just as they are by their partner. Marriage brings to the fore all one's childhood yearnings for unconditional acceptance. A successful marriage includes the freedom to regress, to enjoy a childlike dependency. But in marriage a couple also tend to reenact early relationships with parents who failed to give them enough love. This is particularly hard on those with the emotional vulnerabilities of the narcissist."

Narcissists tend to surround themselves with people who will laud them rather than give criticism when it is needed. They are drawn, too, to the trappings of success, such as automobiles or houses that announce their worth to the world. Narcissists find in such status symbols support for their grandiose sense of themselves. Narcissists are typically charming and friendly. Their own intense need for adulation makes them sensitive to the same needs in others, and so they are particularly adept at ingratiating themselves. But there is a hidden agenda to their friendliness: they are interested only to the extent that their own self-importance is fed by it.

Those narcissists whose hard work has paid off in making them leaders in business, for instance, tend to try to make their organization re-create the childhood they long for, with themselves at the center of a loving world, according to Howard S. Schwartz, a professor of management at the Oakland University School of Business Administration in Rochester, Michigan. Writing in the *Journal of Management,* Mr. Schwartz proposes that many business organizations tend to reward the narcissistic fantasies of those at the upper echelons, who receive the adulation of those below. When the process becomes pathological, the organization becomes totalitarian, with those below fearing to do or say anything that does not fit with the idealized view of those at the top. At its pathological extreme, the narcissist's intense drive to inflate his sense of grandiosity is seen by psychoanalysts as protecting him from a deep feeling that his life is empty and that underneath it all he is worthless. The main theories of narcissism have been propounded by the late Heinz Kohut, a Chicago psychoanalyst, and by Otto Kernberg, a psychiatrist at New York Hospital–Cornell Medical Center.

Some psychoanalysts trace the roots of narcissism to difficulties during the period from eighteen months old to three years, when a child's sense of having a self, independent of others, emerges. When parents need their child to seem perfect, as often happens with narcissistic parents, the child encounters barriers to expressing separateness and establishing an individual identity. Whenever such children try to express themselves in ways that do not fit with the parental needs, they are attacked, criticized, or ignored. This leaves them feeling inadequate and impaired and believing that they can be loved only if they fit the parents' image of their perfection.

The child's resulting fear that no one loves him just as he is leads to a deep belief that there is something repugnant or disgusting about him. It is to protect himself from these feelings that the narcissist builds a facade of grandiosity. When people disappoint narcissists in their search for adulation, they are prone to turn on them with rage. When life deals a setback, narcissists plunge into depression.

This makes narcissists particularly hard to treat in therapy, for at some point the therapist will have to deflate the narcissist's grandiosity, if only to help him or her find a firmer reality. It is at that point that the therapist risks becoming the target of the narcissist's rage.

One of the new treatment approaches to narcissism, developed by Dr. Ornstein and his wife, Dr. Anna Ornstein, a child psychiatrist, revolves around what they call the "curative fantasy."

"Every patient comes to therapy with certain ideas, both conscious and unconscious, about what will cure them," Dr. Paul Ornstein says. "That fantasy will figure in the cure; if the therapist doesn't respond to it early on, the patient may quit, feeling he'll never get what he wants from this therapist." Those fantasies have particular prominence in the mental life of narcissists. "You see the same sort of fantasy in everyday life," Dr. Ornstein says. "The idea that if I just get this job, or if only my wife would treat me in a certain way, or whatever, then my problems would be solved."

In Dr. Ornstein's approach to therapy, he views self-defeating habits such as drug taking and promiscuity as a sign of the patient's curative fantasy. "These are often failed attempts at self-healing," he says. "If you can acknowledge this, and show him that you, too, are trying to solve his problem, then the patient will feel understood and affirmed. But when the patient feels the therapist doesn't understand, he experiences a lack of attunement. Such periods will come and go during the course of a successful therapy with narcissists. It's part of a reparative process, an emotional equivalent to the early love the patient didn't get. It eventually helps them build a sturdier self."

One of the more novel approaches to healing the narcissist's wounds is through meditation. Writing in the *Journal of Transpersonal Psychology*, Mark Epstein, a psychiatrist in New York City, proposes that certain experiences in meditation offer a natural corrective to the narcissist's fragmented sense of himself. Dr. Epstein proposes, for instance, that the intense experience of delight that sometimes occurs during meditation can reassure the narcissist that he can find a sense of well-being within himself and so does not need to depend on others for it.

The narcissist's need for the best of all things can, however, be a problem when seeking treatment. Dr. Masterson tells of having been quoted in an article in *The New York Times* as an expert on the subject. Within days, a dozen people called him who had read the article and thought they needed treatment for the problem. Each came to him for an evaluation, and he recommended they be treated. But when Dr. Masterson said that he was too busy to take them on himself and suggested that they be treated instead by one of his associates, not one of the twelve returned for treatment. [DG]

WHY CYNICS DIE YOUNG

Angry, cynical people are five times as likely to die before reaching fifty as people who are calm and trusting, a psychiatrist has found. He says the finding is more evidence that people with so-called type A behavior—fast-talking workaholics who are always in a hurry—are not more prone to premature death from heart disease. Instead, he says, his work is the most definitive evidence so far that people who are hostile are more likely to suffer premature death from all causes.

Such toxic personalities can be traced to biological differences that are likely present from birth, according to the psychiatrist, Dr. Redford B. Williams, a professor at Duke University Medical Center in Durham, North Carolina.

During the 1970s, Dr. Meyer Friedman and Dr. Ray Rosenman, cardiologists in San Francisco, described "type A" men, a group they said were twice as likely to suffer heart disease as their more easygoing "type B" counterparts. But in the early

1980s this finding was shaken by a series of studies that failed to find that type A personalities predict heart problems.

"We can now state with some confidence that of all the aspects originally described as making up the global type A pattern, only those related to hostility and anger are really coronary prone," Dr. Williams says. Dr. Williams based his findings on numerous studies, including a twenty-five-year study of 118 lawyers. Those who scored high on hostility traits in a standard personality survey taken during law school were five times as likely to die before the age of fifty as classmates who were not hostile.

Personality traits such as paranoia ("People out there are picking on me") and social avoidance ("I'd rather cross the street than meet that person") did not correlate with early death, Dr. Williams says, nor did neurotic behaviors. But personality traits that reflected cynical mistrust ("People lie to get ahead"), anger, and acting out of anger ("I often have to get rough with people") were strong predictors of premature death.

THE PAIN OF OTHERS

ON seeing another child fall and hurt himself, Hope, just nine months old, stared, tears welling up in her eyes, and crawled to her mother to be comforted—as though she had been hurt, not her friend. When fifteen-month-old Michael saw his friend Paul crying, Michael fetched his own teddy bear and offered it to Paul; when that didn't stop Paul's tears, Michael brought Paul's security blanket from another room.

Such small acts of sympathy and caring, observed in scientific studies, are leading researchers to trace the roots of empathy—the ability to share another's emotions—to infancy, contradicting a long-standing assumption that infants and toddlers were incapable of these feelings. And in some of the most surprising findings, researchers have identified individual neurons in primates that respond primarily to specific emotional expressions, a response that could be a neural basis for empathy. These findings are opening a new research area in which scientists are searching for the specific brain circuitry that underlies the empathic impulse.

The scientific interest is spurred, too, by the critical role of empathy in many facets of life, from management and sales to friendship and parenting, to compassion and moral awareness. The absence of empathy is also telling. Its lack is seen not just in criminal sociopaths, who have no concern for their victims. A deficiency in empathy also marks disorders like autism and chronic schizophrenia, "where the impoverishment of skills like empathy is a major deficit," says Leslie Brothers, a psychiatrist at the California Institute of Technology who is conducting brain studies on empathy.

Among the strongest spurs for research on the neurology of empathy have been experiments showing that newborn babies will cry in response to the cries of another infant. In the studies, newborn babies cried more loudly when they heard the sounds of other babies crying than they did when they heard a computer simulation of infants' cries and other sounds that were equally loud and startling. "Virtually from the day they are born, there is something particularly disturbing to infants about the sound of another infant's cry," says Martin Hoffman, a psychologist at New York University. "The innate predisposition to cry to that sound seems to be the earliest precursor of empathy."

Researchers cannot know for sure, of course, that the newborn's cries reflect empathy rather than, say, a reflexive tendency to respond upon hearing another's cry of alarm. Still the responses of infants like Hope clearly go far beyond a reflex, scientists say. These and other findings concerning infants and toddlers contradict an influential view offered several decades ago by Jean Piaget, the Swiss developmental psychologist, who contended that children could not feel empathy until they had achieved cognitive abilities that allow seeing things from another person's perspective. These abilities, he believed, developed around age seven or eight.

However, researchers are finding that the sympathetic distress of infants occurs long before they acquire the sense that they exist apart from other people, which takes place late in the first year. From a few months after birth through the first year of life, studies have shown, infants react to the pain of others as though it were happening to themselves. On seeing another child get hurt and start to cry, they themselves begin to cry, especially if the other child cries for more than a minute or two.

But around one year of age, infants begin to realize that the distress is being felt by someone else. "They realize it's the other kid's problem, but they're often confused over what to do about it," Dr. Hoffman says. For instance, one boy of that age brought his mother over to comfort a crying friend, even though the other child's mother was also in the room. During this phase, toddlers often imitate the distress of someone else—apparently, researchers say, in an effort to better understand what the other person is feeling. This kind of imitation, called "motor mimicry," was the original meaning of the word *empathy*, which was coined by a psychologist early in this century.

The theory was that empathy was based on physically imitating the distress of another in order to better understand what that person was feeling. The word was coined to distinguish it from sympathy, a feeling of concern for another person that may not necessarily be based on sharing what that person himself feels. "From around fourteen months to two or two and a half years, you see children feel their own fingers to see if they hurt when someone else hurts their fingers," says Marion Radke-Yarrow, chief of the Laboratory of Developmental Psychology at the National Institute of Mental Health, where much of the work on altruism in children has been done. "By two and a half, though, toddlers clearly realize that someone else's pain is different from their own, and know how to comfort them appropriately."

The next major landmark in the evolution of empathy occurs around age eight, according to Dr. Hoffman. "At that age you can empathize with a person's overall life situation, not just their immediate circumstances. For instance, you could be sad for someone with a life-threatening disease, even if she seemed happy at the moment."

Beginning in early childhood, differences emerge in the degree to which people are empathic. "From around two or three, you begin to see children develop their own style of empathy, with some showing increasing awareness of other people's plight, while others seem to turn away from such concerns," says Dr. Radke-Yarrow. In studies with Carolyn Zahn-Waxler, Dr. Radke-Yarrow found that children were more empathic when their mothers disciplined them for the distress their "being bad" caused another child, saying, for example, "Look how sad you've made her feel."

Differences in empathy, in Dr. Brothers's view, reflect both how children are raised and, to some degree, biological differences in the brain. By adulthood the differences in empathy affect people's moral awareness, Dr. Hoffman says. For instance, people who are highly egoistic and presumably lacking in empathy keep their own welfare paramount in making moral decisions like how or whether to help the poor. For those who are highly empathic, on the other hand, the welfare of others is paramount in their moral judgments.

Empathy plays different roles in various realms of life, though. "When you encounter someone else in need, empathy triggers the urge to help," Dr. Hoffman says. "But empathy can also be a tool of manipulation in, for instance, trying to make a sale, particularly when the salesman has convinced himself about the worth of what he's selling."

The developmental changes in empathy through childhood seem to reflect, in part, the growth of the central nervous system in those years. "Empathy matures with the growing brain," Dr. Brothers says. It is data from infants and from observations of empathy in other species that have led Dr. Brothers to seek the brain circuits that are essential to the emotional response.

For many years scientists were skeptical that animals could display empathy. But Dr. Brothers points to one classic study conducted in the 1950s that showed they could. In the study monkeys were trained to avoid an electrical shock by pressing a lever after they heard a certain sound. Once trained, the monkeys were separated, but connected through closed-circuit video. Only one could hear the sound, while only the other could pull the protective lever. Whenever one monkey heard the sound, the other monkey saw its distressed face and promptly pulled the lever.

"The monkey that controlled the shock prevented the other monkey from getting it on the very first trial, the moment it saw the distressed face," Dr. Brothers says. Such findings seem to indicate that primates can be suitable subjects for brain studies of empathy.

Neurological data from brain damage in humans have helped focus the brain research on certain neural areas. For instance, injury to certain areas of the parietal lobes, at the back of the brain, produces an inability to understand the emotions expressed through tone of voice. And patients with injuries in some parts of the brain's right hemisphere complain that, though they feel emotions, they cannot convey them in their tone of voice or gestures. Dr. Brothers's research has focused on the visual cortex, the area of the brain that registers the meaning of what is seen, and its connections to the amygdala, part of the brain's limbic system, which responds to emotions. Studies by several groups of researchers, including one led by Edmund Rolls, a neuroscientist at Oxford University, have discovered that specific neurons in that area of the brain respond mainly to facial expressions or to movements that have social meaning in the world of primates, such as crouching.

Dr. Brothers, with her colleague Michael Hasselmo, a neurophysiologist, reports finding specific neurons in macaque monkeys that respond to certain emotional expressions. In their study some of the neurons became active only when the monkeys saw another open its mouth without showing its teeth, a threatening gesture; other neurons activated only in response to a grimace indicating fear. "The evidence seems to suggest that there are specific brain circuits for social response to emotional signals," Dr. Brothers says. "This is the type of neural activity that should lie at the base of empathy." [DG]

PLEASE TOUCH

The experience of being touched seems to have direct and crucial effects on the growth of the body as well as the mind. Touch is a means of communication so critical that its absence retards growth in infants, according to researchers who are for the first time determining the neurochemical effects of skin-to-skin contact. Their work focuses on the importance of touch itself, not merely as part of, say, a parent's loving presence. The findings may help explain the long-noted syndrome in which infants deprived of direct human contact grow slowly and even die. Psychological and physical stunting of infants deprived of physical contact, though otherwise fed and cared for, had been noted in the pioneering work of Harry Harlow, working with primates, and the psychoanalysts John Bowlby and Renee Spitz, who observed children orphaned in World War II.

The research suggests that certain brain chemicals released by touch or others released in its absence may account for these infants' failure to thrive. The studies on the physiology of touch come against a backdrop of continuing research on the psychological benefits of touch for emotional development. In some of the most dramatic findings, premature infants who were massaged for fifteen minutes three times a day gained weight 47 percent faster than others who were left alone in their incubators—the usual practice in the past. The massaged infants also showed signs that the nervous system was maturing more rapidly: they became more active than the other babies and more responsive to such things as a face or a rattle. "The massaged infants did not eat more than the others," says Tiffany Field, a psychologist at the University of Miami Medical School, who did the study. "Their weight gain seems due to the effect of contact on their metabolism."

The infants who were massaged were discharged from the hospital an average of six days earlier than premature infants who were not massaged, saving about $3,000 each in hospital costs, according to Dr. Field.

MAKING THE MOST OF MIDLIFE

A FAR more positive image of middle age is emerging as researchers look beyond their concern about the midlife crisis to examine psychological growth during the forties, fifties, and sixties. For many, middle age is the most fruitful phase of life, a time when intense preoccupations with marriage and career have faded and the inevitable deterioration of the body is yet to come. This positive slant on midlife stands in stark contrast to earlier work by psychologists who had focused on the turmoil that often results when people pause in middle age to reevaluate their lives. The emerging view of middle age accommodates both sides: while there is a reordering of priorities for most people in their forties and fifties, that reassessment often leads to a more compassionate attitude, a richer emotional life, and a deepening of personal relationships.

The newest studies show, for example, a sharp rise in people's altruism at midlife, a key sign of new priorities. Psychoanalysts now see this increased caring as the flowering in middle age of emotional development that begins in childhood. "Most people experience midlife mainly as a period of caring rather than crisis," says John Pollack, a survey researcher with New World Decisions in Princeton, New Jersey, who conducted a national poll on attitudes toward middle age. "It's a time when people look forward to increasing closeness and compassion, as opposed to, say, developing new relationships or making great career changes."

Not all researchers agree on what constitutes middle age; most of them think it is the years from forty-five to sixty-five, but others construe it to cover ages thirty-five to seventy. The survey found that those responding were inclined to define middle age in terms of life's landmarks, rather than years. The survey of 1,200 men and women found that the vast majority perceived middle age as a time of intense deepening of relationships and acts of caring. For example, 84 percent of Americans agree that "at middle age, a person becomes more compassionate to the needs of others," and 89 percent see middle age as a time of becoming closer to friends and family.

Many researchers see the findings as refining speculations by Erik Erikson, dating to the 1950s, that "generativity," a concern with nurturing others or contributing to the general well-being, is the hallmark of psychological growth in midlife. Those who did not find a way to nurture or somehow contribute to the future were left in a state of psychological stagnation, according to the theory. Although the theory was influential among psychoanalysts, only now are researchers beginning to find scientific evidence of its validity. In a study by Elizabeth Midlarsky, a psychologist at the University of Detroit, 1,380 people passing by in a shopping mall were asked by a pregnant woman to donate to fight birth defects. There was a general increase in generosity with age: the older the person, the more likely he or she was to donate. For example, 52 percent of those 15 to 24 years old donated, and 66 percent of those 25 to 34. But there was an even larger jump with middle age: for those 35 to 44, there was a 19-point rise in the percent who gave, to 85 percent. The rate of giving hovered there through age 65. There was another 10 percent jump in the rates of giving among those 65 and over.

This quickening of generosity in middle age fits observations by psychoanalysts who are investigating emotional shifts during those years. "You find people doing things at forty-five they just weren't bothering with before, getting more concerned about other people," says Dr. George E. Vaillant, a psychiatrist at Dartmouth Medical School. "As one man put it, In my twenties I learned to get along with my wife, in my thirties to get ahead in my job. In my forties I worried about other people's lives.' "

One of the few systematic studies of psychological development during middle age is from research on 204 men from the Harvard classes of 1940 to 1942 who have been assessed at regular intervals since graduation. Dr. Vaillant analyzed data from the Harvard men to evaluate their psychological maturation. His analysis, which spanned data from the men's college years to their sixties, examined the emotional defenses each man favored at different points of life. Some defenses, such as handling anxiety by losing oneself in fantasy, are considered immature by psychoanalysts. The most mature reactions to anxiety and psychic conflict include humor, creativity, and altruism.

As adolescents, these men were twice as likely to use immature defenses as mature ones, Dr. Vaillant found. As young adults, though, the men were twice as likely to use mature defenses as immature ones. But by midlife, they were four times more likely to use mature defenses. Dr. Vaillant reported the findings in *The Middle Years: New Psychoanalytic Perspectives,* an edited volume published by Yale University Press. This increase in altruism during middle age emerged in data analyzed by Dr. Vaillant from two other studies in which the same people have been assessed by researchers from childhood to old age. One was of 456 boys from Boston's inner city who were selected in 1940 at age fourteen; the other involved more than 1,200 California boys and girls who were identified by Stanford psychologists in the 1920s as having high intelligence.

The caring and deepening of relationships that blossom in middle age can take a wide variety of forms. Dr. Vaillant says, "It can be getting more involved with your grown children or delighting in your adolescents and encouraging them to flourish, coaching Little League, getting a church group off the ground, being the spirit behind a growth company, or being a dynamic headmistress of a school."

Often, nurturing in middle age takes the form of acting as a mentor. "It might be a scientist who starts worrying less about his career ambitions, tenure and grants, and starts bringing along younger scientists. One prototype is the baseball player who turns into a coach," says Dr. Vaillant. Middle-age comforts may be at play in the increase in nurturing, said Robert Michels, chairman of the psychiatry department at the Cornell Medical School. Writing in *The Middle Years* with Elizabeth Auchincloss, a New York City psychoanalyst, Dr. Michels cites several studies showing that as people approach midlife they tend to experience fewer stressful events, such as losing a job. When they do go through such a major change, the studies found, middle-aged people see it as much less upsetting than do younger people. A life that is less eventful and less stressful than in youth, Dr. Michels says, allows people in middle age to face the central crisis of later life, the approach of death, "with the armor of wisdom, experience, and stability." From the more pressing confrontation with death that occurs in middle age, Dr. Michels says, a new view of one's life emerges. The result is often a shift in people's ideals and hopes from centering on their personal aspirations to "symbolic extension of the self," such as one's family, community, or a cause. "You gain a kind of symbolic immortality by furthering a group you belong to or a cause you identify with," says Dr. Michels.

Dr. Michels cites as an example Eugene Lang, a successful New York City businessman who, at age sixty-two, broke from his planned remarks to the sixth-grade class at New York's Public School 121 to announce he would pay for the education of every student in the class who was accepted into college. Dr. Vaillant says that those who are the most giving at middle age are generally the same people who threw themselves most intensely into developing their careers earlier in life, and that those who had consolidated their careers and found a stable marriage were most able to be giving in middle age. "It's hard to make a contribution to others if you haven't first done so for yourself," he says. "You can't give something away until you have something to give. But longitudinal studies find that people at forty-five are more community-oriented and altruistic than they were at twenty-five."

There have been earlier findings suggesting that middle age is a period of emotional blossoming, but they were largely obscured by the intense research focus on the crises and trials of middle age. The opening that middle age brings differs for men and women, according to studies by David Holmes, a psychologist at the University of Kansas. "Women say the happiest period of their lives is when their children grow up and leave home," Dr. Holmes says. "It opens up all kinds of possibilities for them; they report being more mellow, and yet more assertive. For men in their forties, though, the big change is that they discover relationships. They turn from a focus on their

careers to one that includes the people in their lives; they want both."
Dr. Michels says: "If people are reasonably mature in middle age, they
no longer experience their ideals and ambitions in a limited personal
sense, but in a larger perspective. They are less concerned with their
own striving, and more interested in the meaning of their lives and in
touching the lives around them."

Dr. Vaillant adds: "Whether you can express your altruism depends on your circumstances. At midlife you're at the height of your
feeling of control: you have both good health and good earning
power. By your sixties, you may already be pulling back, and feeling
less in control." [DG]

ON GOLDEN POND

The secret of emotional health among older men is not a successful career, a happy marriage, or a stable childhood, new
findings suggest. It lies instead in an ability to handle life's
blows without passivity, blame, or bitterness.

The findings, which contradict widely held theories about
the importance of early life for emotional well-being in adulthood, are among recent conclusions of a study of men who
have been scrutinized at five-year intervals since they gradu-
ated from Harvard in the early 1940s.

The project, known as the Grant study after the W. T.
Grant Foundation, which initially supported it, is one of a
handful that have intensively assessed people at regular intervals through their adult years. Such studies are particularly
valuable for the understanding of psychological development
because they allow researchers to see what factors matter, for
better or worse, later in life.

The researchers defined emotional health at sixty-five as
the "clear ability to play and to work and to love" and a feeling
of satisfaction with life.

These were among their findings:

• Pragmatism and dependability are particularly important.

- Many factors in early life, even devastating problems in childhood, had virtually no effect on well-being at sixty-five.

- Being close to one's siblings at college age was strongly linked to emotional health at sixty-five.

- Severe depression earlier in life caused problems that persisted.

- Traits that were important at college age, like the ability to make friends easily, were unimportant later in life.

The latest data were collected by George E. Vaillant, a psychiatrist at Dartmouth Medical School.

The men hardly represent a cross section of Americans. All were Harvard undergraduates, white, and in good mental and physical health when selected. The researchers say that by avoiding complicating factors like sex, economic status, and race, they were able to focus on more subtle factors that propel one person forward while another lags.

The Pursuit of Health

THE CODEBREAKERS

A S the human genome project drives steadily forward, the vast new effort to delineate all 3 billion chemical building blocks of humanity's genetic material is arousing alarm, derision, and outright fury among an increasingly activist segment of the biomedical community. The critics argue that the human genome project has been sold on hype and glitter, rather than its scientific merits, and that it will drain talent, money, and life from smaller, worthier biomedical efforts.

Matching the pitch of the criticism is the scale of the project and the intensity of support among its promoters. As biology's first foray into big-ticket science, the $3 billion, fifteen-year human genome project is designed to do nothing less than decipher the complete code of the 50,000 to 100,000 genes that are the genome, the blueprint for a human being. Researchers hope to attain their goals by sketching out biochemical maps and sequences of genes.

Its backers insist that a large-scale study of human deoxyribonucleic acid (DNA) will benefit all biomedical researchers, greatly accelerating the search for the genes that cause many diseases, from rare disorders like Huntington's disease to common ailments like cancer. And they argue that the new technology's spinoffs will benefit other experiments. "Our project is something that we can do now, and it's something that we should do now," says Dr. James D. Watson, a Nobel laureate who heads the National Center for Human Genome Research at the National Institutes of Health (NIH). "It's essentially immoral not to get it done as fast as possible."

But opponents of the project argue that the effort is intellectually questionable. They contend that even if scientists manage to finish the genome project, it will have generated enormous reams of uninterpretable and often useless data, essentially a computerized catalogue of genes, subunits of genes, and long stretches of filler material, with few clues about how any of that genetic material works or can trigger disease. "The human genome project is bad science, it's un-thought-out science, it's hyped science," says Dr. Martin Rechsteiner, a biochemist at the University of Utah. Some critics have begun aggressive letter-writing campaigns, urging colleagues who harbor similar sentiments to write Congress.

Opponents say that the battle has shaped up as a fight between prominent scientists (including Nobel laureates like Dr. Watson and Dr. Walter Gilbert of Harvard University) who support the genome project and less renowned scientists who usually run smaller labs on modest budgets and see the genome project as a threat. Although critics doubt that they will derail the project soon, they hope that their constant hounding will influence the course of the project and assure that it does not drain resources from other research.

Despite its unifying rubric, the human genome project is not really one project, but a series of related projects that are now being performed by scientists at universities and institutes around the country. It is jointly orchestrated by Dr. Watson's office at the NIH and an office at the Department of Energy. Congress first allocated a formal budget of $27.9 million to genome-related research in 1988. In 1990, the NIH and the Energy Department dispensed $87.4 million in genome grants, with the NIH controlling about two-thirds of that money. Together the offices may receive as much as $200 million annually for the remainder of the project. That budget, says Dr. Watson, represents a mere few percent of the current $7.5 billion budget of the health institutes, and he stresses that the genome money is newly won money, not funds that have been diverted from the research coffers of the health institutes.

Although now staunchly behind the genome project, scientists at the health institutes initially scorned the whole notion. The idea for a huge DNA analysis project first arose at Los Alamos and Lawrence Livermore laboratories, the Energy Department's national weapons research centers. Scientists there were trying to determine whether the offspring of the survivors of the Hiroshima bomb had mutations in their DNA as a result of their parents' exposure to radiation. "We realized that we didn't have any methodologies sensitive enough to detect the mutations if they were there," says Dr. Benjamin J. Barnhart, manager of the Human Genome Program at the Energy Department. The scientists decided that the only way to pick up the mutations would be to sequence, or spell out, all the 3 billion subunits that make up the human genome, and then look for individual errors in the subunits. The weapons laboratories also were seeking novel ways to use their huge data banks and technical resources for civilian projects that would keep them busy in the event military research declined in a more peaceful era.

Thus galvanized, Energy Department experts presented the idea of a national mass-sequencing effort at a scientific conference in Colorado in 1984. However, the great majority of scientists dismissed the original proposal with hostility or indifference. They argued that current sequencing technologies are laborious and expensive, costing a biomedical researcher between $3 and $5 to spell out a single DNA subunit, which, multiplied by 3 billion subunits, would amount to up to $15 billion just to sequence the entire genome. And they declared that most of the money would be wasted. Scientists suspect that up to 95 percent of human DNA has no function.

Nevertheless, a few prominent scientists continued pushing the idea of a genome project. They significantly modified the aims of the project along the way, largely to meet the fierce objections of scientific luminaries like Dr. David Baltimore, a Nobel laureate who is currently president of Rockefeller University in New York. Among the biggest of the adjustments was to postpone the idea of sequencing all 3 billion pieces of DNA in favor of first sketching out a good "road atlas" of the genome. That genetic map will feature between fifteen hundred and three thousand genetic markers, or distinct biochemical patterns, evenly spaced up and down the chromosomes. Scientists obtain such markers by subjecting the DNA of large families to chemical manipulations and then painstakingly comparing the resulting patterns of the various family members in search of distinct hallmarks. With those hallmarks distributed around the human chromosomes, researchers can use the patterns as trailblazers, pointing them in the direction of genes that they wish to isolate and study. Such a map, many researchers concur, will be a magnificent resource.

As the genetic mapping proceeds, other researchers will strive to create another sort of map known as a physical map, consisting of the human chromosomes chopped into giant pieces; each long piece is then inserted into a yeast cell or other cell that can be manipulated to reproduce the piece of chromosome. The individual pieces will be lined up end to end in proper order, as they exist in human cells. This operation will essentially create a single photocopying machine for the entire human genome. In theory, a researcher will employ the genetic map to locate a gene of interest and then turn to the physical map to actually pull out the gene and copy it until he or she has a quantity that can be analyzed.

Before undertaking the final phase of the project, other scientists will attempt to make sequencing technology ten times more efficient than it is today. Then they will attempt to spell out the entire human genome to determine if the supposed "junk" DNA harbors important insights about human evolution.

In another crucial concession to early critics, genome proponents agreed that some laboratories should study in parallel the DNA of favorite experimental organisms, like worms, fruit flies, and mice, comparisons that could allow researchers to better understand the DNA of humans. Thus mollified, a number of early doubters have rallied behind the project. "I have no objection to the program as it's currently organized," says Dr. Baltimore. Dr. David Botstein, a former vice president of Genentech who now heads the genetics department at Stanford University Medical Center, says that he initially viewed the genome project with "extremely negative feelings," but that when genome backers agreed to emphasize mapping and technological development for sequencing, he changed his mind.

Grousing about the genome enterprise might have continued to wane had it not been for recent budget limitations on new grants at the health institutes. Between 1987 and 1990, the number of new grants dispensed to young, independent researchers shrank from 6,446 to 4,633, largely as a result of administrative changes in financing policy that had nothing to do with the genome project. But left without support for their struggling labs, many young researchers have attacked the genome project as unfair. "You can't prove it, but I think it is widely felt that the two are in competition," says Dr. Bernard D. Davis, a professor of microbiology and molecular genetics at Harvard Medical School.

Critics have also questioned the need for a new enterprise that receives special consideration in the federal budget. "If this has validity as a scientific project, why not let it compete with all other grant applications to the NIH, instead of saying, this $200 million can only be spent on this particular type of effort?" says Dr. Ry Young, a professor of biochemistry and biophysics at Texas A&M University in College Station. "Why should it be insulated from the fray?"

Opponents wonder whether many of the new converts to the human genome project have been persuaded less by the inherent worthiness of the project than by the possibility that they, too, can win genome-related grants. The project has even greater appeal now that the NIH genome office is offering a new type of grant to support large groups of scientists through so-called genome centers. These centers will receive around $3 million a year for up to five years, which is about ten times the average grant given to a university lab. Eventually, the institutes' genome office intends to distribute about half of its money in giant grants to centers. Young researchers worry that financial devices like the big grants will concentrate ever more resources in ever fewer hands. "I had a brilliant young scientist say to me recently, 'The fat cats are all getting the cream, while I'm sitting here starving,' " said Dr. Rechsteiner.

Beyond money issues, critics of the program quarrel with its medical and scientific claims. They say that the best approach to understanding human disease is not to thwack away randomly at the thick forest of human DNA, as they say genome researchers will do, but to study one specific disease at a time, as scientists traditionally have done. Indeed, they doubt that the human genome project will quickly cure any diseases at all, observing that the mere identification of genes responsible for certain illnesses is no guarantee that researchers can easily learn how the genes work or what can be done to correct their defects. "The gene for sickle-cell anemia has been known for twenty years, and it has yet to lead to a cure for the disease," says Dr. Michael Syvanen, a microbiologist at the Medical School of the University of California at Davis and a stout antagonist of the genome project.

Critics also worry that many jobs necessary to complete the genome project, like identifying biochemical markers on chromosomes, are both so difficult and so numbingly tedious that the students and postdoctoral fellows who will be expected to do the job will rapidly become bored and disillusioned. And though technicians may be able to handle some of the tasks of genome studies, other jobs will require graduate-level training without offering the commensurate intellectual rewards. "I haven't the foggiest idea how I would inspire my students to work on this sort of thing year after year," says Dr. Michael Wigler, a well-known geneticist at Cold Spring Harbor Laboratory on

Long Island, who is not considering genome-related experiments for his group.

But supporters dismiss the complaints as naive and short-sighted. They say that, while identifying a gene is only the first step toward curing a disease—or understanding how the brain works or how a fertilized egg turns into a human baby—it is a vital first step. "Most knowledgeable people and most eminent scientists are solidly behind" the genome project, says Dr. James Wyngaarden, who was in charge of the health institutes when it joined the genome project. "The ones who are critical are journeymen biochemists who may be having a hard time competing themselves." [NA]

THE IMPRESARIO OF LIFE

THE question of how a single fertilized egg blossoms into a complete human infant is one of the magnificent puzzles of biology, and scientists are just beginning to pinpoint the key genes and molecules that direct the intricate unfolding. But in a rush of new experiments, researchers have made the surprising discovery that one of those crucial impresario molecules of life is not some exotic or arcane compound, but retinoic acid, a familiar derivative of vitamin A. Retinoic acid lately has won fame as the active ingredient in Retin-A cream, a treatment for acne and, more debatedly, for the feathery wrinkles of age. The molecule also made news as the main component of Accutane, a more potent anti-acne drug that can cause extremely serious birth defects if women use it during pregnancy.

Now biologists are finding that retinoic acid plays a pivotal role in normal development as well. In a recent flurry of papers published in *Nature, Cell, Development*, and other scientific journals, they report that retinoic acid helps determine the shape and pattern of a broad array of the body's organs, including parts of the brain and spinal column, the face, the limbs, the heart, the skeleton, and the skin. The results come largely from studies of chicks and mice, but researchers believe that they are likely to apply to human development as well. "It's a good bet that this is one of the body's master molecules," says Gregor

Eichele, a cellular physiologist at Harvard Medical School and a leading researcher in retinoic acid.

The molecule appears to work by flicking on whole groups of genes during key moments of development. Those genes are thought to then stir up other batches of dormant genes, thereby setting off a cascade effect that has the collective might to sculpt an undifferentiated blob of cells into a defined organ. The apparent importance of retinoic acid to the growing embryo has impressed and surprised many researchers. Lorraine Gudas, a developmental biologist also at Harvard Medical School, says: "I think this is going to be one of the most exciting molecules in respect to embryogenesis that exist. It's a master regulator that can send a very loud signal at critical points throughout development."

Dr. Gudas and other investigators also believe that retinoic acid remains important in cell control throughout life, particularly in orchestrating the growth and health of epithelial tissue, which makes up the bulk of the skin, the breast, and the lining of the lungs, intestines, and other organs. They hope that by understanding the mechanisms of retinoic acid, they will better comprehend why certain cells go awry, proliferating into tumors, for example, or simply shutting down and dying. "I think a lot of this new basic science research into retinoic acid will be of real value to the clinical treatment of cancer," says Dr. Gudas.

Already the revelations about retinoic acid help explain why drugs such as Accutane produce the harrowing constellation of birth defects that physicians have observed. Since 1984, Dr. Edward Lammer of the California Birth Defects Monitoring Program in Emeryville has studied almost ninety cases of malformations believed to have been caused by Accutane. He has seen children with misformed hindbrains, or clusters of brain cells in parts of the brain where they do not belong. Other children are missing the bones of the middle ear or the entire ear canal altogether. Some young patients have hearts in which the two major arteries never separated. Others lack a functioning thymus, the gland where many white blood cells mature, and so they suffer from immune deficiencies. Some of the children do not survive.

Dr. Lammer notes that nearly all the defects occur in those organs believed to be controlled by the body's own stores of retinoic acid during early stages of embryogenesis, usually within the first month or two of pregnancy. He and other researchers suggest that the extra dose of retinoic acid from the acne drug, which the women took before realizing they were pregnant, probably disrupted the exquisite biochemical precision necessary for normal growth.

The basic research and the clinical observations "all fit together like a nice neat glove," says Dr. Lammer. "Too neatly, unfortunately." For biologists, the severity and the pattern of the birth defects provide striking evidence that retinoic acid modulates development in a profound and specific manner.

All vertebrate animals need vitamin A to survive. They get much of it by eating sources of beta carotene, the compound that makes carrots orange, spinach green, and lobsters brownish-red. Beta carotene is then metabolized in the body into vitamin A and its derivatives, including retinoic acid. Fetuses obtain their vitamin A from the mother, although they probably transform it into retinoic acid within their own tissue.

Scientists were inspired to explore the role of retinoic acid in organ formation because, among other reasons, studies showed that excess vitamin A caused birth defects in laboratory animals. In other provocative experiments, biologists found that when they swabbed retinoic acid onto cultured cancer cells, the cells stopped dividing wildly and took on the features of more normal cells. In an effort to begin pulling apart the exact mechanism of retinoic acid, biologists examined its impact on limb development. They demonstrated that if they applied beads soaked in the molecule to particular spots in the growing limb buds of chick and mouse fetuses, the animals developed distinctively defective limbs: twin wings or twin paws extending from a single limb, the two appendages facing one in another in a bizarre mirror image.

The results indicated that retinoic acid works as a morphogen, a molecule that helps cells to migrate and form patterns characteristic of mature organs. Researchers believe that under normal circumstances the earliest cells of the limb bud release a specific amount of retinoic acid. The signal that dictates the release is not yet known, but the effects of the release are becoming clear. As retinoic acid seeps across the limb bud, different cells receive different concentrations of the molecule. The amount of retinoic acid that reaches a given cell in part dictates its fate. In the case of the growing mouse paw, for example, cells exposed to higher amounts of the molecule migrate outward to form the equivalent of the pinky, while those receiving smaller doses become the equivalent of thumb cells.

In 1987, Ronald Evans of the Salk Institute in La Jolla, California, and Pierre Chambon of the Institute of Biochemistry in Strasbourg, France, isolated the retinoic acid receptor, a protein inside the cell that links up to the vitamin derivative and allows cells to respond to it. Since then, two related retinoic acid receptors have been identified, and researchers believe there may be others. Importantly, the receptors are equipped with a telltale looping appendage at one end, called a zinc finger, which scientists think allows the proteins to clasp onto DNA (where the cell's genetic information is stored) and ignite a burst of gene activity. "The receptor is what takes something with a simple structure, like retinoic acid, and translates that into a complicated signal," says Dr. Evans. "The receptor turns on the stereo system."

Many details of the retinoic acid pathway remain to be fleshed out, but scientists think that the vitamin derivative operates by entering a cell and somehow arousing one or more of its designated receptors. The receptors then glide over the cell's DNA molecule and flick on a battery of genes. The timing and amount of the retinoic acid that infiltrates the cell seem to help determine which genes become activated. Again in the case of the growing mouse paw, those cells in the limb bud that receive a lot of the molecule at just the right time have their genes tweaked to become pinky cells, while the cells that are doused at a later time by less retinoic acid are genetically inspired to form a thumb, though additional signals surely contribute to the process.

With the retinoic acid receptors isolated, biologists have been screening embryonic tissue to see what other budding organs might employ retinoic acid as their sculptor. From studying activation patterns of the retinoic acid receptors, researchers suspect that the molecule works first on cells destined to become part of the spinal column. Soon afterward, it influences the construction of the hindbrain, face, ears, and jaw. Later the vitamin is involved in the dispatch of cells to the heart, liver, circulatory system, and cartilage.

Scientists are also fitting together retinoic acid with other molecules thought to be master orchestrators of development, including proteins known as growth factors and a class of genes called homeobox genes. "The field is still in an embryonic stage, but all our findings are starting to converge," says David L. Stocum, a developmental biologist at Indiana University–Purdue University at Indianapolis. "I expect that in the next two or three years we'll make a lot of headway in understanding just what gives a cell its knowledge of what it's supposed to be." [NA]

WHEN IS A FETUS "VIABLE"?

ALTHOUGH doctors have made great progress in helping severely premature infants survive, they have been unable to overcome a seemingly impenetrable barrier: if a baby is born before twenty-three or twenty-four weeks of pregnancy, experts say, it simply cannot survive. And nothing that medical science can do will budge that boundary in the foreseeable future. Experts say that the overriding problem for these infants, as compared to those born after the forty weeks of a normal pregnancy, is that their lungs are too immature to function, even with the help of respirators. The issue has become important in the abortion debate, where arguments over just when a fetus is "viable"—when it can survive childbirth—have taken the spotlight, especially in court.

Whatever the actual date of viability might be, the vast majority of abortions are performed well before it. About 90 percent of the 1.6 million abortions in the United States each year occur in the first three months of pregnancy, says Susan Tew, a spokeswoman for the Alan Guttmacher Institute in New York, and fewer than 1 percent occur at or after twenty-one weeks of pregnancy.

Advances in technology have improved survival rates for premature infants born after twenty-three to twenty-four weeks. According to a recent report by the Congressional Office of Technology Assessment, only 10 percent of babies weighing less than 2.2 pounds, born before about twenty-eight weeks of pregnancy, survived in 1960. Now half of them live. The report said the advances in saving these babies were most pronounced in those weighing 1.6 to 2.2 pounds, or 750 to 1,000 grams, which corresponds approximately to twenty-five to twenty-eight weeks of gestation. Two thirds of these babies survive today whereas almost all died in 1960.

Dr. Mary Ellen Avery of the Harvard Medical School says that in her own hospital, Brigham and Women's Hospital in Boston, about half of all babies born after twenty-four weeks of pregnancy survive and one tenth of those born after twenty-three weeks survive. But, she added, "the odds of survival become infinitesimal before twenty-three weeks." Dr. Watson A. Bowes, Jr., a professor of obstetrics and gynecology at the University of North Carolina in Chapel Hill, who says that he takes "a strong pro-life position" and who advises the National Right to Life Committee, says he views twenty-three weeks as about the limit of fetal viability. He has seen babies born at twenty-three weeks who survived but has not seen younger survivors. "In our own hospital, the official lower limit of viability, when we will not do pregnancy terminations, is twenty-three weeks. That is sufficiently well documented that we've really taken it as our limit."

Dr. Jack Wilke, a general practitioner who is president of the National Right to Life Committee, says he has seen newspaper accounts of babies who were born at twenty weeks of pregnancy and survived. He says hospital or medical personnel present at the births confirmed the reports. "There have been no survivors that anyone knows of below twenty weeks," he adds, but "it could be that next week we'll save an eighteen- or nineteen-weeker."

But neonatologists, pediatricians, and many scientists say they have yet to see a twenty-week-old fetus that survived. They said they suspect that the length of pregnancy in the cases Dr. Wilke cites may have been miscalculated and that the fetuses he referred to may have been closer to twenty-four weeks. It can be very difficult to determine the length of a pregnancy accurately. Many women do not know exactly when they became pregnant. And even the best methods of estimating a woman's stage of pregnancy are not entirely accurate, particularly late in pregnancy.

The usual way of dating a pregnancy is by counting from a woman's last menstrual period. At best, this method will be accurate within a week for only 80 percent of women, according to Dr. Sheldon B. Korones, director of the Newborn Center at the University of Tennessee. Dr. Korones says that the best way to date a pregnancy is with ultrasound pictures before twenty weeks of pregnancy. But even dates obtained from ultrasound early in pregnancy can err by a week in either direction.

The primary reason that babies younger than twenty-three weeks almost never survive is that their lungs are unable to breathe. During the first twenty-three to twenty-four weeks of fetal life, the lungs and the capillaries that carry blood to the lungs are very immature, Dr. Korones says. Before twenty-three or twenty-four weeks, capillaries have not yet moved close enough to the air sacs of the lung to carry gases to and from the lungs, and the lining of the air sacs is too thick to allow gas exchange. Moreover, the lungs are too stiff to expand and contract effectively.

"What it boils down to is that the blood vessels and the airway have not come close enough together nor have they developed sufficient maturity to function," Dr. Korones says. "Time must pass. Maturity doesn't develop instantly. The way we've got it pegged now is twenty-three or twenty-four weeks. We're going to get better at saving babies that are that small. More will survive. But I don't think we'll be able to do much about babies that are much younger."

A committee of experts appointed by the New York State Task Force on Life and the Law to determine the age at which fetuses can survive outside the womb concluded that "23–24 weeks is the threshold of fetal survival." Technical advances in the foreseeable future will increase the survival of infants above that threshold, the group said, "but will not lower the threshold for fetal extrauterine survival." Babies born at twenty-three or twenty-four weeks also have very immature kidneys that cannot properly excrete urine. "The ability to excrete waste products is extremely limited," says Dr. Arthur E. Kopelman, head of neonatology at the East Carolina School of Medicine, who has done research on the smallest premature infants.

Citing still another difficulty, Dr. Bowes said: "Our microtechnology is at its limit. You have to feed these babies. Their veins are tiny. Their vessels are tiny. Just to get the equipment in, we've reached an impasse." Some researchers have suggested that it might one day be possible to devise an artificial placenta to take the place of a baby's immature lungs and kidneys if he or she is born too early. The artificial placenta would function like a heart-lung machine, recirculating the baby's blood, cleansing it, and adding oxygen.

"In theory, you can imagine how you would do that, but in fact problems such as overwhelming and fatal infections would probably preclude it," said Dr. Kopelman. "There are also problems with hemorrhaging. You have to anticoagulate blood so that it doesn't clot as it goes through the tubing. Tiny babies have a tremendous propensity to hemorrhage into their brains anyway," he said, and if they were put on an artificial placenta machine "I think they would suffer tremendous brain hemorrhages."

Dr. Leo Stern, professor and chairman of the pediatrics department at Brown University, says that to be nearly certain that a fetus would not have been viable at the time of an abortion, the procedure would have to be done before the twentieth week of pregnancy.   [GK]

SMOKING MOTHERS

Two doctors in Sweden say they have found a strong link between smoking and sudden infant death syndrome. They attributed more than one-fourth of the deaths in a three-year study to smoking by mothers during pregnancy. The doctors, Bengt Haglund and Sven Cnattingius, compiled data on nearly 280,000 live births in Sweden from 1983 to 1985. Of those children, 190 died of sudden infant death syndrome, and Dr. Haglund and Dr. Cnattingius said smoking could be blamed for 50 of the deaths. Sudden infant death syndrome, known as SIDS, is a term for deaths with no known cause in infants from one to six months old.

The doctors found that a pregnant woman who smoked moderately, which was defined as one to nine cigarettes a day, was twice as likely as a nonsmoker to lose her infant to SIDS. Women who smoked heavily, which the researchers defined as more than 10 cigarettes a day, tripled the risk of SIDS. For these women, the study found 1.47 infant deaths for every 1,000 live births, as against 0.49 death per 1,000 live births in nonsmokers.

The study also found that among infants who died of SIDS, those whose mothers smoked during pregnancy lived two thirds as long as the others. The doctors said that 15 to 20 percent of women quit smoking during pregnancy, suggesting that the link between SIDS and smoking might be even greater than their study had found it to be. "From a preventive point of view, smoking is the single biggest cause of SIDS," Dr. Haglund said.

But he added that other socioeconomic factors, like the age of the mother, social class, and whether the father lived with the mother and child, affected the results. Dr. Haglund said that while the cause of SIDS was unknown, "if we know the relationship between various factors, at least we can work towards finding a solution." Dr. Haglund said no correlation could be made between the study in Sweden and possible results elsewhere. The study showed that the incidence of SIDS is lower in Scandinavian countries than it is in other industrialized countries like the United States.

PASSIVE SMOKING

AFTER years of questioning the potential health hazards of second-hand cigarette smoke, a growing number of scientists and health officials are becoming persuaded that the dangers are real, broader than once believed, and parallel to those of direct smoke. It has long been established that smoking harms the health of those who do the smoking. Now new epidemiological studies and reviews are strengthening the evidence that it also harms the health of other people nearby who inhale the toxic fumes generated by the smoker, particularly from the burning end of the cigarette. Such indirect, or second-hand, smoking causes death not only by lung cancer, but even more by heart attack, the studies show. The studies on passive smoking, as it is often called, also strengthen the link between parental smoking and respiratory damage in children.

In reports and in interviews, more than a dozen experts say that there is little question that passive smoking is an important health hazard. What has swayed many scientists is a remarkable consistency in findings from different types of studies in several countries with improved methods over those used in the first such studies a few years ago. The new findings confirm and advance two landmark reports in 1986 from the Surgeon General, who concluded that passive smoking caused lung cancer, and from the National Research Council, which said passive smoking is associated with lung cancer. "The links between passive smoking and health problems are now as solid as any finding in epidemiology," says Dr. Cedric F. Garland, an expert in the epidemiology of smoking at the University of California at San Diego. An Environmental Protection Agency (EPA) report concludes that passive smoke causes 3,800 lung cancer deaths each year. The study calls the statistical data firm, not a result of chance, and biologically plausible.

The newer understanding of the health hazards of passive smoking are underscored in a report presented at a world conference on lung health in Boston in May 1990. The report's author, Dr. Stanton A. Glantz at the University of California at San Francisco, estimates that passive smoke kills 50,000 Americans a year, two thirds of whom die of heart disease. Passive smoking ranks behind direct smoking and alcohol as the third leading preventable cause of death, Dr. Glantz says in what experts call an unusually thorough review of all eleven epidemiological studies published on the subject. About 400,000 Americans die from breathing their own smoke each year, the Surgeon General's 1989 report said. About 100,000 die from alcohol, according to a report in 1987 from the National Institute on Alcohol Abuse and Alcoholism.

Donald Shopland of the National Cancer Institute, who has helped prepare the Surgeon General's reports on smoking since 1964, says "there's no question" now that passive smoking is also a cause of heart disease. The evidence for the health hazards of passive smoking is largely statistical and epidemiological; such evidence has often pointed the way for scientists to confirm the findings from laboratory research.

Dr. Mark L. Witten of the University of Arizona has reported on what he says is the first documented damage to animals from passive smoking. In experiments begun when he was at the Massachusetts General Hospital in Boston, Dr. Witten exposed rabbits to sidestream smoke for fifteen minutes a day for twenty days in amounts comparable to those received by children of smoking parents. The animals developed lung damage resembling an asthmatic reaction. Cells in the airways of the rabbits degenerated and their lungs became more permeable, posing a significant risk for the development of serious lung disease by creating an easier entry point for microbes, pollutants, and toxins.

The new findings on passive smoking parallel recent changes in laws and rules that limit smoking in public places. In recent years, all but four states (Missouri, North Carolina, Tennessee, and Wyoming) have passed comprehensive laws limiting smoking in places like restaurants, schools, stores, hospitals, and theaters, according to Action for Smoking and Health, an advocacy group. Airlines have banned smoking on all domestic flights of less than six hours.

Only a decade ago many scientists were skeptical over the initial links between passive smoking and lung cancer. Dr. Garland, the San Diego expert, says: "We were out on a limb when we started, but now we have the kind of replication one would want to see play a role in public policy."

But Walker Merryman, a spokesman for the Tobacco Institute, a trade group, disputes the conclusions, citing reports from a 1989 conference at McGill University in Montreal that was partly sponsored by the tobacco industry. Dr. Joseph M. Wu, a biochemist at New York Medical College in Valhalla, said in his concluding remarks at the meeting that the published data "are inconsistent with the notion that environmental tobacco smoke is a health hazard." He also said, "It appears premature to take any sort of regulatory action with regard to environmental tobacco smoke at this point."

Cigarette smoke consists of more than 4,700 compounds, including 43 carcinogens, the EPA says. Major differences exist in the components of mainstream and sidestream smoke that largely reflect the degree of combustion. Mainstream smoke is inhaled from smoking and consists of large particles deposited in the larger airways of the lung. Sidestream smoke is generated from the burning end of cigarettes, cigars, and pipes during the smoldering between puffs. Sidestream smoke may come from someone else's tobacco or from one's own and is the major source of environmental tobacco smoke. Sidestream smoke is a mixture of irritating gases and carcinogenic tar particles that reach deeper into the lungs because they are small. Scientists say that because of incomplete combustion from the lower temperatures of a smoldering cigarette, sidestream smoke is dirtier and chemically different from mainstream smoke.

Dr. Henry D. McIntosh, a former president of the American College of Cardiology, says he agrees with Dr. Glantz's review, which found a 30 percent increase in risk of death from heart attacks among nonsmokers living with smokers. The risk correlates with the amount of the spouse's smoking; the heavier a smoker, the greater the risk.

Researchers have found that passive smoking makes platelets, the tiny fragments in the blood that help it clot, stickier. The findings were made on ten healthy nonsmokers who sat for twenty minutes in an open hospital corridor beside smokers. Platelets can form clots on plaques in fat-clogged arteries to cause heart attacks and may also play a role in promoting arteriosclerosis, the underlying cause of most heart attacks. Researchers have also shown from animal and human studies that chemicals in sidestream smoke may injure cells on the inside lining of the arteries and thus promote development of plaque and arteriosclerosis.

Dr. Glantz's review, which was done in collaboration with Dr. William W. Parmley, chief of cardiology at the University of California at San Francisco and a former president of the American College of Cardiology, highlights the interplay between such processes. Although platelets do not normally stick to the inside lining of arteries, such sticking is easier when cells in the lining have been damaged. Under such conditions, the platelets can release substances that cause further damage to the artery.

Researchers have also shown that passive smoking adversely affects heart function, decreasing the ability of people with and without heart disease to exercise. "You don't get oxygen to the heart as well," Dr. Glantz says. Passive smoking increases the demands on the heart during exercise and reduces the heart's capacity to speed up. For people with heart disease, the decreased function can precipitate the chest pains from angina. The decreased heart function may reflect impeded enzyme activity within cells, an effect that has been documented with passive smoke in animals, Dr. Glantz says.

Blood tests of adolescent children whose parents smoke show changes that increase their risk of heart disease, according to studies by Dr. William B. Moskowitz's team at the Medical College of Virginia in Richmond. The children, exposed to passive smoke since birth, had increased amounts of cholesterol and lower levels of HDL, a protein in blood that is believed to protect against heart attacks. The researchers found that the greater the exposure to passive smoke, the greater were the biochemical changes.

The effects of passive smoking on heart disease were not part of the EPA study, which relied on the agency's carcinogenicity assessment guidelines to evaluate lung cancer and respiratory illness. The EPA reviewed twenty-four epidemiological studies of passive smoking and lung cancer, eleven more than in the Surgeon General's report in 1986. The newer studies confirm the first thirteen studies. "The evidence is remarkably consistent, even with different methods, and is persuasive," says one EPA official who asked not to be identified. A pioneering report linking passive smoking and lung cancer came in 1981 from a fourteen-year Japanese study by Dr. Takeshi Kirayama. His research methods were criticized at first, but critical review, corrections, and revisions have "failed to discredit the findings," according to the EPA report. Lawrence Garfinkel, an epidemiologist who is vice president of the American Cancer Society, says that he was at first skeptical of Dr. Hirayama's report but that he was convinced by later studies, including his own, that there was about a 30 percent increased risk of developing lung cancer from passive smoking. Mr. Garfinkel adds that the results of an ongoing study of 1.2 million Americans should help clarify the degree of risk from all types of cancer and other diseases.

Dr. Glantz estimates that one third of the fifty thousand deaths from passive smoking were from cancer. In addition to lung cancer, researchers have linked cancer of the cervix to both mainstream and sidestream smoke. Mainstream (but not sidestream) smoke has been linked to cancers of the mouth, throat, larynx, esophagus, urinary bladder, kidney, and pancreas. Sidestream (but not mainstream) smoke has been linked to cancers of the brain, thyroid, and breast. Several groups are trying to do other studies to determine whether such findings are spurious and whether sidestream smoke can cause cancers not caused by mainstream smoke.

The American Academy of Pediatrics estimates that 9 million to 12 million American children under the age of five may be exposed to passive smoke. The nearly thirty epidemiological studies on passive smoking and respiratory disorders published since the 1986 reports strengthen earlier conclusions that passive smoke increases the risk for serious early childhood respiratory illness, particularly bronchitis and pneumonia in infancy. Increased coughing has been reported from birth to the mid-teen-age years in thirteen newer studies of passive smoking and respiratory symptoms.

Researchers say that young children may develop symptoms after only a few months of exposure to passive smoke. The EPA says passive smoke may be a particular problem because of an infant's immature immunologic and respiratory systems. The EPA has also found that passive smoke can lead to middle ear infections and other conditions in young children. The agency has said that asthmatic children are particularly at risk and that lung problems developed in childhood can extend to adulthood.   [LKA]

THE MYSTERY OF THE FEMALE HEART

MEN and women are different at heart, literally. Medical scientists are both perplexed and intrigued by a growing body of evidence that shows differences in the way women and men develop heart disease. The findings suggest there may be important biological distinctions between the sexes in the functioning and development of the heart and cardiovascular system. The distinctions need to be understood because they may provide new insights into the causes and treatment of heart disease in both sexes.

So far, virtually every study of methods to prevent or treat heart disease has been done in men, with the results assumed to apply to women. This is partly because the emphasis has been on reducing the rate of heart attacks among the middle-aged, and most middle-aged heart attack victims are men. But later in life heart attacks are the leading cause of death in women too, and some researchers are now looking into the aspects of heart disease that are unique to women.

The experts all stress that women should continue to follow established guidelines for heart disease prevention such as avoiding smoking and high-fat diets and curbing high blood pressure. "There certainly are differences," says Dr. Peter Frommer, deputy director of the National Heart, Lung, and Blood Institute, who characterizes the new evidence as "tantalizing leads." One of those leads comes in a report that women who have had children tend to have wider coronary arteries, which may help explain their protection from early heart attacks. A study of monkeys points to another difference: on a high-fat diet, female monkeys tend to produce more high-density lipoproteins, HDLs, which carry cholesterol away from blood vessels, than male monkeys do. Yet another recent finding highlights the inadequacy of drawing conclusions about women from studies of men. Researchers found that a high level of fats called triglycerides by itself portends heart disease in women, even in the absence of high cholesterol levels. This is not true of men.

The most obvious difference is that women tend to develop coronary heart disease far later in life than men. Heart disease becomes the number one killer of women in their sixties, but it is the number one killer of men at the age of thirty-nine. That disparity points to the long-assumed protection offered to premenopausal women by their hormones. But scientists now believe it may be too simplistic to expect the hormone estrogen alone to explain differences in heart disease. Dr. Antonio Gotto of the Baylor College of Medicine in Houston believes there are cellular and biochemical differences in female blood vessels. "I feel confident that female blood vessels have some inherent protection," Dr. Gotto says. He speculates that female blood vessels may be better able to repair early damage from high blood pressure, for example, preventing the development of artery-clogging plaques.

Women have different symptoms of heart disease than men. Women often have pains on and off for a long time before they have a heart attack; in men, chest pains are more often a sign that a heart attack has begun. Women who have a heart attack are twice as likely as men to die within sixty days. Diabetes seems to eliminate the protection accorded women in their earlier years, whereas it seems to have less effect on men.

Contributing to the sense of mystery is the finding that women do only half as well as men in bypass surgery. Although a relatively simple reason is suggested—women have smaller hearts, more difficult to operate on, and they tend to come into the operation older and sicker than men do—not everyone is convinced that this is the whole story.

The aim of the heart researchers is not just to improve the prevention and treatment of heart disease in older women but to learn what protects younger women from heart disease and perhaps to use the knowledge to help men. Dr. William Castelli, director of a long-range heart study in Framingham, Massachusetts, bemoaned the lack of attention being given to heart disease in women. "Their signs and symptoms are not being taken seriously because of the myth that women don't get heart attacks," he says.

The Framingham study is a major source of data on women and heart disease because it follows both men and women throughout their lives, correlating such factors as blood chemistry and life-style with incidence of heart disease. Two similar federally funded studies, in Tecumseh, Michigan, and Evans County, Georgia, and a study in Rancho Bernardo, near San Diego, are among the few to include women. Framingham and Evans County data indicate that women are at particular risk of heart disease if they have high levels of triglycerides in their blood, said Dr. Millicent Higgins of the heart institute.

In men, according to Dr. Castelli, triglycerides do not predict heart disease risk independent of other factors, such as blood cholesterol.

"Triglyceride," says Dr. Castelli, "is the chemical name for what most of us think of as fat. Corn oil is pure triglyceride." He explains that triglycerides in the blood are converted by the body to low-density lipoproteins, or LDLs, which carry cholesterol to the blood vessels, and may contribute to artery-blocking plaques in other ways too. Dr. Castelli said he does not know why triglycerides should be so much more dangerous in women than in men. According to Dr. Castelli, the importance of triglyceride levels is usually overlooked in women.

The Framingham data and data from the Rancho Bernardo study also point to the importance of diabetes as a risk factor for women. As far as heart disease goes, says Dr. Castelli, "Women with diabetes are no longer like women. They are like men." And the Framingham data indicate that "there are different initial symptoms of heart disease in men and women," he said. Women tend to have chest pains or a feeling of pressure in the chest as their first sign, whereas men tend to have heart attacks.

Of 1,600 women studied in Framingham, only 6 had heart attacks before menopause, Dr. Castelli says. Although it is often assumed that female sex hormones are what protect premenopausal women against heart attacks, experts say the exact link has never been firmly established. "It is very attractive to think that female sex hormones are good for you and male sex hormones are bad for you, but in point of fact, the data are mixed," says Dr. Elizabeth Barrett-Connor of the University of California in San Diego. "Most of us don't believe it anymore."

In support of estrogen's purported beneficial effects are several reports indicating that women who take estrogen supplements after menopause are less likely to have heart attacks. But the Framingham data did not show a protective effect and there are many unanswered questions. The heart institute was planning a study of female hormones and heart disease. The study will involve about one thousand postmenopausal women who will take various combinations of estrogen and another female sex hormone, progesterone.

One hypothesis is that estrogen protects against heart disease by increasing the concentrations of the beneficial HDLs and decreasing the concentrations of harmful LDLs. But there is a trade-off, said Dr. Basil Rifkind of the heart institute. Estrogen also increases the tendency of blood to clot, which in itself can increase the chances of a heart attack. In the late 1960s the heart institute studied the use of estrogen supplements in men and concluded not only that the men were not protected against heart disease but that they may have had an even higher death rate from blood clots when they took the female hormone. [GK]

Q&A

Q. *Why don't people who take nitroglycerin for a heart condition explode?*

A. Nitroglycerin, the highly explosive compound of carbon, hydrogen, nitrogen, and oxygen developed in the nineteenth century, acts directly on the walls of the blood vessels. But the amounts used as medicine are not large enough or concentrated enough to cause any sort of explosion, explains Dr. Thomas Robertson, chief of the cardiac diseases branch of the National Heart, Lung, and Blood Institute, part of the National Institutes of Health. The drug "is diluted with filler in the tablet," Dr. Robertson says. "By the time it is absorbed in the body it is in minute concentrations. It dilates the vessels, which both increases the blood supply to the heart and reduces the work of the heart by reducing blood pressure," he said.

HETEROSEXUALS AND AIDS

HETEROSEXUAL transmission of AIDS continues to advance slowly in the United States as a whole, but in some inner cities the virus is beginning to move more rapidly through heterosexual intercourse. The AIDS epidemic has moved in three waves in the United States over the last decade. The disease emerged first among homosexual men, then spread widely among intravenous drug users and later, in the first wave of heterosexual transmission, to the sex partners of drug addicts and bisexual men.

The disease has not yet spread significantly from them into the general heterosexual population but it is moving into at least part of that group, people with multiple sex partners and rampant venereal disease. "Most people want to know, is there going to be a widespread heterosexual epidemic or not?" says Thomas C. Quinn, an AIDS researcher at the National Institute of Allergy and Infectious Diseases, part of the National Institutes of Health. "The answer is not a yes or a no. Heterosexuals will get infected, but they will not be your everyday person. It will be the people already at risk for syphilis, gonorrhea, chlamydia, and with life-styles that include risky sexual partners. Among those, it will be an epidemic." As the 1990s begin, only about 5 percent of the full-fledged AIDS cases in this country have been officially attributed to heterosexual transmission, but there are indications in some cities that heterosexual transmission is increasing, exacerbated by an epidemic of other sexually transmitted diseases that facilitate transmission of the AIDS virus.

Dr. Timothy Dondero, chief of the seroepidemiology branch of the AIDS program at the Centers for Disease Control (CDC), says that the explosive epidemic of syphilis in the inner cities has raised concern about heterosexual transmission of AIDS and that the widespread use of crack may be a special problem because it is facilitating very active, imprudent heterosexual activity. The ultimate question about heterosexual transmission, Dr. Dondero says, is, "Will it remain primarily an offshoot of other transmission? Is IV drug use the engine that keeps the epidemic going? Or will it now be able to run on its own? We don't have the answers yet." In a paper delivered at an AIDS conference at Johns Hopkins University in Baltimore, Dr. Quinn and his colleagues described syphilis as one of the key factors in spreading AIDS among heterosexuals in the inner city.

They studied 4,863 patients in two inner city clinics in Baltimore that treat sexually transmitted diseases. Among heterosexuals who denied using intravenous drugs, those who had syphilis were seven to nine times more likely to have AIDS than other patients at the clinic, suggesting that heterosexual intercourse was involved in transmitting both syphilis and AIDS to these patients. The sores caused by syphilis and other venereal diseases are believed to facilitate the entry of the AIDS virus into the body. Dr. Quinn said that the research of Dr. Francis Plummer in Kenya showed that only 2 percent of men who are circumcised and have no genital ulcers acquired AIDS from an infected woman. But 52 percent of uncircumcised men with genital ulcers became infected. Among the intravenous-drug and crack users and their partners in this country, similar conditions now occur: a combination of promiscuity and untreated disease. More alarming were data from Baltimore for those clinic patients under twenty-five years old. Seventy-two percent of the women infected with the AIDS virus in 1988 and 46 percent of the infected men denied engaging in any activity other than heterosexual intercourse that would put them at risk for AIDS, a sharp increase from a similar survey a year earlier. "This disease in some places is now following the pattern of Africa and moving on to those who bear the burden of all sexually transmitted diseases—the minorities and lower socioeconomic classes," says Dr. Edward Hook, a specialist in sexually transmitted diseases at the Johns Hopkins Medical Institutions. "We have created Africa" in the inner cities of America, says Dr. Sten H. Vermund, an epidemiologist at the National Institute of Allergies and Infectious Diseases. "The crack houses have done that."

Experts believe that many crack users are already infected with the AIDS virus, which can be spread heterosexually as people at crack houses trade sex for drugs. A study published by the New York City Health Department says that about 7 percent of the New York cases are the heterosexual partners of drug addicts or other high-risk individuals. Heterosexual transmission beyond such immediate sex partners has occurred "relatively infrequently," the study said. There has been very little documented heterosexual transmission of the AIDS virus from infected women to men in New York City. Of 634 heterosexual contact cases reported in the city through late 1989, 627 were women who acquired the virus from men and only 7 were men, all of whom had sexual contact with female intravenous drug users.

The low number of men may result from city health officials' extremely vigorous efforts to question them and their acquaintances about their habits. If there is any chance they contracted the virus through drug use or homosexual encounters, they are not listed in the heterosexual category. Further, the seven cases are men who presumably became infected an average of ten years ago. Studies in the city's clinics for sexually transmitted diseases showed an increase between 1986 and 1988 in the number of people infected with the virus who had no risk beyond heterosexual intercourse. In women it rose from 0.4 to 3.5 percent of AIDS infection cases in the clinic, and in men it rose from 1.2 percent to 4.6 percent. "There are some trends there that are a little alarming," says Dr. Mary Ann Chiasson, assistant director of the city's AIDS Research Unit. "It's too early to tell how widespread the problem will be because we don't know how big the population is," referring to those engaging in frequent sex in places like crack houses. Dr. Chiasson says work in city clinics has showed a change in the female-to-male transmission pattern. In couples in which one partner was a drug user and the other was not, the transmission of AIDS was about equal from men to women and women to men, with 10 to 13 percent becoming infected.

The transmission of the human immunodeficiency virus by heterosexual contact has been controversial throughout the epidemic. The first cases reported were considered suspect for years. There was some fear that the epidemic would move quickly from homosexual men to the heterosexual population at large, especially after it was reported in the early 1980s that two cases of AIDS infection were found in a single Minnesota singles club of fifty members. But the fear of widespread outbreak failed to materialize. A book published in 1990 called *The Myth of Heterosexual AIDS* argues that the media and some researchers have created fear and misgiving among heterosexuals when there is actually very little risk. It points out that the virus is not very contagious by ordinary heterosexual intercourse, especially from women to men.

But each year the number of cases of heterosexually transmitted AIDS has become larger and increased faster than the epidemic as a whole. And these cases of AIDS are the result of infections that took place eight or ten years ago. Many researchers believe that the virus is already spreading further heterosexually than the official case count indicates, especially among the poor. "The heterosexual epidemic is no myth," says Dr. Jerome Groopman, head of the AIDS program at Harvard's New England Deaconess Hospital. "It is real." The share of all cases that are heterosexual remains small. Overall in the epidemic, about 128,319 cases of fully symptomatic AIDS have been reported from 1981 through March 1990. Of those, the Centers for Disease Control estimate that about 6,231, or 5 percent, have been cases in which the disease was transmitted by heterosexual sex.

But cases attributed to heterosexual transmission are growing faster than any other category of AIDS cases. For example, the CDC reports that from 1988 to 1989, cases caused by heterosexual transmission jumped 36 percent. To compound the misery, many women infected with the AIDS virus heterosexually then have babies that are also infected. About 60 percent of the heterosexual transmission cases reported through March 1990 involved people who had sex with drug users, bisexuals, people infected by a blood transfusion, or others at high risk of AIDS. About 10 percent got the disease from other heterosexuals who are believed to fit none of the risk groups, but simply caught it themselves heterosexually. And about 30 percent were born in countries in Africa or the Caribbean where the disease is chiefly spread through heterosexual intercourse.

The official statistics may understate the true extent of heterosexual transmission of AIDS. The CDC and the New York City Health Department both designed reporting systems that were "hierarchical," meaning that AIDS cases were attributed to homosexual sex, intravenous drug use, and all other risk categories first. Only those that did not fit into these categories in any conceivable way were put in the heterosexual transmission category. Thus a person with AIDS who is an intravenous drug user as well as a heterosexual is deemed to have acquired the infection through sharing drug needles rather than through heterosexual intercourse. In 1988 and 1989 a majority of all cases of AIDS reported in New York were among heterosexuals if one counts the heterosexual drug users. Because male drug users pass the virus efficiently to wives and girlfriends, 26 percent of all the women who get AIDS now in New York get it through heterosexual intercourse. The number is climbing steadily. "Up to now," says Robert Rolfs, a specialist in syphilis at the CDC, "heterosexual transmission has been a relatively unimportant mode of transmission compared to homosexual transmission and intravenous drug use. Heterosexual transmission appeared relatively inefficient. But now we may see an explosion in the inner cities; HIV infection is certain to increase in that population but how much and how fast we don't know yet."

Dr. Quinn, the AIDS researcher at the National Institute of Allergies and Infectious Diseases, notes that there is a rapid increase in heterosexually transmitted disease in several cities, especially those of the East Coast, including Newark, Philadelphia, Baltimore, Washington, and Miami. [PJH]

AIDS: THE DEADLY ROLE
OF SCAVENGER CELLS

YEARS into the AIDS epidemic, some researchers have been shifting their focus, concentrating on what they believe might be the crucial, perhaps overriding role of scavenger cells of the immune system in the development of the disease. Scientists studying macrophages, white blood cells that are present everywhere in the body, are finding answers to puzzling questions about how the AIDS virus invades the body and causes disease.

Most scientists had been paying greatest attention to another type of white blood cell, the T-4 cell, in their effort to understand AIDS. The T-4 cells are often invaded and killed by the AIDS virus, and their depletion in patients has been associated with the onset of disease. But the study of those cells has left unanswered many questions about AIDS. "We definitely have to zero in on the macrophage," says Dr. Peggy Johnston of the National Institute of Allergies and Infectious Diseases in Bethesda, Maryland.

Dr. Jay A. Levy of the University of California School of Medicine at San Francisco says, "I go to meetings now and all I hear is the macrophage, the macrophage, the macrophage." Dr. Levy believes that T-4 cells and other body cells also play important roles in the disease, but that the macrophage is "a pivotal cell."

For several years, some researchers, including Dr. Robert C. Gallo of the National Cancer Institute, a discoverer of the AIDS virus, have suggested that macrophages were important targets of the AIDS virus, and could pass it on to other immune system cells. But only more recently have researchers had the technical ability to grow macrophages with relative ease and study them in the laboratory. Now it appears that macrophages may be the first, and sometimes the only, cells invaded by the AIDS virus. The new findings mean that some people who had been declared free of the virus by customary tests may actually be infected, the virus hiding in their macrophages. Studies have been attempting to determine how often this happens and to see whether many people need to be retested.

It appears that infected macrophages may be responsible for the dementia that sometimes accompanies AIDS. With macrophages so clearly important, drugs against AIDS will have to be tested to see if they can curb the virus's action in these cells. There was uncertainty, for example, over whether azidothymidine, or AZT, deters the virus in macrophages as effectively as it does in T-4 cells. Macrophages are present in the blood, the brain, mucous membranes, semen, and cervical fluid. When they are in the blood, they are traditionally called monocytes.

These macrophage cells, whose name is derived from the words "big eater" in Greek, help fight infection by ingesting invaders, such as bacteria and protozoa. They normally do this after being signaled by T-4 cells that these organisms have infected the body. Macrophages also directly activate other immune system cells to attack disease-causing organisms in other ways, such as by producing antibodies. But even when macrophages are not fighting infections, they are crucial to health. The cells secrete hundreds of substances that keep body tissues alive and growing and that stimulate other immune system cells to fight disease.

Researchers had focused on T-4 cells in their studies of AIDS because these are the cells most obviously destroyed by the virus. Although the depletion of T-4 cells is still considered central to the development of AIDS, investigators cite evidence that this alone is not sufficient to explain the disease. Dr. Jacques Leibowitch of the René Descartes University in Paris cites two indications that T-4 cell destruction is neither necessary nor sufficient for severe immune system damage to occur. First, he says, people can have secondary infections associated with AIDS while their T-4 cells remain normal in number, usually in the early years after invasion by the virus.

Then, "at the other end of the story," when people are in the later stages of the disease and have virtually no T-4 cells left, "you can go without an infection for months, or even as long as a year. If the song of the T-4 orchestra were true, you would have an infection every other second," he adds.

Another mystery had been the difficulty in finding T-4 cells containing the AIDS virus, even in patients who had antibodies to the virus and who had symptoms of AIDS. Doctors have estimated that as few as one in a million T-4 cells are infected, which led some to ask where the virus hides. The emerging picture is that the virus goes to the macrophage first and spreads from there to T-4 cells. The virus kills T-4 cells and thereby prevents these cells from signaling macrophages to fight certain infections. At the same time, the infected macrophages do not function properly to fight diseases even if they are signaled by T-4 cells. As a consequence, patients become vulnerable to organisms that normally would never make them ill.

In addition, scientists believe, infected macrophages do not properly release the substances that keep other tissues growing and healthy. One result, in some patients, may be neurological symptoms of AIDS and widespread destruction of brain cells. Macrophages serve as a reservoir for the AIDS virus because the virus multiplies in them but does not kill them, as it kills T-4 cells. In T-4 cells, the virus is released from the cells as it reproduces in a process known as budding. In macrophages, the virus buds inward, remaining in the cell rather than being released. Macrophages become "like beanbags, filled with hundreds of viral particles," according to Dr. Monte S. Meltzer of the Walter Reed Army Institute of Research in Washington.

Infected macrophages can transmit the virus to other cells, possibly by touching the cells. But the infected macrophages may bypass the body's normal immune defenses so that they never trigger the production of antibodies against the AIDS virus. This may explain, scientists said, mysterious cases in which patients developed AIDS without ever having antibodies against the virus in their blood. A growing number of researchers are looking to macrophages for an explanation of why AIDS patients are plagued with certain infectious diseases but not others. They reason that the infections typical of AIDS mostly involve organisms that invade and kill body cells. These are exactly the types of organisms normally killed by macrophages.

"In the beginning, all the diseases that AIDS patients get are intracellular parasites," notes Dr. Jeffrey C. Laurence of Cornell University School of Medicine in New York. "And the major way the body attacks intracellular parasites is with macrophages." Dr. Levy said macrophages normally engulf and destroy the parasite that causes Pneumocystis carinii pneumonia, the major killer of AIDS patients. But macrophages that are filled with the AIDS virus "just sit there."

Dr. Howard E. Gendelman of the Walter Reed Army Institute also has evidence that infected macrophages may cause AIDS dementia. In laboratory experiments, Dr. Gendelman showed that infected macrophages release a substance that kills brain cells, while healthy macrophages release substances that nourish brain cells.

Researchers looking at macrophages are now finding that the AIDS virus is there in abundance, even when it cannot be isolated from T-4 cells. Dr. Meltzer of Walter Reed isolated AIDS virus from the macrophages of three homosexual men who had been exposed to the virus on many occasions but who had no antibodies to it and who had no detectable virus in their T-4 cells.

Once people have antibodies to the AIDS virus, they almost always have the virus in their macrophages, too. And the virus is far more prevalent in macrophages than it is in T-4 cells. Dr. Suzanne Crowe, Dr. Michael S. McGrath, and Dr. John Mills of the University of California in San Francisco found that they could isolate the AIDS virus in from 3 to 9 percent of the macrophages of people who are carriers of the AIDS virus but have no symptoms of disease, and that they can isolate the virus in from 10 to 20 percent of the macrophages of people who are ill with AIDS. Moreover, Dr. Crowe says, these are just the macrophages that are actively releasing the virus. Many more will release it if they are grown for a week or so in the laboratory. Thirty percent of all macrophages from people who are AIDS virus carriers or AIDS patients release the virus under these conditions. [GK]

THE HELPING VIRUS

Viruses have been described as pieces of bad news wrapped in protein, but a study suggests that a virus infection can be life-saving good news, at least in mice prone to diabetes. There also may be lessons in the research for treatment of human disease, according to a report in the journal *Science*. The study involved non-obese diabetic mice, a breed known to develop fatal diabetes in their first half year of life. These mice are often used as an animal model for human type I, insulin-dependent diabetes. The animals suffer from an autoimmune process in which some of their own defensive white blood cells, a type called helper T-cells, attack the insulin-producing cells of their own pancreas glands. The result is diabetes caused by a lack of insulin.

In the studies reported by Dr. Michael B. A. Oldstone of the Research Institute of Scripps Clinic in La Jolla, California, the diabetes-prone mice were protected against this self-destruction by a carefully selected virus infection. The scientist in fected some of the animals with a relatively benign form of lymphocytic choriomeningitis virus that primarily infects the helper T-cells. The result was a lifelong infection with the virus, but no diabetes. Ordinarily, nearly all of the mice would have died of diabetes by the time they were a year old.

The evidence from the experiments indicated that prevention of fatal diabetes was most likely a result of inactivation of the self-damaging T-cells by the virus infection. Dr. Oldstone notes that viral genes and their products have profound effects on cells, and that it should be possible to obtain and use such viral products to treat specific cells of the body that are the virus's natural targets. "Viruses and, presumably, their products can be developed to be beneficial and may have potential as a component for treatment of human diseases," Dr. Oldstone's report concludes.

PAIN, PAIN, GO AWAY

A REVOLUTIONARY approach to pain control, pioneered at research centers over the last fifteen years, is gaining rapid acceptance at hospitals across the country and should spare patients from considerable misery. Leading the revolution is a new type of doctor, a pain specialist, who brings to the bedside an ever-improving understanding of the mechanisms that cause pain and of new technologies to fight it. "We have changed the way doctors think about pain," says Dr. Richard Patt, director of the pain treatment center at the University of Rochester. Until recently, physicians were trained to accept some discomfort and to use powerful pain medications sparingly. Children were treated with special caution and infants often were not treated for pain at all. But "we are now much more aggressive," Dr. Patt says.

The two main advances for people in acute pain are patient-controlled analgesia, in which a patient can regulate pain by pressing a button that delivers powerful narcotics through an intravenous line, and epidural narcotic administration, in which a physician places tiny doses of morphine directly into the space just outside the spinal cord. Although successes in the treatment of chronic pain have been less dramatic, people with conditions like bone cancer have been significantly helped by tiny morphine pumps implanted in the abdominal wall and improved long-acting narcotics that are taken only twice a day. The Alza Corporation has filed a new-drug application with the Food and Drug Administration for a bandage-like patch that delivers painkiller through the skin.

To oversee the new techniques, pain treatment services have been springing up in hospitals "all over the place," according to Dr. L. Brian Ready, chief of the nation's first acute pain service, which opened in 1985 at the University of Washington in Seattle. These services are generally run by anesthesiologists or sometimes neurologists. Gone are the days when pain doctors were widely considered the snake oil salesmen of the profession. "Treating pain was sort of quackery, mixed up with hysterical women and acupuncture," recalls Dr. Kathleen Foley, a neurologist who is director of the pain relief service at Memorial–Sloan Kettering Cancer Center in New York. "Scientific studies and advances in neuroscience have moved us out of that realm and into the realm of respectability."

"I think you can argue that until very recently pain didn't get treated—or get treated well," Dr. Foley says. In the last decade a number of studies have shown that doctors and nurses routinely undertreated patients' discomfort, fearing that they would become addicted to painkillers like morphine and that the drugs could dangerously lower their breathing rate. "The typical pattern was the doctor ordered less than was helpful and the nurse gave less than was ordered," Dr. Ready says. Research indicated that 35 to 75 percent of patients did not get adequate pain relief after surgery.

Experts say recent experience has proved these worries unfounded. Narcotics given for pain control produce only temporary physical dependence and are not terribly addictive, says Dr. Patt. Even cancer patients who are on morphine for weeks are readily weaned from it when they enter remission, he says. "In fact, we have to warn them not to stop the narcotics suddenly, because most can't wait to get off." (The drugs often have unpleasant side effects, like grogginess and constipation, and patients are eager to shed them and their association with the disease.)

Two discoveries in neuroscience further paved the way for new treatments. Although morphine, an opiate, has been used to quell medical discomfort for more than a hundred years, it was not until the mid-1970s that scientists first discovered how opiates work to relieve pain. The receptors that recognize the drugs, the researchers found, are concentrated in the spinal cord and in a small region deep within the brain. Scientists also isolated natural opiates called endorphins and enkephalins, substances made by the body but closely resembling painkilling drugs. "Morphine couldn't be bad if we had a natural substance like it in our bodies," reasons Dr. Foley.

The patient-controlled analgesia pump is a concrete symbol of how philosophies have changed. The pump was first marketed in 1984; now an estimated 46 percent of hospitals have at least one, though many are still inexperienced in using them and few offer them routinely. The pump, which contains a solution of morphine or another narcotic, is hooked up to an intravenous line in the patient's arm and set so that pressing a button delivers a premeasured dose into the bloodstream. The patient may press the button as often as needed, but a computer in the pump sets some safety limits. A pump for an adult might be set to deliver one-milligram doses at least eight minutes apart, with a total dose of no more than thirty milligrams in four hours, Dr. Ready says. Even before these safeguards come into play, "most patients who press more than they should just get drowsy and fall asleep" from the morphine, and "no harm done."

Experts who have used the technique say it avoids many pitfalls of earlier techniques. There is no waiting for a nurse to answer pleas. Different patients need different levels of medication, and the pumps can generally take these needs into account. And the drug is delivered directly into the bloodstream, so it works instantly. "Patients react with great enthusiasm, and the enthusiasm is greatest from patients who've had surgery before," according to Dr. Ready, who has also used the pump to treat pain stemming from a variety of conditions, from bone marrow transplants to sickle cell disease. "They say this is dramatically different."

Studies show that the new technique does not control pain much more effectively than does properly administered conventional treatment. But Dr. Foley says patients are much more satisfied with "the sense of control," and "the patients can tolerate more pain, because they know they can have pain meds whenever they want them." In terms of pure power to conquer pain, experts say nothing beats the epidural narcotics. Once scientists discovered clusters of opiate receptors in the spinal cord, it was a short next step to see if infusing opiates into this region would prevent discomfort. The spinal cord is the common pathway for nerves from all over the body; doctors have long injected anesthetics into the region, but only in the last decade have they tried to inject narcotics specifically to relieve pain without the general numbing and paralysis caused by anesthesia.

Dr. Tony Yaksh of the University of California at San Diego first experimented with the technique in 1979, and within the last five years it has become widely used. With the patient in the fetal position used for spinal taps, a physician threads a delicate catheter through a needle into the epidural space, which surrounds the spinal column. The needle is then removed, leaving a capped end of the catheter protruding from the skin; morphine can be delivered through it directly to the spinal receptors. Since the morphine is confined to a small space, a small dose often lasts twenty-four hours and only tiny amounts enter the bloodstream. There are none of the side effects that usually result from narcotic painkillers. The most common side effect is mild itching.

"This technique produces previously unknown levels of pain relief," according to Dr. Ready. And with fewer side effects, "a day after surgery, people are much more awake and functional." When President Ronald Reagan had colon surgery in the mid-1980s, he had an epidural narcotic catheter in place so he could resume working in his hospital bed the next day. The catheters remain as long as significant pain persists, and patients can move about freely. Cancer patients with longstanding pain often live at home with a modified version that they can inject themselves.

At the University of Washington, where virtually all major surgery is followed by care from a pain specialist, about half the patients receive epidural narcotics and half receive the patient-controlled pumps.

In one German study, when patients were given the choice of using a patient-controlled device, 20 percent refused and 25 percent said they did not like it. "Pumps are not good for the elderly," Dr. Patt says, or for patients with mental impairments or those "who want to be cared for."

One of the great advances of the last two years has been the discovery that these techniques could be applied to children, whose pain had been largely neglected. A University of Iowa study in the mid-1970s found that fewer than half of a group of surgical patients aged four to eight received any painkillers during their hospital stay, even though they had undergone major operations, including amputations and open heart surgery. Until last year, almost no children under two ever got narcotics, according to Dr. Charles Berde, director of the pediatric pain service at Children's Hospital in Boston. One reason was fear that they would be more vulnerable to morphine's side effects, particularly its effect on breathing. Another was a debate over whether young children, who cannot say what hurts, could experience or later remember pain.

"There is a greater realization that children do feel pain," says Dr. Berde. "We are more familiar and comfortable giving them drugs." At Children's Hospital, children frequently receive epidural narcotics and patients as young as seven are routinely offered patient-controlled pumps and use them well.

Although some advances in the field of acute pain have spilled over to benefit patients in chronic pain, experts agree that lasting pain is the next great challenge. Traditional morphine pills last only a few hours, requiring continual pill-popping and producing wide swings of pain. Newer long-acting opiates have somewhat eased the plight of such patients. But when drugs fail or lead to too many side effects, experts turn to mechanical and surgical techniques, none of which has proved widely successful.

For example, electrodes may be placed over a muscle that hurts and wired to a small tuning box, which produces a mild tingling that distracts the nerve that carries the pain. The units are harmless and can be worn continuously, but experts say that at best they reduce pain and rarely eliminate it. When all else fails, chronic pain patients are often treated by injecting alcohol or phenol into deep nerve bundles coming from a painful region. But since most bundles carry fibers that control muscle movement and sensation as well as pain, these injections can produce handicaps.

Dr. Foley hopes that the rapidly improving understanding of the pain pathways will rescue chronic pain patients. Experts now feel that the opiate receptors in the central nervous system, along with enkephalins and endorphins, are part of a human nerve network that inhibits painful stimuli racing toward the brain. They are hoping to beat pain by bolstering this natural defense.

In experiments researchers have implanted electrodes into patients' brains. This massages the region, releasing quantities of endorphins and enkephalins, easing pain without the sedation of morphine. Experts say the natural opiates seem to produce pure pain relief, with no side effects.

Schering-Plough has made a compound that prolongs the normally short life of enkephalins in the body and is now testing the drug in humans. Dr. Foley's group is working with preparations that promote the release of these natural opiates, she says, "so you wouldn't have to give drugs like morphine at all." [ER]

Q&A

Q. *Do people with arthritis suffer more joint pain in damp weather?*

A. "The answer, of course, is yes and no," says Dr. Michael Belmont, a rheumatologist who is clinical instructor of medicine at New York University School of Medicine. Many patients report that they feel better when the climate is dry and warm and many report that their joints ache when it is about to rain or is raining. But these reports are not scientifically controlled and could have a strong psychological component, he said.

There are two major kinds of arthritis: inflammatory arthritis, such as rheumatoid arthritis or gout, and osteoarthritis, or "rheumatism," also described as wear-and-tear arthritis, which is less inflammatory. "The consensus is that weather does not affect the inflammatory diseases," Dr. Belmont says, "but when it comes to osteoarthritis and some of the other soft-tissue rheumatisms, there is contradictory information." There have been studies in which patients kept diaries and researchers tallied their ratings of how bad they felt on a given day against reports of local temperature, humidity, and barometric pressure. "Different studies came to conflicting conclusions," Dr. Belmont says, "and none of them was large enough, well enough designed, with homogeneous patient populations, or long enough to know the answer." There is "perhaps some consensus" that changes in barometric pressure, rather than extremes themselves, may be a cause of discomfort in noninflammatory arthritis, he says.

"However, there are a hundred and one different causes for arthritis, and the main determinant of how bad you feel is probably not climate, but the nature and course of the disease itself," he continued. "That is the reason we don't tell all our patients to move to Arizona."

KICKS

AS reports of cocaine-related deaths steadily accumulate, doctors are gaining a sharper awareness of the dangers of the drug. Increasingly, the doctors realize they must add cocaine abuse to their checklist of considerations when they are confronted with certain medical emergencies—in infants as well as adults. Among the most striking of cocaine's dangers are the potentially devastating effects on the cardiovascular system. The reports have included the following:

- A thirty-three-year-old man from North Carolina used cocaine for the first time and suffered a heart attack a few hours later.
- A nineteen-year-old man's chest ached nearly every time he used cocaine. Finally, he had a heart attack. At his New England home five weeks later, he resumed cocaine use and had a second heart attack. A month later, he suffered his third heart attack, again after taking cocaine. Four months after his first attack, the man again complained of crushing chest pains, collapsed, and died.
- Three men in Detroit, two in their twenties and one in his forties, used crack, the potent, smokable form of cocaine generally sold in pellets. All three had paralyzing strokes.
- A healthy forty-five-year-old man puffed for several hours on a pipe containing cocaine, then died suddenly from a ruptured aorta, the main artery leading from the heart.

Cocaine's dangers have not always been recognized. The drug was used for centuries by South American Indians and since 1884 in medicine as a local anesthetic. Sigmund Freud even used it in the belief that it would be a "cure-all" before he recognized the threat of addiction. Since Freud, doctors have learned of more dangers, such as seizures, delirium, fever, loss of the sense of smell, and death of cells in the lining of the nose, which leads to holes between the nostrils.

Nevertheless, a belief persisted that cocaine was not particularly harmful until the recent epidemic of recreational use and abuse. The government has estimated that 5 million Americans are regular users of cocaine and 30 million have tried it at least once. With the increase in use, doctors have discovered that cocaine can trigger such things as heart attacks, strokes, angina, destruction of the liver, and other complications.

Cocaine causes blood vessels to tighten, the heart rate to quicken, and blood pressure to rise suddenly. It can also trigger potentially fatal heart rhythms. Any of these effects can increase the heart's need for more oxygen-rich blood to nourish its muscle cells, thus making cocaine a hazard to anyone whose coronary arteries are narrowed by fatty deposits due to atherosclerosis. If the oxygen shortage is mild and brief, the person may suffer angina but no heart attack. If the spasms of chest pain last long enough to kill cells, they can cause heart attacks.

Cocaine is also causing angina and heart attacks in younger, healthy people who have normal coronary arteries. The danger to these people is believed to result from spasms that reduce or shut off the flow of oxygenated blood that nourishes the heart. The sudden clamping of an artery in the brain can cause a stroke, and some people have suffered pains in the abdomen when the spasms struck arteries that feed intestinal cells.

Although the walls of the aorta have a degree of elasticity, a sudden rise in blood pressure can exert a great force on walls that have been weakened by atherosclerosis. Such a spurting pressure is thought to be capable of rupturing the aorta. Other cocaine users have drowned from the sudden accumulation of fluid in the lungs, a condition known as pulmonary edema, that was somehow triggered by the drug.

In a February 1987 report in *Human Pathology*, Stanford researchers said they had found that cocaine can painlessly and permanently damage heart muscle, leaving red streaks called contraction bands.

The lesions in the cells make the cells useless and block the normal electrical pathways of the heart, making the beat dangerously irregular. The Stanford team said it suspected that cocaine users who have not previously had heart problems could damage their hearts over time. Of concern is that when these individuals grow older and develop atherosclerotic heart disease, their already damaged hearts may be unable to withstand the added stress.

Cocaine can be dangerous whether it is snorted, smoked, or injected into the skin, muscles, or veins. Seizures, for example, can result from a single dose and can be difficult to treat. Also, because much of the cocaine sold is impure and dealers often mix it with less expensive drugs such as amphetamines, the reactions may be due to several drugs, not just cocaine. Moreover, cocaine can endanger babies of users, both before they are born and, apparently, after birth. Three hours after a breast-feeding mother in Chicago snorted cocaine, her two-week-old infant developed symptoms of intoxication, researchers said. The baby's heart rate sped up, its blood pressure rose, its breathing became more rapid, and it became irritable and had tremors. The symptoms persisted until the mother excreted all the cocaine and the biochemical breakdown products made in her liver.

Cocaine use during pregnancy can cut off oxygen to the fetus, killing it or causing such complications as premature delivery or the detachment of the placenta from the womb. Mothers who use cocaine during pregnancy often deliver babies of low birth weight. Other babies suffer from cocaine's powerful effects on the nervous system: neurobehavioral abnormalities with tremors, mood swings, and irritability.

Despite its reputation as an aphrodisiac, cocaine can actually take away from sexual performance. Some men who have used cocaine for a long time, particularly in high doses, have reported difficulty in maintaining an erection and in ejaculating. Some women have had difficulty achieving orgasm.

Some of the most drastic examples of the dangers of cocaine have involved so-called mules or body packers. "Mule" is the name given to those who transport cocaine from one country to another. "Body packers" stuff cocaine into condoms, balloons, or plastic bags and, when trying to elude authorities, swallow them or conceal them in the rectum or vagina. Packets have broken open, overwhelming the body with massive amounts of cocaine. In other instances, holes in the condoms or bags have allowed cocaine to leak or fluid to be drawn in, causing the bags to burst. In such cases, the victims have suffered seizures and died. As a result, when doctors detect bags of cocaine in X rays of these people, they consider emergency surgery to remove the bags before they rupture. [LKA]

COCAINE CRAVING

Scientists believe they have discovered a new chemical clue to what makes cocaine so attractive that the user feels a need for it again and again. Cocaine can produce a feeling of euphoria, freedom from fatigue, and related effects. Now scientists have established the chemical basis for these reactions and, particularly, the process that makes the user want the drug again.

Cocaine interferes with nerve cells' uptake of a particular nerve-signaling substance, dopamine. It was assumed that the drug must attach itself to a particular receptor on nerve cells to which dopamine would otherwise bind. A report in *Science* notes that cocaine has previously been found to bind itself to several different receptors on nerve cells of the brain. But the new research showed that binding to only one of these would "initiate the effects that lead to cocaine addiction."

Scientists of the Addiction Research Center of the National Institute on Drug Abuse in Baltimore say the key receptor was one found on a substance called the dopamine transporter that takes dopamine out of circulation as it leaves one nerve cell, preventing it from attaching to the adjacent nerve cell that is its normal target.

The scientists discovered this by training rats to ask for cocaine by pressing a lever, a standard research technique, and matching the avidity of the rats in doing this with the potency of the drug and its uptake by various receptors in the brain. Identification of the receptor's role might be useful in future studies of cocaine addiction and prediction of the abuse potential of other drugs.

CALLING ALL ORGANS

AS organ transplantation becomes increasingly successful, surgeons are facing a new series of thorny practical and ethical issues. The fundamental problem is that there are far more potential patients than donors. At least seventeen thousand Americans are now waiting for an organ. As a practical matter, doctors are trying innovation and improvisation to expand the pool of donors. For instance, they are moving to relax age restrictions on donations. They have even used organs from seventy-year-old adults in children.

On the ethical issues, however, doctors are finding that improvised solutions are either inappropriate or impossible. With the number of multiple-organ transplants growing, for example, society must decide whether limits should be set on the number of organs one person receives. In reconsidering the age restriction on donations, doctors are charting new biological territory. Physicians have been reluctant to transplant older organs that might be damaged or might not function long. But the existing standards reflect arbitrary limits. For kidneys it is about sixty years, while for livers it is fifty-five. For hearts, it is forty-five for women and forty for men. The age differences generally reflect the relative incidence of organ damage from clogged arteries.

"The chronological cutoff doesn't make a lot of sense because someone who dies at seventy-five can be biologically younger than someone at fifty," says Dr. Calvin R. Stiller, the head of transplant surgery at the University of Western Ontario in London and a leader in transplantation. Dr. William J. Wall from the Western Ontario team has taken the liver from a seventy-year-old man and given it to a ten-year-old boy. At least initially, the liver seemed to function well.

That success has led the surgeons to begin a research project to answer a fundamental question about biology: how long can an organ live? The project involves liver transplants in rats. It is tedious work that requires microsurgical techniques unnecessary in transplanting the larger human livers. Dr. Stiller said he is working with a team of three surgeons from Shanghai, China, who are skilled in the techniques needed for animal liver transplants. In the experiments, the surgeons allow litters of rats to live into old age. As the rats near the end of their lifespan of two and a half years, the surgeons transplant the rats' livers into young rats. The recipient and donor rats are from inbred strains to avoid the problems of rejection, so that the doctors can focus on the longevity of the organ. The crucial factor in using an older organ is the condition of the blood vessels. "If they are open or easily reparable, we use the organ," Dr. Stiller says.

In one unusual case, a Western Ontario transplant team headed by Dr. Neil McKenzie removed a heart from a fifty-eight-year-old man who died of a stroke. Finding the arteries blocked by fatty deposits, the surgeons performed a bypass operation on the heart and then transplanted it into a patient dying of heart disease. The Western Ontario team has performed sixteen other heart transplants with organs donated by individuals aged forty-six to fifty-eight.

Doctors at Western Ontario are also testing the use of human hearts as bridges to keep patients alive while they seek a better match or stronger heart. This runs parallel to the efforts in many medical centers to use artificial hearts as bridges. Dr. Stiller's team has used human hearts as bridges in several cases. The hypothesis is that a human organ involves less risk and cost, even if it does not achieve peak performance. The surgeons also know that if the heart fails, they can remove it and implant an artificial heart.

In three of the cases, the temporary human heart transplant did the job until a better heart was found after about a week. In others the "temporary" heart has become a permanent, well-functioning transplant. Doctors are also turning to living donors for organs that they had not expected to be able to use. In the past, doctors were limited to removing one of two kidneys from a living donor, or taking a portion of the pancreas, which lies deep in the abdomen. But recently doctors in four countries have taken a portion, or lobe, of the liver from a parent and have given it to a child with fatal liver disease. Two such transplants were done at the University of Chicago in late 1989.

Members of the transplant team were optimistic, believing that by using a living family member for the donation, they would not have to wait as long to find a good match with a cadaver liver. As a result, the tiny patients might be healthier when they receive the permanent transplants. In the first procedure in Chicago, both the recipient and the donor suffered complications. The infant had several additional operations to stop bleeding after the transplant. And the infant's mother had to have her spleen removed because doctors accidentally nicked it during the operation on her liver.

The potential for such complications have made many people uneasy about using the liver or pancreas because there are no backups. In contrast, if a kidney fails, a patient can turn to dialysis. If living-donor liver transplants succeed, they are bound to raise questions about the extent of a parent's obligation to take such risks in donating an organ for a dying child. Such concern compelled the Western Ontario team to seek alternatives that would not pose such a risk. "There are donors currently not being used because of age," Dr. Stiller says. "Better we find out whether they are usable and, if necessary, take the lobes out of those and put them into small children."

But with so many patients waiting for an organ, the move toward multiple-organ transplants raises anew the question of whether people who have already received one organ should get another when those who have never had a transplant die while waiting for one. Yes, says the United Network for Organ Sharing. "If there is a reasonable chance of success, persons whose previous grafts failed should be given equal access to another," the organization has said. The network is a federally financed group that keeps track of organ transplants.

Further success with multiple-organ transplants will surely exacerbate the shortage of organs. Dr. George J. Annas, professor of health law at Boston University School of Public Health, says that "with a permanent shortage of organs, we must pay more attention to how the line is formed and who is next." A complicating factor is that the number of hospitals doing organ transplants has grown rapidly. That can lead to pressure for keeping organs for use in single-organ transplants at local centers, even though the same organs could have been used for multiple-organ transplants elsewhere, says Dr. Arthur L. Caplan, who heads the biomedical ethics program at the University of Minnesota. But should three organs be used to save three lives or one? asks Dr. James F. Childress, a professor of religious studies and medical education at the University of Virginia in Charlottesville.

Equity issues could become especially important if multiple transplants are done on a larger scale, Dr. Starzl says, adding that in such cases, "If Solomon were around, I'd turn the matter over to him." [LKA]

THE CAUSES OF OBESITY

RECENT findings on the causes of obesity and the metabolic consequences of "yo-yo" dieting are forcing weight-reduction specialists to reconsider both their methods and the goals of treatment. The studies show, for example, what many obese people have been saying for years: they get fat or stay fat on a caloric intake no greater than (and sometimes less than) the amount consumed by people of normal weight.

The dieter accused of "cheating" when losses grind to a halt has also been vindicated. Low-calorie diets, long the mainstay of treatment, are now known to have limited effectiveness in many people because their metabolic rate drops to "protect" them from starvation, sometimes falling low enough to prevent further weight loss on as little as one thousand calories a day. And while obesity that runs in families had long been blamed almost entirely on household gluttony and sloth, a major study of people who were adopted showed that genetic factors seem to predispose many people to gain weight easily, especially in a land of perpetual plenty where there is little need for physical exertion.

One by one, obesity experts are concluding that many, if not most, people with serious weight problems can hardly be blamed for their rotund shape and that, given the effects and effectiveness of current methods of weight reduction, some would be better off staying fat. Only about one dieter in ten achieves lasting success, and many obese people who manage to lose significant amounts of weight may have to exist in a semistarved state indefinitely to maintain the loss. "At least half of obese people—those who are more than 30 percent overweight—who try to diet down to 'desirable' weights listed in the height-weight tables suffer medically, physically, and psychologically as a result, and would be better off fat," says Dr. George Blackburn, an obesity specialist at Harvard Medical School.

But while these emerging conclusions may seem depressingly fatalistic to obesity specialists and the people they hope to help, the new understanding may eventually lead to safer and more effective treatments for obesity than having to subsist on very-low-calorie diets or resorting to dangerous drugs or surgical procedures. "For the last five to eight years, I was really in the doldrums," says Dr. Jules Hirsch, an obesity specialist at Rockefeller University in New York. "Whatever we tried had the same grim results: people could lose half their body weight, but they'd be miserable in the reduced state and in two to five years, they'd gain it back." But Dr. Hirsch adds: "Prospects opened up by new techniques in biology have really raised my spirits. For example, we are now trying to clone the gene that makes mice obese. In less than ten years, we should know how the obesity gene acts, whether people are different from mice, and whether there are multiple types of obesity. I think, too, that we will better understand the biological factors that regulate body fat and find ways to manipulate them with drugs."

More immediately, some of the recent discoveries can be applied now to improve the health and fitness of obese people and to help those with lesser weight problems, most of which are environmentally induced, to shed unwanted pounds permanently without really dieting. New studies indicate that for many obese people, relatively small weight losses—often only 10 percent of body weight—can correct a tendency toward diabetes or high blood pressure. Thus, major health risks associated with obesity might be countered with modest losses of ten to twenty-five pounds that are easier to maintain. "The whole premise that the goal of weight reduction should be to reach 'desirable' weight is the major flaw in weight-loss strategies," Dr. Blackburn says. "It's the first ten percent of weight loss—not the last ten percent—that's important."

For people already consuming a normal number of calories, such losses can often be achieved through an hour a day of physical exercise, with little or no change in caloric intake and with a more lasting reduction than that achieved through dieting alone. For example, at Stanford University Dr. Peter Wood put one group of men whose weight averaged 220 pounds on a diet that reduced caloric intake by 300 calories a day. A similar group of men were instructed to eat as usual but to run or walk ten to twelve miles a week. At the end of a year, the exercisers had lost an average of 9 pounds, all in body fat, and the dieters had shed 15 pounds, 12 of which were fat. However, two years later, the dieters had regained half their lost pounds but the exercisers had kept off all the weight.

Even if no weight is actually lost, Dr. Hirsch says, exercise can improve the health of overweight people by reducing their percentage of body fat and their risk of developing a life-threatening illness.

Furthermore, the popular motivational principle of "if at first you don't succeed, try, try again" may not apply to weight reduction. Rather, the new studies indicate, the dieter's motto should be "Get it right the first time," according to Kelly Brownell, a psychologist at the University of Pennsylvania. He showed that yo-yo dieting—regaining weight and losing again—increases body fatness and may ultimately result in an inability to lose weight even on a very low caloric intake.

Some women attending the university's obesity clinic failed to lose weight when eating only 800 or 900 calories a day, Dr. Brownell says, adding that "they seemed to be the ones who'd been on the most diets." In a study of dieting rats, he showed that at first it took the animals twenty-one days to lose a specific amount of weight and forty-six days to regain it when they returned to a normal caloric intake. But in the next diet cycle, the same diet took forty days to accomplish the weight-loss goal but the animals regained the weight in only fourteen days. At the same time, their bodies got progressively fatter because in losing weight, they lost both muscle and fat but they gained back proportionately more body fat than they had lost.

Dr. Brownell found that yo-yo dieting increased the activity of lipoprotein lipase, an enzyme that promotes the storage of body fat. And since fat tissue is metabolically less active than muscle, with each diet cycle the animal's daily caloric needs dropped and they gained weight on fewer calories. The psychologist concluded that yo-yo dieting increases the body's efficiency in using food for fuel and may ultimately make weight loss impossible. In past centuries, he and others have suggested, this genetically programmed ability to conserve calories improved survival chances in periods of food scarcity. But today it is maladaptive. Dr. Brownell warns: "Don't start a diet unless your motivation is high and you adopt a good program of life-style changes that promote permanent weight loss. If the time isn't right to diet, wait."

There is already evidence that certain foods are better than others in promoting weight loss. Contrary to long-held assumptions, all calories are not equal. A calorie of fat counts more to the body than a calorie of starch. Dr. Elliot Danforth of the University of Vermont in Burlington explains that dietary fat is the only nutrient that can beat a direct path to the body's fat depots. Only 2.5 percent of the calories in fat are needed to accomplish this. Starches, on the other hand, "cost" about 25 percent of ingested calories to be stored as fat, and only about 1 percent of ingested carbohydrates end up as body fat. Thus, simply switching from a high-fat diet to one high in carbohydrates, without actually lowering total caloric intake, can result in a net caloric loss to the body. In addition, switching from simple carbohydrates—sugars—to complex carbohydrates—starches—and increasing dietary fiber can reduce the high insulin levels often found in fat people. Since a main role of insulin is to promote the storage of body fat, lowering insulin levels should facilitate fat loss.

It might also help to divide caloric intake into as many as six mini-meals a day and to avoid consuming concentrated sweets between meals. Both large meals and sweet snacks trigger an outpouring of insulin and may increase body fatness beyond their strict caloric contribution. Still, obese people may have difficulty achieving a normal level of the fat-storing enzyme, lipoprotein lipase. Dr. Robert H. Eckel of the University of Colorado Health Sciences Center reported to the American Diabetes Association that obese people, in comparison with people of normal weight, produce too much of the enzyme and that even after weight loss, their enzyme activity did not fully return to normal. This suggests that obese people who lose weight may have to continually fight a biochemical tendency to load fat into their cells.

Dr. Robert Schwartz of the Veterans Administration Hospital in Seattle found that people who had maintained a large weight loss for eight or more years still produced too much of the lipase enzyme. But as soon as those obese people who have lost weight start regaining it, their enzyme level drops. Other studies at Rockefeller University showed that obese people may never be able to maintain a reduced state on the caloric intake consumed by those who have never been fat.

Dr. Rudolph Leibel and Dr. Hirsch found that a group of obese people, whose weight averaged 334 pounds, consumed about 3,650 calories a day, whereas people of normal weight who averaged 138 pounds ate 2,300 calories a day. After six months on a 600-calorie diet, the obese people had dropped to an average of 220 pounds, but to keep the weight off they could eat only 2,200 calories a day. The group never reached "normal" weight, but if they had, the researchers say, they would have to consume only 1,700 calories a day indefinitely to maintain the loss. Such biochemical differences between obese people and those of normal weight are most likely genetic in origin, according to studies by Dr. Albert Stunkard, a psychiatrist at the University of Pennsylvania who showed that adopted adults in Denmark were much more like their biological parents than their adoptive parents in body weight.

In another study of identical twins, siblings and nonrelatives, Claude Bouchard of Laval University in St. Foy, Providence, Quebec, showed that weight gain in response to the consumption of excess calories also seems to run in families. Other studies showed that usual levels of activity might be an inherited trait. For example, Dr. Danforth says, a tendency to fidget, which can use up as many as eight hundred calories a day, seems to be inborn and possibly inherited. "More people in the obesity field are now looking at metabolism as a biochemical phenomenon that is derived from inherited traits," says Dr. Theodore B. Van Itallie of St. Luke's–Roosevelt Hospital Center in New York. For example, he explains, fat cells have two kinds of receptors on their surface, one that promotes the breakdown of fat and the other that favors fat accumulation. "People might differ genetically in these receptors," Dr. Van Itallie says. "Such findings are beginning to explain why people living in the same environment vary in fatness."

At the least, the fat-cell receptors seem to explain why people often cannot lose weight in particular trouble spots. Dr. Leibel at Rockefeller University found that the fat cells on women's thighs and hips predominantly contain the receptors that accumulate fat. One woman with a pear-shaped body was shown to have almost none of the fat-releasing receptors on her thighs and hips; when she lost weight, she lost it everywhere except where she wanted to lose it. There may also be an inherited difference in the ability of people to generate body heat from food, which uses up some of the calories a person consumes.

Researchers at the University of Lausanne in Switzerland showed that women who had been obese since childhood generated significantly less body heat from a meal than did women of normal weight. Even after weight loss, the obese women showed a deficit in heat production. Still, Dr. Van Itallie says: "The growing evidence that fatness is inherited should not discourage physicians about the treatability of obesity. Body fatness responds to environmental conditions. As members of a sedentary and food-laden society, obesity-prone persons who wish to control their weight must learn to maintain a high level of physical activity and eat defensively." [JEB]

Q&A

Q. *I notice I lose about five pounds overnight but gain it back during the day. Why?*

A. "Assuming a person is overweight, a three- to five-pound overnight weight loss is not so unusual," says one weight expert, Professor M. R. C. Greenwood, chairman of the biology department at Vassar College and president of the North American Association for the Study of Obesity. The processes of metabolism continue overnight, while the sleeping person is fasting, and depending on the person's total body weight, "that much weight could be lost as the result of digestion, the loss of heat through metabolism, and therefore the loss of calories," she says. The weight loss could occur even if there were no fluid loss through urination. The overnight digestive process could cause measurable weight loss for a person of any weight, she said, but for most normal-weight people, a weight change of much more than a pound overnight would be unusual.

EAT LESS; LIVE LONGER?

A FLIMFLAM ad writer could not have invented more outrageous claims. Here is a diet that extends lifespan by 50 percent or more. It prevents heart disease, diabetes, and kidney failure, and it greatly retards all types of cancer. It eliminates or forestalls many of the usual banes of aging, including cataracts, gray hair, and feebleness. Adherence to the diet keeps the mind supple and the body spry to an almost biblical old age.

On a microscopic level, the diet protects the genes against environmental insults, keeps important enzymes operating at peak efficiency, and cuts back on dangerous metabolic by-products in the body. And, oh yes, the dieter stays slim. Very, very slim. These claims are not mere snake-oil phantasms, but the results of astonishing studies that lately have captured broad attention among scientists in the fields of aging, toxicology, oncology, and other disciplines.

In laboratory experiments, investigators have discovered that animals raised on a meal plan containing all the necessary vitamins and other nutrients, but only 60 to 65 percent of the calories of the animal's normal diet, will live significantly longer than expected. By nearly all measures, from the health of the creature's organs and the robustness of its immune system to the lustrous appearance of its fur, the animal on the restricted diet maintains the vigor of youth long after the well-fed control animals in the experiment have become weak, sluggish, and grizzled, indeed, long after the controls have died.

Laboratory mice fed the restricted diet, for example, have lived to fifty-five months. The average life span of lab mice eating a normal diet, in which they consume as much as they want to, just as most laboratory rodents do, is about thirty-six months. "The outcome of caloric restriction is spectacular," says Richard Weindruch, a gerontologist at the National Institute on Aging in Bethesda, Maryland. "Gerontologists have tried many things to extend life span, but this is the only one that consistently works in the lab."

Much to their surprise, researchers have found that it does not matter whether the sharply restricted diet is composed largely of fat or of carbohydrates. As long as the animal receives a minimum amount of protein and enough vitamins and minerals to prevent malnutrition, the creature survives to the same venerable old age.

But researchers warn against people undertaking an ascetic regimen too hastily. They stress that experimental animals are fed carefully measured and planned menus that are difficult to translate into human fare, and that it is easy to become malnourished. "At this point, I definitely would not recommend a calorie-restricted diet for people," says Dr. Angelo Turturro, a biologist at the National Center for Toxicological Research, in Jefferson, Arkansas, a division of the Food and Drug Administration that is studying caloric reduction. "There are still too many unknowns about its physiological effects that we have to sort out."

Nevertheless, gerontologists are excited about the insights they will glean by studying caloric restriction. Initial observations that an extremely low-calorie diet extends life span in animals date back to the 1930s, but they were long shrugged off as mere laboratory curiosities. Only now is the study of dietary restriction receiving wide attention and extensive financial support. "It's been a sleeping giant," says Dr. Weindruch. "But the giant has awoken." As a measure of the new enthusiasm surrounding the field, the National Institute on Aging was spending about $3 million in 1990 on studies related to the effects of calorie restriction on longevity, compared with less than $1 million for 1987. A report on the ability of a low-calorie diet to suppress the growth of breast tumors in lab mice is the lead article in a recent issue of the *Proceedings of the National Academy of Sciences*, a prominent science journal.

And four hundred researchers from around the country and abroad flocked to a recent meeting held in Washington on the physiological consequences of dietary restriction. "Six years ago you would have had ten people show up," says Dr. Roy L. Walford, a professor of pathology at the University of California at Los Angeles School of Medicine and one of the pioneers in the field.

Thus far researchers have studied only rats, mice, and species even lower on the phylogenetic scale, including fish, spiders, worms, water fleas, and protozoa. All creatures fed a restricted diet have had greatly extended life spans. Studies are now under way to examine the effects of a low-calorie menu on two species of primates, squirrel monkeys and rhesus monkeys, which scientists hope will be applicable to another sort of primate, the human being. The projects were begun only in the late 1980s, and results are too preliminary to draw any firm conclusions. But scientists say that they will not have to wait for the monkeys' deaths—even with a normal diet, a rhesus monkey lives about thirty-five years—before they will begin to see the impact of caloric restriction on primate health.

Most researchers suspect that the monkeys will reap in years what they are losing in calories, and that, by analogy, reduced food intake probably would prolong the life span of humans as well. "One argument in its favor is that so far dietary restriction has worked in lower animals across the board," says George S. Roth, a molecular geneticist at the National Institute on Aging who heads one of the primate trials. "If it works in that direction evolutionarily, we should be able to extrapolate out in the other direction, to humans, equally well."

But Dr. Roth offers an alternative explanation of how caloric restriction might work for lower species but be less effective for animals already endowed with relatively long lives. By this theory, short-lived species such as insects and rodents could possess a built-in mechanism that allows them to withstand a couple of famine years and still survive long enough to reproduce. But longer-lived species, such as primates, are fertile for so many years that they may not need to have a life-prolonging mechanism set in motion when confronted by a couple of lean years. To resolve the issue, said Dr. Roth, "we really have to wait and see what we get from the monkeys."

The most zealous researchers contend that people need not wait for any more data before considering a curtailed diet. Dr. Walford says he believes that humans could live to an extraordinarily advanced age if they were to limit their caloric intake. "Right now, the maximum human life span is about a hundred and ten years, and only a few people live to that age," he says. "But if what is true for other species is true for man, then with a sufficiently vigorous caloric restriction, the maximum life span could be extended to about a hundred and seventy." Basing his theory on animal studies, Dr. Walford thinks that caloric restriction would have a salubrious effect regardless of how old the new dieter was, although the earlier the better. "This may be too broad a statement, but caloric restriction will halve the rate of aging at whatever time you begin," he predicts. "If you start at fifty, and you were genetically set to die at eighty, then you might live another sixty years rather than thirty."

Dr. Walford himself, who is sixty-five, has been on a low-calorie diet for about five years. He eats between 1,500 and 2,000 calories a day, compared with an average daily intake for men of about 2,500 calories (for women the figure is about 2,100). The bulk of his calories come from vegetables, grains, and other foods considered healthful by most nutritionists. Some of Dr. Walford's colleagues, however, think he may be going too far. "I wouldn't do what Roy is doing," says Dr. Turturro. "He seems pretty healthy, but he chooses his clothes to conceal his boniness."

Scientists in the calorie-restriction business have mapped out many of the effects of diet on animals, mostly mice and rats. They have found that animals on a normal diet often die of kidney disease or destruction of heart muscle tissue. Those that survive a bit longer fall prey to tumors, and all die by the age of thirty-two months. Rats fed a tightly restricted diet almost never contract kidney or heart disease, and though they, too, develop cancer, they do so at a much later age.

They live to almost fifty months and sometimes die of no discernible cause. "When we look inside them, they're completely clean," says Dr. Edward Masoro, a physiologist at the University of Texas Health Science Center in San Antonio who is considered another leading figure in the field.

Whereas free-eating rodents that had white coats as youths often turn gray and oily by two years of age, the restricted-diet rats keep their shiny white fur, sometimes for more than forty months. Rats on restricted diets are able to run mazes more successfully than the well-fed rats at comparable ages. They suffer far less diabetes and fewer cataracts, and their immune system remains vigilant well into old age. Even when they are exposed to serious carcinogens, diet-controlled animals often fail to develop tumors. And rodent strains that are specially bred to be prone to cancer, hemolytic anemia, or autoimmune diseases gain some protection against such disorders if they are assigned a low-calorie diet. "Any kind of screwed-up animal seems to benefit" from caloric restriction, says Dr. Weindruch.

Researchers have now turned their attention to the puzzle of how dietary restriction works its apparent magic. Dr. Masoro and his colleague in the physiology department, Dr. Byung Pal Yu, are examining the metabolic machinery of the calorie-restricted rats compared with that of animals allowed to feed at will. Dr. Masoro has focused on the role of blood glucose levels in health and longevity. All animals rely on glucose, which is metabolized from carbohydrates in the food, as a primary source of fuel for the body's tissues. Dr. Masoro has found that animals fed a restricted diet burn as much glucose per gram of tissue as do the plumper control animals. However, the restricted rats have a significantly lower concentration of glucose circulating in their bloodstream than do the controls.

Dr. Masoro believes that the result of a lowered blood glucose level is globally beneficial to the body. He points out that free-floating glucose can promiscuously interact with many important enzymes and proteins in the body, distorting their shape and function. Dr. Yu has examined the contribution of oxygen to the aging process. As a by-product, food metabolism creates so-called free radicals, highly reactive oxygen molecules that combine with and can damage many parts of the body, particularly the slippery, fat-studded membranes that surround cells. Dr. Yu has discovered far less oxygen damage to cell membranes in food-restricted animals than in the controls. He and other researchers also have found that a liver enzyme designed to detoxify free radicals is 50 to 70 percent more active in the dieting animals than in the controls.

Some scientists have detected evidence that in calorie-restricted animals, the enzymes that repair damaged DNA are more robust than in the free-eaters. They say that because mutations in DNA, the cell's information molecule, can lead to cancer, the ability to promptly fix DNA mishaps could explain the delay in tumor growth among dieting animals. Other researchers believe that dietary restriction works by slightly lowering the body's temperature. "The whole system in a food-restricted animal cools down," says Dr. Turturro. "And as the fires of metabolism die down, the amount of damage from metabolism is decreased."

Researchers admit that their understanding of how diet affects lifespan remains embryonic, and a subject of much dispute. "There are as many theories of aging as there are gerontologists," says Dr. Weindruch.   [NA]

EATING WELL IS THE BEST REVENGE

A new study of more than one thousand Wisconsin residents suggests that frequent childhood consumption of fresh vegetables—especially broccoli, cauliflower, and salad—is associated with a reduced risk of developing colon cancer later in life. As expected, high consumption of "hidden fats"—burgers, luncheon meats, and fried foods, for example—was linked to an increased cancer risk. Previous studies in animals have shown that dietary fats and cholesterol enhance the effects of chemicals that cause colon cancer. However, all fats may not have an ill effect.

The study, by Theresa B. Young, an assistant professor of preventive medicine at the University of Wisconsin Center for Health Sciences, suggests a protective effect from childhood diets rich in peanut butter, milk, and cheese, which are also high-fat foods, and the study found no particular risk from the use of fats like butter and salad dressing that are added to foods. Ms. Young explains that added fats are not necessarily safe. Rather, she says: "The protective effect of salad consump-

THE BURDEN OF PROOF • 223

tion is almost shocking. Thus, the use of added fats, which include salad dressings, may actually indicate a healthier diet, because lots of salad is being eaten." The study, conducted among white men and women from the ages of fifty to ninety, revealed some benefit to switching to a better diet later in life. But it is still unknown whether such a switch can fully compensate for a poor diet in childhood.

THE BURDEN OF PROOF

AFTER years of advising many Americans to lower their blood cholesterol levels to prevent heart attacks, federal health officials drew up plans in 1989 for a new study to determine whether that advice holds true for women, who make up the bulk of the population, and the elderly of both sexes, who suffer the vast majority of heart attacks. When the National Heart, Lung, and Blood Institute described the plan for testing, the agency focused new attention on a broad debate over the wisdom of the sweeping National Cholesterol Education Program, begun in 1985. The program seeks to make all Americans aware of their cholesterol levels, to treat those whose cholesterol levels put them at particular risk of heart disease, and to institute a national dietary change that would diminish the average American's cholesterol levels by 10 percent.

The biochemical mechanism by which cholesterol can, in some cases, cause heart attacks is poorly understood. Researchers believe that cholesterol becomes trapped inside the walls of arteries, where it initiates the formation of plaque, the artery obstructions that can eventually block the flow of blood to the heart, brain, or other vital organs and cause heart attacks or strokes. Although the body makes most of its own cholesterol, foods that are rich in saturated fats or in cholesterol itself can also drive up the level of cholesterol in the blood. Critics of the education program have asserted since it began that the costs and benefits of lowering cholesterol for all but middle-aged men with extremely high cholesterol levels is sketchy, and the evidence that lowered cholesterol level extends life is tenuous.

They have questioned whether results from studies of middle-aged men taking powerful cholesterol-lowering drugs should be used to advise older people, in whom high cholesterol levels are not as good a predictor of heart disease, and women, who have never been explicitly studied in a clinical trial of reducing cholesterol. They have also questioned whether even middle-aged men with only moderate cholesterol levels need worry much about reducing their cholesterol. Heart attack rates for women start rising after menopause. But before menopause their risk is low.

Many experts argue forcefully that cholesterol reduction does no harm to the individual and might well prevent heart attacks in many people. They also believe the recommended dietary changes would be healthful for a variety of reasons, not just cholesterol reduction. But critics point to puzzling results in some studies suggesting that cholesterol reduction might cause health problems, and they worry about the long-term side effects if patients resort to a lifetime of cholesterol-lowering drugs. The effort to reduce cholesterol levels in tens of millions of Americans has been endorsed by some of the most prominent mainstream health organizations, including the American Heart Association, the American Medical Association, a panel of experts assembled by the National Institutes of Health, and the National Heart, Lung, and Blood Institute, among others. But some skeptics have been challenging the importance of cholesterol reduction from the start, and as the national effort to reduce cholesterol has swung into higher gear, so has the volume of questions and criticism. For example:

- A group of medical experts convened by the Congressional Office of Technology Assessment recommended that Medicare not pay for cholesterol screening in people over age sixty-five. The group said that it was not established that cholesterol reduction in older people saved lives or reduced the chances of having a heart attack.
- In a report in the *New England Journal of Medicine*, Dr. Allan S. Brett of New England Deaconess Hospital in Boston said studies of people who have lowered their blood cholesterol levels with drugs indicate the benefits are much less than portrayed.
- In an article in the *Atlantic Monthly* and in a subsequent book, Thomas J. Moore, a journalist, argues that the dangers of high blood cholesterol have frequently been exaggerated and that the

national campaign to reduce cholesterol levels in more than a quarter of the adult population will be expensive and misguided.

- Prominent individuals, including the heart surgeon Dr. Michael DeBakey, who is chancellor of the Baylor College of Medicine, and Dr. Thomas N. James, a past president of the American Heart Association and president of the University of Texas Medical School at Galveston, continue to question the recommendation that old people with high cholesterol levels try vigorously to lower them. And Dr. James questions the basis for recommending cholesterol-lowering diets for children and others not at high risk of heart disease.
- The American Council on Science and Health, a policy analysis group financed largely by industrial sources, argues that the role of dietary cholesterol in causing heart disease is being overstated and distorted.

Although some proponents of the cholesterol-lowering campaign have bristled at the critics, they readily concede that not every aspect of their advice is based on irreproachable clinical data. "We reached a consensus about cholesterol, but naturally there were questions," says Dr. Scott Grundy, director of the Center for Human Nutrition at the University of Texas Southwestern Medical School at Dallas and a past chairman of the American Heart Association's nutrition committee. "The excitement was so great when we decided to start the program, there was a lot of enthusiasm on the part of the press. There was bound to be oversimplification. Undoubtedly, people have taken it to be less of a complex issue than it really is."

Dr. Basil Rifkind of the heart institute added that the nagging questions about cholesterol reduction for all members of the population are what prompted the institute to undertake its new study. "I do not think the case for cholesterol reduction has been proved to the degree we all would prefer," Dr. Rifkind said. In particular, 75 percent of all heart attacks occur in people over age sixty, but previous studies have specifically excluded people sixty and over.

Dr. Grundy says he still advises people to follow low-fat diets and lose weight if they are overweight because this advice is generally considered to be at worst harmless and almost certainly beneficial. He adds that the value of treating old people and women for high cholesterol and the value of diets (as opposed to the use of drugs) in reducing cholesterol are going to be questioned, and rightly so. "The debate was there all along, but because of the tremendous interest in cholesterol, these issues were put on the back burner. But I think a more sobering view is inevitable. All of these questions that were put in the background will have to be reexamined. Most of us realize that that's inevitable."

But Dr. Daniel Steinberg, a cholesterol researcher at the University of California at San Diego, wonders what the critics expect policy planners to do. "We could have said, 'We don't have the data, so don't do anything about cholesterol,' " says Dr. Steinberg, who headed a federally sponsored panel that recommended lowering cholesterol for the nation. "Meanwhile, people are dying of heart attacks at the rate of five hundred thousand a year. We would have felt derelict if we had not made the recommendation."

When the National Cholesterol Education Program was initiated in the mid-1980s it represented the consensus of leading heart disease experts. These experts reviewed a huge body of research, dating back many decades, that showed to almost everyone's satisfaction that the higher a person's cholesterol level, the greater the risk for heart disease. The data also showed that people could reduce their cholesterol levels by an average of 10 percent if they followed a diet low in saturated fat, according to Dr. Rifkind.

But it was not enough to find that people were more likely to have heart attacks if their cholesterol levels were high. Before doctors recommended that the nation reduce its cholesterol levels, they had to show that lowering cholesterol levels prevented heart attacks. That evidence came in 1984. A heart institute study, the Coronary Primary Prevention Trial, showed that when middle-aged men with very high cholesterol levels reduced those levels with the drug cholestyramine, they had significantly fewer heart attacks than a control group of men who did not take the drug. The drug is manufactured by Mead Johnson Laboratories, a division of the Bristol-Myers Company. The study found, in general, that for every 1 percent drop in cholesterol levels, there was a 2 percent reduction in heart attacks.

After the national education program had begun, researchers got what they viewed as further confirming evidence for their public health stand. In 1987, another study of cholesterol-lowering drugs, the Helsinki Heart Study, showed similar results. About the same time, the University of Southern California reported on a study of men who had already had a coronary bypass operation for severe atherosclerosis, then followed strict cholesterol-lowering diets and took cholesterol-lowering drugs. The study found that the growth of the artery-blocking plaques was slowed or even reversed, in some cases. In addition, researchers looked back at data from men who took the generic drug nicotinic acid to lower their blood cholesterol during a five-year study in the 1960s. The follow-up data on the men nine years later showed they had a lower death rate than men who had not taken the drug.

There were some nagging unanswered questions, however. One disappointment, Dr. Rifkind says, was that the major cholesterol-reduction studies, like the Helsinki study and the Coronary Primary Prevention Trial, did not show that the men who reduced their cholesterol levels lived longer. In both studies, for unknown reasons, the participants who lowered their cholesterol were more likely to die violently or in accidents. In the Helsinki study, which relied on the drug gemfibrozil to get cholesterol levels down, the lowering of cholesterol was also associated with an excess number of operations to treat gastrointestinal problems. The drug is made by the Parke-Davis Group, a division of the Warner-Lambert Company.

Dr. Paul Meier of the University of Chicago says the failure to demonstrate a reduction in mortality "is a real worry." Dr. Meier, a statistician who has been part of the team that analyzed numerous clinical trials in heart disease, and Dr. Thomas Chalmers of the Harvard University School of Public Health brought up the troubling drawback of the studies at a heart institute conference held in 1985 to reach a consensus on the value of lowering cholesterol for the entire population. "We were as welcome as ants at a picnic," Dr. Meier recalls. Dr. Myron Weisfeldt, head of cardiology at Johns Hopkins University School of Medicine and president of the American Heart Association, and others believe that a mortality advantage will emerge as time goes by and the trial participants get older. But Dr. Brett, commenting in the *New England Journal of Medicine*, wrote that "although one may legitimately speculate that further years of treatment could convert favorable trends into statistically significant ones, is it not conceivable that the same reasoning may hold for unfavorable trends such as gastrointestinal disease and violent death?" Another question that critics of the national cholesterol education program frequently raise is whether older people should be concerned about cholesterol. The studies that showed the benefits of lower cholesterol levels involved middle-aged men with very high cholesterol levels. Although cholesterol levels are good predictors of heart disease risk in middle-aged people, they are considered less reliable in older people. And cholesterol levels tend to rise as people get older, although no one knows why. "If you believe there is a benefit to cholesterol lowering in older people, it is based on inference," Dr. Weisfeldt says. "There are no data on intervention in older people that are of comparable magnitude to the data on intervention in younger people. Many clinicians, including myself, are certainly more cautious in the use of cholesterol-lowering drugs in older people." He notes, however, that cholesterol-lowering diets are safe and he would recommend them to older people.

Dr. James, a past president of the American Heart Association, disagrees: "One of the saddest things is to see patients who are in their seventies or eighties and who are terrified by what they are eating. One of the first things they want to know is what their cholesterol level is." He says there are even examples of malnutrition in the elderly because they were so afraid that their favorite foods would raise their cholesterol levels. The obsession of the elderly with diet and cholesterol "is a national tragedy," he says.

Another question has to do with long-term use of cholesterol-lowering drugs. Those studied the longest, like cholestyramine, are so unpleasant to take that many patients give them up. In the Coronary Primary Prevention Trial, participants had to drink a gritty solution of the drug. One trial participant said the men in the study complained that taking it was "like drinking Miami Beach." The drug caused nausea, heartburn, and bloating, and most of the men did not take the full dose.

A new class of drugs, like lovastatin, made by Merck and Company, is now on the market and much easier to take. But these drugs have not been tested for long-term effects. Dr. Weisfeldt says he would probably avoid putting a man under age forty on such a drug, unless the man had a strong family history of heart disease and had very high cholesterol levels. "These drugs are potentially toxic," Dr. Weisfeldt says. [GK]

IS ALCOHOLISM INHERITED?

RESEARCHERS studying sons and daughters of alcoholics are detecting specific biochemical and behavioral differences in their responses to alcohol that may be a key to why these children are prone to becoming alcohol abusers themselves. For years, scientists have been reporting that a tendency to become an alcoholic can be inherited. Expanding on recent studies, and with new findings appearing almost monthly, researchers are identifying some inherited physiological differences among children. The differences may indicate a predisposition to alcoholism, researchers say.

The newest studies reflect the resourcefulness required in facing one of science's most elusive challenges: identifying genetic factors in human behavior. One key, much-discussed finding is that college-age sons of alcoholics tend to have better eye-hand coordination and muscular control when they drink. They also tend to have a lower hormonal response to alcohol and, as compared to young men whose parents are not alcoholic, to feel less drunk when they drink too much. Another group of researchers has shown that college-age daughters of alcoholics exhibit most of the same traits as the sons. And young boys who do not drink themselves but whose fathers are alcoholics tend to have the same unusual brain wave patterns seen in alcoholics, another research group finds.

What researchers strongly suspect but have not yet proved is that the children who are born with these various traits are more likely than others to actually become alcoholics. But there are strong reasons for suspecting this is so, the scientists say. According to Dr. Enoch Gordis, director of the National Institute on Alcohol Abuse and Alcoholism, there are 10 million "full blown alcoholics" in this country and an additional 7 to 8 million alcohol abusers. There is wide agreement that a tendency to alcoholism can be inherited. "There were more than one hundred studies published this century indicating a familial basis of alcoholism," says Dr. Henry Begleiter of the State University of New York Health Science Center in Brooklyn.

Strong evidence indicates that this reflects genetic as well as social factors. Recent studies of adopted children of alcoholics indicate that 30 to 40 percent become alcoholics themselves, regardless of the drinking habits of their adoptive parents. In contrast, 10 percent of the general population is dependent on alcohol.

Among the first to study the children of alcoholics was Dr. Marc Schuckit of the University of California at San Diego, who began recruiting college students and examining their responses to alcohol thirteen years ago. By the late 1980s he had studied 400 men; half had alcoholic fathers and none were alcoholic themselves at the time of the study. Dr. Schuckit and all the other researchers restricted themselves to children of alcoholic fathers to exclude the possibility that an alcoholic mother could have affected her child by drinking during her pregnancy.

Dr. Schuckit invented an apparatus with which he could give alcohol or nonalcoholic beverages to the volunteers without letting them know which they were drinking. All the drinks, real as well as sham, had the odor and taste of alcohol. He then gave the men up to the equivalent of four to five drinks and kept track of their blood alcohol levels. As a group, the sons of alcoholics said they felt less drunk than the sons of nonalcoholics and they performed better on tests of hand-eye coordination even when their blood alcohol concentrations were identical to those of the sons of nonalcoholics. In addition, Dr. Schuckit reports, the sons of alcoholics swayed less when they walked and, finally, they had less pronounced changes in those hormones whose levels rise in response to alcohol. The hormones are prolactin, cortisol, and adrenal cortical trophic hormone.

Forty percent of the sons of alcoholics showed decreased sensitivity to alcohol in terms of perception of drunkenness, performance after drinking, and hormone levels, Dr. Schuckit found. That was true of less than 10 percent of the control group. Although Dr. Schuckit says he did not know what the hormonal findings might mean in terms of physiology, the hormone test was "less volitional, more biological" than the other tests, and so it emphasized his conclusion that there are real differences in the sons of alcoholics.

Dr. Jack Mendelson and Dr. Barbara Lex of McLean Hospital in Boston repeated Dr. Schuckit's experiments, but this time with daughters of alcoholics. After about fifty women were studied, the results, according to Dr. Lex, were in general agreement with Dr. Schuckit's.

Women had not been studied previously because hormonal changes during their menstrual cycle can change their responses to alcohol. Dr. Mendelson and Dr. Lex overcame that obstacle by making sure, with blood tests, that all the women in their study were at the same hormonal stage in the menstrual cycle when they were tested.

Dr. Begleiter, also displaying ingenuity and determination, was attacking the question of inheritance from a different perspective. He began, he says, more than twenty years ago by studying the brain waves of alcoholics. To do this, he developed a method of examining how the brain works. Dr. Begleiter learned that he could measure electrophysiological patterns of the brain while subjects were asked to think, anticipate, or remember. For example, he would show subjects a series of photographs, including pictures of the subjects' own family members, public figures, and strangers. He would ask the subjects which photos they recognized and would note how their brain waves changed when they saw a familiar face in a picture.

Dr. Begleiter decided to use the method to study the brains of abstinent alcoholics to determine whether drinking had damaged their mental abilities. He found deficits, he said, and so he wondered whether the problems would clear up in time. Some did clear up, Dr. Begleiter found, but others did not. Even when he studied former alcoholics who had not had a drink for as long as five years, some of the brain deficits remained. "In the majority of cases, you wouldn't know anything was wrong until you did the test," he says, but because the deficits persisted, "we got the idea that maybe some of the deficits that did not recover were not consequences of alcoholism but anteceded it." So he decided to study young children who had had no exposure to alcohol but whose fathers were alcoholics.

Dr. Begleiter found that as many as 30 to 35 percent of the sons of alcoholic fathers had the deficits typical of alcoholics, whereas less than 1 percent of the boys in a matched control group did. Five other groups have recently replicated this finding, Dr. Begleiter says, and one group replicated the results on two separate occasions.

Dr. Begleiter was intrigued by the finding because a large study of adoptees in Sweden had indicated that about a third of the sons of alcoholic fathers become alcoholics themselves. Dr. C. Robert Cloninger, an investigator in the Swedish study, has proposed that there are subgroups of alcoholics and that inheritance is more pronounced in a group that, Dr. Cloninger suggests, use alcohol because it releases their inhibitions.

When Dr. Begleiter looked specifically at sons of alcoholic fathers who fit this particular subgroup in Dr. Cloninger's classification scheme, he found that 89 percent had the deficits on the brain wave test. "We were very surprised," Dr. Begleiter says.

The next step for all three groups of investigators is to follow the children of alcoholics and see whether the sons and daughters who the researchers expect, on the basis of the tests, are most likely to become alcoholics do, in fact, become dependent on alcohol.

And the scientists would like to learn how what they are measuring relates to the process of becoming dependent on alcohol. Dr. Schuckit speculates that if many children of alcoholics do inherit an increased ability to tolerate alcohol, that may set the stage for alcohol dependence by making it difficult for these people to learn to stop.

Young men, in particular, often are encouraged to drink heavily, Dr. Schuckit says. If they do not have the inherent protection of feeling and acting drunk, they may begin "drinking regularly enough and heavily enough so that it becomes a problem." [GK]

THE DISEASE THAT JUST WON'T GO AWAY

WHEN officials in Hartford barred spectators from a North Atlantic Conference basketball tournament one recent season because they feared an outbreak of measles, it was a stark reminder that the disease, which seemed on the verge of disappearing only a decade earlier, remains a public threat. The highly contagious malady lingers despite concerted efforts to eliminate it; three thousand to four thousand cases are reported across the United States, in scattered outbreaks, in a typical year.

This is a small fraction of the 500,000 cases a year that afflicted Americans before a vaccine was licensed in 1963. But federal health officials warn that the common nine-day measles, also called rubeola, is still a dangerous disease. In a renewed attempt to contain it further, a panel of experts who advise the Public Health Service has strengthened its recommendations on vaccinations, specifying who should receive shots now and under what conditions.

Measles can cause hearing loss and encephalitis. It opens the way to streptococcal infections and pneumonia. And it can kill people. Among the world's children, measles is the second most prevalent cause of death after diarrheal diseases. Even in the United States, children sometimes die from its complications. Most who die are under two years old. "It's not a simple, mild, benign disease," says Dr. Rebecca Meriwether, who heads the communicable disease control branch for the health department of North Carolina, one of the states that have had measles outbreaks.

In one of the most dramatic examples of the measles threat, sixty unvaccinated children contracted measles in 1989 in the Samoan community in Los Angeles. Six died of complications, mainly pneumonia. Moreover, nearly a quarter of all measles sufferers in Los Angeles County in 1988 were hospitalized with secondary infections or dehydration brought on by diarrhea. "This emphasizes that it is a serious disease," says Dr. Stephen Waterman, the chief of the county health department's acute communicable disease control unit.

The symptoms of measles include fever, a running nose, a hacking cough, inflammation of the eyes, sensitivity to light, the familiar rash, and usually mild itching. Diarrhea is not uncommon, and secondary infections can strike the ear and the lungs. Federal health officials say that 98 percent of all children are now vaccinated by the time they enter school. But they estimate that fewer than half of the nation's preschoolers are. (Precise figures are not available.) The most vulnerable groups include children who were vaccinated before the age of fifteen months. Before then, measles antibodies transferred to the child from the mother before birth can block the vaccine's ability to confer full immunity.

Also among the vulnerable are some people born after 1956 and before 1980, even though they were vaccinated in that period. Officials at the federal Centers for Disease Control (CDC) assume that Americans born before 1957 have had the disease and are immune. But a small percentage of those vaccinated before 1980 received ineffective doses because the vaccine then was less stable, weakening in the presence of light and heat. More stable vaccines were introduced in 1980.

In line with these vulnerabilities, the CDC said, those contracting measles this year fall into two broad groups: preschool children, most of them in inner cities, and young people in junior high school through college.

Local health officials have tried to control outbreaks in several ways. For instance, when measles hits a school, all children who have not been vaccinated are sent home until they receive a shot. The incubation period is about ten days, and immunization can protect a person early in that period. Local officials in recent years have also closed or canceled basketball games, curtailed seasons, and prohibited other gatherings at scattered high schools and colleges. The officials fear that measles would spread rapidly in the confines of an indoor arena or auditorium.

In fact, the transmission of the measles virus has been traced to such enclosed settings in some instances, says Dr. Sonja Hutchins, an epidemiologist in the division of immunization of the CDC. She says that the experts who advise the Public Health Service have not specifically recommended that games be canceled or closed to the public, but because the spread of measles at such events has been documented, some local health officials have decided to do so.

New York health officials have proposed a bill in the legislature that would require students to prove immunity to measles before they can enter college, according to Bill Fagel, a spokesman for the state Department of Health. The same requirement applies throughout the country for children entering elementary school.

Within four years after a vaccine was licensed in 1963, the number of cases reported in the United States plunged 87 percent, from roughly 500,000 a year to 63,000. A decade later, only 27,000 cases a year were reported, and the CDC undertook a program intended to eradicate the disease by 1982. It adopted an aggressive policy of encouraging children to be vaccinated at fifteen months. It also sought to improve surveillance and reporting and to control outbreaks by zeroing in on specific homes where one member of the family was infected to prevent other members from becoming so.

Almost simultaneously, a more stable vaccine was introduced, and in 1983 the number of reported cases dropped to 1,500. Since then, the totals have risen to 3,000 to 4,000 in most years. The post-1983 peak occurred in 1986, when more than 6,000 cases were reported.

"We came very close to eliminating measles," Dr. Hutchins says, "and we're still very close, but outbreaks are still occurring." None of them reaches anything like epidemic proportions, but they cause worry nevertheless, particularly when they involve poor residents of cities, whose infants are less likely to be vaccinated and who are more vulnerable to complications.

New recommendations, issued by the Public Health Service's Immunization Practices Advisory Committee, call for children living in areas of continuing outbreaks among preschoolers to be vaccinated at the age of nine months as a precaution and again at fifteen months for lasting immunity. Continuing year-to-year outbreaks occur primarily in inner cities.

Many of those now in junior high school, high school, and college have no natural immunity. About 5 to 10 percent in this group may be vulnerable because they received the less stable vaccine, the CDC estimates. For this group, the expert panel recommends that in schools where there are outbreaks, all those vaccinated before 1980 be vaccinated again, along with their siblings.

Of the two risk factors, vaccination before 1980 and vaccination before the age of fifteen months, the latter appears to be more important, says Dr. Meriwether. In a junior high school that was in the "epicenter" of North Carolina's recent outbreak, students who had been vaccinated before the age of fifteen months were more than twice as likely to contract measles as were schoolmates who had been vaccinated after fifteen months.   [WKS]

THE TRUTH ABOUT COMAS

MIRACULOUS recoveries from prolonged comas are the stuff of movies and fairy tales. But although medical advances have opened the way for more coma patients than ever to survive, most linger in a limbo state of unconsciousness, somewhere between life and death. Few of these patients, even if they can be kept alive indefinitely and even if they appear to awaken and sleep regularly, have any hope of full recovery.

By one recent estimate, at least ten thousand Americans are in irreversible comas at any given time. Each case means an extended period of anguish for loved ones, who hang on to a thread of hope, and for the doctors and nurses who must give daily care to patients with hopeless prognoses. Deciding how to treat comatose patients with little chance of recovery poses wrenching and unresolved ethical and economic issues.

Doctors have long recognized the coma as an advanced state of brain failure in which a person lies in a sleeplike state with eyes closed.

But recent research has shown that continuous sleeplike comas seldom last more than a month. Those who survive that long usually proceed into a condition described as a persistent vegetative state. In this condition, the mind is dead but the brain is not.

Life continues in this state because the brain stem activates the vegetative, or autonomic, nervous system, to carry on the vital mechanical functions governing breathing, heart pumping, blood pressure, and elimination of wastes. The patient is unresponsive, often appearing awake but giving little or no evidence of awareness of the environment or ability to express thoughts. Many of these patients resume normal cycles of sleeping and waking. Their eyes may open spontaneously, and they may reflexively blink when menaced. "It is hell for people to see a loved one in a sleep-wake cycle, moving their eyes, and expecting—falsely—that the individual understands and will recover," says Dr. Fred Plum of the New York Hospital–Cornell Medical Center, an expert on comas.

The longer someone lives in a coma, the less likely he or she will recover. Virtually no one comes out of a vegetative state that has persisted as long as six months. But researchers hope to learn what allows the rare individual to emerge from a coma after being in that state for a few weeks. They hope that such knowledge could be used to help others. The problem of long-term comas looms over doctors and families every day in every hospital. Although accurate statistics about the numbers of comatose Americans are lacking, experts believe the numbers are increasing, largely as a result of the introduction of intensive care units in hospitals and new treatments for once-fatal conditions.

Some coma patients are victims of heart attacks and strokes who were saved only to suffer brain damage. Many victims of Alzheimer's disease eventually become comatose. And some coma patients are victims of drug overdoses or automobile accidents. In recent years, researchers have used the new technology of PET scanners—for positron emission tomography—to measure biochemical actions in the brain.

Such tests have confirmed the reliability of the conventional appraisals doctors use to measure cognition and brain damage, such as flicking fingers in front of the eyes and pouring cold water in the ears to test the direction of eye movement and other reflexes. The PET scans have shown that the persistent vegetative state is comparable to the deepest stages of anesthesia and that such patients do not feel pain, exerting only reflex responses when pinched or otherwise stimulated. "What was learned is extremely important because it allows physicians to deal more humanely with families," Dr. Plum says. With a more scientific basis for diagnosis and prognosis, families' uncertainty can be reduced.

Yet the coma still has its mysteries. One is why some autopsies have found such a striking disparity between the limited extent of structural brain damage and the total devastation of the mind. Another concerns those who suffer from prolonged comas and the persistent vegetative state. What selectively kills the brain cells? Does the injured brain produce poisons that create even more brain damage? Would transplants of brain cells help?

The bleak outlook for people in the vegetative state underscores the crucial decisions that doctors and families must make about a patient's clinical care in the first hours of a coma. In this period, neurological signs may be more important predictors of the patient's future than the actual diagnosis of the underlying illness or injury. This could help physicians and families decide whether there is any point in taking extraordinary life-saving measures.

Although the American Medical Association has said that it is ethical for doctors to withhold all means of life-prolonging medical treatment, including food and water, from people in irreversible comas, not everyone agrees. Families rarely get enough counseling in making the crucial decisions about a loved one's care and in dealing with any feelings of guilt that may develop, according to Dr. Gerald Steinberg, director of the Western Massachusetts Hospital in Westfield, who treats many coma patients.

Dr. Sheldon Borrel, a rehabilitation medicine specialist at San Francisco General Hospital, surveyed health care workers and determined that those who were farthest removed from the coma patients' bedsides found it easiest to approve withholding life support. "The closer you are to being the one who has to remove the tube," he says, "the more difficulty you have with the decision." [LKA]

Our Troubled Environment

THE THREAT OF THE NEW

THE boatman called Mohammed said the river was his life, as it had been his father's before him, and would in time become his son's. The Nile, he said, does not abandon its children. Yet, along the river's banks, where humanity has thrived across the millennia, a discordant note is being sounded by those concerned with old Egypt's dwindling ability to withstand the encroachment of the new. Monuments that have endured thousands of years, surviving even polluted air and a flood of tourists, many specialists say, are under new assault from rising ground water and pollutants that threaten even unexcavated relics once thought to be shielded by the sand that buries them. "Our belief that an unexcavated monument was a safe and protected one, whose pristine state would be enjoyed by later generations, has also been found to be wrong," Kent Weeks, an American Egyptologist at the American University of Cairo, said in an address. "Even undug sites are suffering from these insidious forces of destruction."

For several years, Egyptologists have acknowledged the threat to excavated antiquities. But they have clung to the belief that sites left underground could safely be left for excavation and discovery by future generations, allowing today's specialists to concentrate on the vast array of archaeological treasures already unearthed. But according to several experts, a body of evidence from new excavations shows environmental damage to delicate sandstone and other structures, even before excavation took place. That adds a new sense of urgency to combating the problems afflicting the legacy of ancient Egypt, the specialists say.

As the scope and magnitude of problems emerge, conflict between Egyptians and foreigners and political infighting within Egypt's own antiquities department have all but paralyzed the effort to take action to avert a possible disaster. A sense of urgency has been building since the Soviet-built Aswan High Dam, upstream from here, brought a fundamental change to the environment of the Nile Valley, the slender ribbon of fertility that covers 3 percent of Egypt's surface area but houses most of its 55 million people.

When the dam filled in the 1960s, the seasonal flood and ebb of the river was stemmed and harnessed, enabling farmers to plant twice a year and fill irrigation channels all year round. Previously the annual flood dictated the planting of a single crop each year as the flood waters ebbed. At the same time, with more predictable food supplies, enhanced health care, and protection from both flood and drought, Egypt's population has been growing. By the year 2000, according to Egyptian and Western estimates, the population will swell to more than 70 million people, but the land available for traditional farming and housing cannot expand to meet that growth.

For Egypt's great trove of archaeological treasures—the world's most extensive—the changes have brought calamity, from the Sphinx in the north to the Luxor Temple in the south. "We are like a good mother who has one hundred children," says Sayed Tawfiq, the director of the Egyptian Antiquities Organization in Cairo. "All of them need to be cared for. There are more than two thousand tombs, a lot of monuments, pyramids, obelisks. If you gave every tomb two years of restoration, that would be four thousand years, which we don't have."

Ground water beneath the monuments has risen: the air is more humid because the irrigation canals never empty; salts in the soil are drawn through ancient facades, peeling them away from the rock below; sewage has tainted the soil. And, as visitors throng this resort and visit the sites of the City of the Living and the City of the Dead on each bank of the Nile, tour buses vibrate over the tombs and perspiration fills chambers once left in seclusion. Each day, some specialists say, perspiration in the Tomb of Tutankhamen is the equivalent of six gallons of liquid released in its claustrophobic confines, damaging old wall-paintings. Some pillars at Luxor Temple are corseted with wooden scaffolding because they shifted in the changing environment. A chunk of limestone simply fell off the Sphinx, offering a seven-hundred-pound reminder of antiquity's decay.

American researchers working at Karnak Temple recently discovered mystifying sand deposits during excavation and concluded that they had once been sandstone blocks that were eroded underground by changes in the soil around them. In early 1988, a new tomb was opened for the first time in the Valley of the Kings, across the Nile, but the stench of sewage from a broken pipe was so great that no work could be done for six months. "There is not a single site in Egypt free of the threat," says another American Egyptologist, who asked not to be identified by name.

As they grapple with issues that affect the future of some of humanity's great benchmarks, both Egyptian and foreign Egyptologists say there is a pressing need for an international effort to preserve the monuments. "Certain types of monuments are just disintegrating before our eyes," says a Western specialist. "But restoration is a slow process and it is going to take a lot more time than Abu Simbel," he said, referring to the major effort that went into relocating the temple of Abu Simbel as the waters rose behind the Aswan Dam.

The question, however, seems to be how Egyptians and foreigners define their relationship as they contemplate renewal. "It is like a man who has had a heart attack," observes Professor Tawfiq, the director of antiquities. "No one tries to help because no one knows the remedy. Nobody tries to undertake studies for restoration [of the Sphinx]. The reason is that it's difficult and if you fail, it's your last chance."

He says foreigners are drawn to Egypt in hopes of making a discovery that will rank them with the academic equivalents of Indiana Jones. "If you make an excavation, you will get results. You'll be famous. But you don't get a doctorate for restoration." Some outsiders attribute such comments to what one American called a wave of xenophobia among people who resent foreigners for what they regard as an attitude that casts Egyptians as inferior. Most foreign Egyptologists will not discuss the issue with reporters unless their names are withheld because, they say, it is delicate.

Some American and other Western specialists acknowledge that foreign Egyptologists have not always been scrupulous or even scientific in unearthing a major find. "The idea was that if you could find something dramatic, you didn't bother too much about the environment you found it in," an American expert said. But the foreigners also blame political infighting for a marked tightening in the bureaucratic procedures required to gain permission to work as an archaeologist in Egypt. Some paperwork that used to take two months now requires five or six.

Polish archaeologists who worked for more than two decades on the restoration of the funerary Temple of Hatshepsut were ordered in 1989 to suspend their activities, because, as Professor Tawfiq asserts, "the temple has lost its dignity." Professor Tawfiq says the Polish restoration had distorted the temple's sense of antiquity. "They have built a new temple, in my opinion," he says. "It is beautiful and wonderful because it is ancient." But he believes it should have been restored without imposing a vision of how the completed temple might have looked when it was originally built.

The tourists provide a source of dispute within the many competing Egyptian bureaucracies that have a say in how the country's legacy is preserved and developed. "There is a constant battle between the Ministry of Tourism, which is trying to promote a thriving industry, and the Ministry of Culture, which is trying to protect a priceless heritage," a Western expert says. Underpinning the debate is a perception, common to both Egyptians and foreigners, that no one has ever tackled a problem on this scale. In recent years, for instance, people have experimented with ways to protect the four-thousand-year-old Temple of Karnak from rising underground water by digging a trench around it to drain surplus water away. But the trench itself filled with water and compounded the problem. Some tombs have simply been closed to stem their deterioration.

"The enormity of the problem is one of the major considerations," observes an American specialist. "The second consideration is that we don't know what to do because no one has ever done this before." [AC]

YOU DON'T NEED A WEATHERMAN

A cylinder of ice nearly a mile and a half long, painstakingly drawn from the Antarctic icecap by a Soviet and French scientific expedition, has documented 160,000 years of the earth's climate, revealing a surprisingly precise connection between changes in temperature and in the amount of carbon dioxide in the air. "Fortunately, nature has been taking continuous samples of the atmosphere at the surface of the ice sheets throughout the ages," the scientists report in the journal *Nature*. The Soviet-French data reach back past the ice age that ended 10,000 years ago to the ice age before.

As carbon dioxide concentration falls, so does temperature. As carbon dioxide concentration rises—and it is rising with

unprecedented speed now, because of human activity—temperature rises, too. In what is known as the greenhouse effect, atmospheric carbon dioxide traps heat that would otherwise be radiated out to space. Although the researchers stop short of claiming to have demonstrated a cause-and-effect relationship, they have provided strong evidence of a link between climate and carbon dioxide. They suggest that carbon dioxide, produced and consumed by biological processes on land and in the oceans, is part of a complex feedback loop, sharply amplifying small changes in the amount of sunlight that reaches earth.

LAKE 302

DESPITE its utilitarian name, Lake 302, ringed by bedrock outcrops, by jack pine and spruce, is as beautiful as any of the thousands of lakes that dot western Ontario. But no one is flying a floatplane in to fish here or buying waterfront property for a cottage. The lake's water is acidic and its ecosystem is in rapid decline. Acid rain is not the problem here, for the rain in northwestern Ontario is relatively pure. Instead, the lake has been systematically acidified by government technicians trickling sulfuric and nitric acids into the propeller wash of a research boat.

In some of the boldest experiments into lake ecology ever conducted, scientists have turned Lake 302 and others nearby into the equivalent of huge, real-world test tubes. Since the 1960s the lakes have been intentionally filled with pollutants so that scientists can observe and chronicle the progress of chemical and biological damage. The sacrifice of the lakes has produced "the best science ever conducted" on lake acidification, says Eville Gorham, an ecologist at the University of Minnesota who in the 1950s in England was a pioneer researcher on the topic. The studies have virtually undone the position, voiced by some industry experts and others, that damage observed in acidifying natural ecosystems could not be conclusively linked to acids in the rain.

The isolated Experimental Lakes Area, operated by Canada's Department of Fisheries and Oceans, has provided a rare opportunity for scientists to experiment with real food chains and real ecosystems, says Dr. David Schindler, a freshwater biologist who has led much of the research here for twenty years. In the case of acids, the approach has allowed researchers to monitor the intricate patterns of ecological change that occur as a lake becomes more and more acidic. By manipulating the ecosystem in controlled experiments, scientists can establish cause and effect far more convincingly than before.

Among the key findings of the studies are these:

- Biological destruction can occur in a lake even in the very early stages of acidification. Such subtle change would have been impossible to prove at already-acidified lakes because of natural lake-to-lake variations in the types and numbers of species.
- The disappearance of adult fish from a lake is not an accurate gauge of damage to fish populations. Reproductive failure and loss of prey species occur earlier.
- Lakes can at least partially recover soon after acid input ceases, but it might take dozens or hundreds of years before the original ecosystem fully reestablishes itself.
- Wetlands can serve as filters for acid-forming compounds and thus help to protect lakes.
- Acidification causes a dramatic shift in the kinds and number of species that can survive in a lake, although it might not reduce the biomass, or total amount of cell matter, of the organisms in the lake.

Dr. Schindler first conducted "whole lake" experiments here in the late 1960s, studying not acid rain but more conventional pollutants. A scientific and political debate was then raging about whether controlling phosphorus in detergent and sewage effluents would control lake eutrophication, the "rapid aging" process that turns a clear lake murky and chokes it with algae. The detergent industry contended that some other nutrient in waste water, such as carbon or nitrogen, was the chief factor in the explosive growth of undesirable algae. Traditional laboratory experiments were inconclusive.

The newly established Experimental Lakes Area, made up of forty-six isolated, pristine lakes five hours by road from Winnipeg, provided an ideal laboratory for testing the various hypotheses. In the most striking of a series of experiments, Dr. Schindler and colleagues segmented the two basins of an hourglass-shaped lake with an impermeable curtain. Both basins received equal amounts of carbon and nitrogen, but one also received doses of phosphorus. Within months, the basin with phosphorus turned into a pea soup of algae while the other remained clear. That and a series of related experiments showed that a lake could obtain carbon and nitrogen, but not much phosphorus, from the atmosphere, and that phosphorus was usually the nutrient that triggered premature lake eutrophication. The research disproved arguments that phosphorus control would not help Lake Erie and other polluted waterways.

The Experimental Lakes Area has become an established, if out-of-the-ordinary, research center. The headquarters is a rugged hour away from the Trans-Canada Highway down a treacherous dirt logging road. At road's end is a ramshackle village of cabins, house trailers, radio towers, and laboratories, as well as toys, dogs, and playing children. In the summer, as many as fifty scientists, graduate students, and their families live and work here. Dr. Schindler says the region offers what may be the best setting in North America for whole-ecosystem studies on lakes. The air is relatively unpolluted, and the ecosystems in the small lakes are very similar. There is virtually no interference from development or tourism.

"It's about as remote as you can get," he says. Acidification experiments began on Lake 223 in 1976, just as concerns in Canada over possible acid damage to its lakes and streams were beginning to mount. Even sensitive lakes have some capacity to neutralize acid. Thus in the first two years of the experiment, there was little obvious change, although chemical analysis showed a steady decrease in acid-neutralizing compounds. Researchers also were surprised to discover that bacteria in the lake provided additional buffering by consuming or converting some of the acids.

But by the third year, with the neutralizing capacity used up, acidity increased from the lake's normal pH level of 6.5 to a very mildly acidic 6.1 (On the pH scale, a reading of 7 is neutral; progressively lower numbers indicate increasing acidity. A pH reading of 5 is 10 times more acidic than pH 6.) In the fourth through the seventh years, acidity increased steadily, and so did damage to the ecosystem. Freshwater shrimp, an important food source for trout, declined.

Later, fathead minnow reproduction failed. However, in the absence of competition, the pearl dace, another minnow species that is more resistant to acid, temporarily flourished, providing continuing food for the trout. The shells of crayfish began to harden more slowly.

Trout reproduction eventually failed. Then, as the pearl dace also began to decline with increasing acidity, the trout began to show signs of starvation. By the seventh year, when the pH reached 5.0, white suckers, the most acid-hardy of the lake's fish, were no longer able to reproduce. Crayfish numbers declined sharply and, again, reproduction ceased. By the autumn of the eighth year, with the pH still at 5.0, all fish reproduction had stopped, and crayfish, leeches, and mayflies had vanished. Adult trout began to look almost like eels. Shortly afterward the researchers allowed Lake 223 to begin a recovery.

Contrary to earlier speculation by many scientists, the actual biomass of organisms in the lakes did not decline at these acidity levels. Instead, there was a striking shift in the diversity of species. For example, in Lake 223, the number of aquatic insect species, originally about seventy, was cut in half, even though the number of individual insects appeared to remain about the same. Peering through a glass-bottom boat into Lake 302, one can see where much of the biomass there is now concentrated. A form of dark, sticky, filamentous algae floats just off the lake floor in grotesque clumps the size of beach balls. Dr. Schindler says the algae began to appear in clumps at about pH 5.6, and at higher acidity levels they began to dominate, even as a diverse range of other forms of algae began to vanish. At Lake 302 in 1988, acidity was increased to about pH 4.5, a level found in some of the most acidified lakes in Scandinavia and eastern North America. Dr. Schindler says preliminary data suggested that the total biomass finally begins to decrease at such extreme levels. Indeed, the lake may be in the early stages of dying.

Across another lake, Dr. Suzanne Bayley has set up a complex of irrigation pipes and spray heads on an eight-acre peat bog. The system regularly irrigates the bog with artificial rainwater containing sulfuric and nitric acid. Because the bog lies above impermeable bedrock, Dr. Bayley can monitor chemicals in outflowing water. Her work suggests that such wetlands can act like filters, reducing the sulfuric acid to sulfur and water and converting the nitric acid into a nutrient form of nitrogen that bog plants use as a fertilizer. Thus, wetlands may play an important role in buffering some lakes against even more extensive acidification.

One of the research team's most significant findings is that lakes will indeed recover some biological diversity if acidification slows or stops. Beginning in 1984, Dr. Schindler's team decreased acid injections into Lake 223, allowing the pH to recover gradually to about 5.6. As a result, two fish species, the white sucker and the pearl dace, have begun to spawn in the lake and adult lake trout have regained lost body weight. The diversity of plant plankton in the lake has again increased and the amount of the algal balls has declined. However, several of the small fish and invertebrates, including crayfish, fathead minnows, and freshwater shrimp, still had not returned. There were sketchy indications that lake trout eggs are hatching but that the young trout are dying.

"The good news is that there can be a substantial recovery in acidified lakes if acid deposition slows," Dr. Schindler reports, adding that it may be possible, in time, to restock some species in lakes that have adequately recovered. "The bad news is that even if we could turn the tap completely, it could be hundreds of years before we would see a full recovery in an acid-damaged lake." [JRL]

DAMAGE CONTROL

Until now, chance and the irresistible forces of nature have overwhelmed the comparatively puny efforts of humans in determining whether a major oil spill becomes an ecological disaster or spares the environment. Even when attempts to contain the spillage are undertaken promptly and well, experts say, factors beyond human control usually decide a locality's ecological fate.

Will the technology for cleaning up spills ever be good enough to do the job on its own? The question gained new urgency in the wake of the *Exxon Valdez* tanker accident, which brought ecological ruin to Prince William Sound in Alaska, and other major spills. Researchers are working to develop a variety of techniques for dealing with oil spills, ranging from relatively mundane methods of burning the oil or simply sopping it up with absorbent material to sophisticated ways of tracking the myriad threads and tendrils of spilled oil. But in practical application, many experts say, the technology has not fundamentally advanced in two decades, although there have been refinements and improvements. It can deal with some kinds of spills, but experts say that under the worst of conditions, with the biggest spills, the best of it is all but useless.

"There may be no technological fix for big spills in adverse conditions," says Richard S. Golob, the director of the Center for Short-Lived Phenomena in Cambridge, Massachusetts. The center keeps track of oil spills, and Mr. Golob is a recognized independent expert on the subject who has studied it for the last fifteen years. "The public has always believed that in an oil spill, we should be able to contain and recover a vast majority of the oil spilled. Historically, that is just not the case. It's not just a problem of organization and available resources. It's a problem of technology and our ability to deal with winds and waves. We are dealing with some of the largest forces of nature. In major spills, when there hasn't been serious damage, the reason is that Mother Nature has been kind to us."

In 1976, the New England coast was spared a major ecological catastrophe from an oil spill, not because of anything humans did, but because of what nature did. The tanker *Argo Merchant* ran aground on Nantucket Shoals off Massachusetts. It spilled nearly 8 million gallons of No. 6 fuel oil, which is heavy, viscous, and long-lived. If the wind had been blowing toward land, Cape Cod, Martha's Vineyard, and Nantucket would have experienced a disaster. Instead, the wind blew the oil out to sea. By contrast, onshore winds blew oil onto the beaches of Brittany after the grounded tanker *Amoco Cadiz* spilled 68 million gallons into the English Channel in 1978. The result was widespread damage to whole species and communities of wildlife.

Existing techniques are usually effective in containing spills of about a hundred thousand gallons or less, Mr. Golob says, and they can be successfully used to prevent spilled oil from reaching marshes or sheltered coves. But "there are limits to what the equipment and technology can do." Under adverse conditions in the open sea, when waves and winds are high, containment and clean-up attempts can be futile, Mr. Golob says. "Most experts will say that if you can recover 10 to 20 percent of the oil, you're doing well. Many spill experts will say privately that sometimes it is totally ineffective to try to respond to a spill out at sea, and yet that's a very impolitic attitude, because it's giving in. Inaction is perceived as bad." So they proceed to try anyway, he says, while the public meanwhile has "a misperception of what is possible."

OUR VANISHING WETLANDS

A NEW and contentious phase has opened in the struggle over the future of the millions of acres of marshes, swamps, bogs, prairie potholes, and bottomlands that not only provide an irreplaceable habitat for wildlife but also help replenish the nation's water supplies and control its floods. Despite longtime efforts by the federal government to arrest the steady shrinkage of this natural resource, an estimated 200,000 to 400,000 acres of wetlands continue to vanish each year, most in freshwater drainages in inland areas, where 95 percent of all wetlands lie. Largely untouched by federal and state regulatory efforts, millions of acres of these freshwater wetlands have disappeared since the mid-1950s, overwhelmingly because farmers converted them to cropland.

Efforts to preserve wetlands have focused mainly on coastal areas, but now environmentalists, politicians, and regulators are engaged in a campaign to control the attrition of inland wetlands for the first time and ultimately to reverse it. They have declared their near-term objective to be no net loss of wetlands. Under the no-net-loss principle, conversion of wetlands would be permitted only if the user restored previously converted wetlands or created new ones.

But despite widespread acceptance of the basic principle, early attempts to put it into practice have encountered stiff and angry resistance. Opposition comes especially from farmers who see regulatory efforts as a confiscation of their assets. "I bought my farm, and if the government wants it, they should acquire it the good old American way—buy it," Rick McGown, a Missouri farmer, said in hand-written testimony that he presented to Congress. And on Maryland's Eastern Shore, where most nontidal wetlands lie on farms and the issue has become especially heated, enforcement of the no-net-loss policy has discouraged banks from lending money to farmers because of the uncertain status of their lands. It has also stopped development by preventing farmers from selling excess lands to developers, a move many had counted on to provide for their retirement. "The farmers are very upset," says Samuel Q. Johnson, an Eastern Shore legislator who is a member of the Maryland House of Delegates' committee on the environment. "And not only the farmer. This thing is spreading to the banking community, to the real estate community. I know people who have been turned down on loans for land. It has nothing at all to do with their credit or ability to pay. It is simply the unknown factor about whether this was nontidal wetlands." Two real-estate developments worth $22 million together have shut down because of uncertainty over the situation, he says.

Most nontidal wetlands are on private property, and as the wet-lands debate heats up, policymakers are searching for ways to head off conflict and confrontation, allay concern, and safeguard both public and private interests while making the no-net-loss principle a reality. Proposals have emerged in Congress to offer tax incentives for wet-lands conservation and even to pay farmers to reconvert land, which could then be placed in a wetlands conservation bank. Underlying the action and argument is a shift in perception about wetlands. Through most of American history, they have been considered wastelands to be eliminated and converted to productive use. "For a long time, we regarded swamps as sources of pestilence that should be filled in as quickly as possible," says James Leape, a wetlands expert at the Con-servation Foundation, a Washington-based organization that spon-sored an influential national study of wetlands loss. The study was carried out by the National Wetlands Policy Forum, a broadly based group that included environmentalists, industrialists, government of-ficials, developers, and farmers. Its report, issued in late 1988, first proposed the no-net-loss goal. The report has largely shaped the renewed argument and is the focus of the congressional hearings.

Wetlands today are seen increasingly not as nuisances, but as in-valuable resources. Scientists say that they protect shorelines from erosion by ocean waves and storms and reduce the severity of inland floods by slowing and storing floodwaters. They remove pollution

from waters that flow through them and recharge underground aqui-
fers. They store water in the wettest parts of the year and release it at
a constant rate to maintain regular stream flows. They provide critical
breeding and nesting habitats for a wide array of fish and wildlife,
including migratory waterfowl. They support many commercial fisheries, provide an array of commercial products from cranberries and
timber to fish and shellfish, and offer widespread opportunities for
hunting, fishing, boating, nature study, and photography. And because of the gases they emit and absorb, like methane, nitrogen, and
carbon dioxide, they help maintain the planet's atmospheric balance.

Wetlands may be marshes extending across thousands of acres,
like the vast Okefenokee Swamp in Georgia and Florida; or the
squishy borders of inland rivers and lakes; or wet spots smaller than
a room. They include not only Alaskan tundra but also the Everglades
and the bottomlands of the Mississippi Delta. Ecologically speaking,
the loss of wetlands can be disastrous. In one example, the conversion
of prairie potholes of the upper Midwest to agricultural use has been
"extremely devastating" to certain species of ducks, says Joseph S.
Larson, an expert on wetlands who is the chairman of the department
of forestry and wildlife at the University of Massachusetts.

The potholes, large depressions in the prairie created when huge
chunks of ice were buried in the last ice age and then melted, are the
nesting grounds for "probably 80 percent of the ducks in North
America," according to Dr. Larson. But so many of the potholes have
been converted from wetlands to wheat fields, he said, that there has
been a tremendous loss of habitat, and species like the redhead and
canvasback ducks are in serious decline as a result. In addition, a
number of communities have experienced serious flooding after wetlands were drained and replaced by streets, parking lots, and other
surfaces that allow water to run off more freely.

Of more than 200 million acres of wetlands that existed in the
contiguous forty-eight states when Europeans arrived in North America, an estimated 95 million remain, according to the Environmental
Protection Agency. They are found across the country but are concentrated especially in the Southeast, the upper Midwest, and the
Northeastern coastal states. Another 200 million acres remain in
Alaska in the form of tundra, which is frozen wetland. Government
studies by the Fish and Wildlife Service have calculated that from the
mid-1950s to the mid-1970s, 11 million acres of wetlands vanished, a
net loss of about 550,000 acres a year, in the contiguous forty-eight

states. About 80 percent of the loss resulted from the draining and clearing of inland wetlands by farmers. The remainder was attributed to such activities as urban and suburban development (which accounted for about 6 percent), dredging, mining, and the discharge of pollution. Natural phenomena like erosion, storms, and subsidence accounted for about 5 percent of the loss.

How much additional loss has occurred in the last fifteen years is a matter of sharp contention. The Congressional Office of Technology Assessment estimated in 1984 that the rate was about 240,000 to 360,000 acres a year. Whatever the scope of the problem, spokesmen for farmers and developers minimize their constituents' contribution to it. "I don't think agriculture has played that much of a part" in the last ten years, says Mark Maslyn, an official of the American Farm Bureau Federation, who took part in the Wetlands Policy Forum study. "I think that was just a time and point in history that passed." But environmentalists and some government officials insist that interests like agriculture and real estate are responsible for the continuing attrition. "We're losing wetlands by small increments," says Hope Babcock, general counsel for the National Audubon Society, who also took part in the Wetlands Policy Forum study, "and when you start nibbling away, pretty soon you're destroying a whole ecosystem."

The struggle over the issue sharpened as the no-net-loss principle was incorporated into an agreement between the Environmental Protection Agency and the Army Corps of Engineers over the enforcement of wetlands protection under Section 404 of the Clean Water Act. This section, which regulates the filling of wetlands, is considered by environmentalists the most important tool available for controlling wetlands loss. Some environmentalists say that the authority of the Corps of Engineers in some ways does not extend far enough because it does not apply to the excavation, clearing, or draining of wetlands. Such activities are responsible for most wetlands loss, especially inland. While coastal wetlands are reasonably well protected, said the 1984 report by the technology assessment office, freshwater wetlands away from the coasts "are generally poorly protected." In an attempt to help plug this gap, in 1985 Congress prohibited farmers from growing crops on any tracts converted from wetlands after that year. If they do, they are supposed to lose all their federal farm program benefits, including subsidized loans, crop insurance, and price supports.

In the end, many conservationists and officials are convinced, wetlands protection will have to look beyond regulation to economic solutions. "It's unrealistic to expect private land owners to do what's good for society just out of the goodness of their hearts," says Ralph Heimlich, an analyst in the Economic Research Service of the Agriculture Department who has studied the wetlands issue extensively. Outright purchase of land for a wetlands conservation bank is one strategy being pursued in Florida, where officials planned to buy and restore more than a hundred thousand acres of freshwater marshland that were drained and converted into orange groves and truck farms. Other strategies being proposed involve economic aid for not developing or otherwise converting ecologically important wetlands, as well as tax breaks for donating them to a conservation reserve. "You have to begin to tilt the economic playing field," says Ms. Babcock.   [WKS]

BAD NEWS FOR BEACHES

USING computerized measuring techniques, scientists are tracking with new precision the erosion that is eating away 90 percent of the nation's coasts. Their findings are bad news, especially for the 295 barrier islands that lie like beads on a necklace along the coast from New England to Texas. The islands, long sandy strips of land immediately off the coast, are home to many of the nation's shore resorts and coastal towns. The sea has been rising relative to the land for at least a hundred years, geologists say, and most agree that erosion is going to accelerate as global warming melts the polar ice packs, sending sea levels even higher.

The barrier islands are adapting to the rising sea by slowly migrating to higher elevations inland; as their seaward dunes erode, new dunes rise just behind them. But in the process, structures newly exposed to the sea are often undermined. When heavy storms hit, dozens of houses can be destroyed, as happened on the Outer Banks of North Carolina in March of 1989. Over the last one hundred years, the Atlantic coast has eroded an average two to three feet a year, according to Stephen P. Leatherman, director of the Laboratory for Coastal Research at the University of Maryland. The average yearly erosion on the Gulf Coast is four to five feet, and in some places it is much worse. The south shore of Martha's Vineyard is losing about ten feet a year, while Cape Hatteras, North Carolina, is losing twelve to fifteen feet, and Louisiana is losing thirty to fifty feet a year.

"It's no secret anymore what's happening on the coast," Dr. Leatherman says. "Ten years ago we didn't have good data on erosion rates. Now we do, and we know about 90 percent of the coast is eroding." The threat has driven many communities to seek refuge behind seawalls and jetties, only to find that these costly engineering projects often make the erosion worse, either for them or for neighbors down the coast. Other resorts are spending vast sums to pump new sand onto their beaches, only to see the sand disappear again within a few years. Many engineers maintain that seawalls, jetties, and replenished beaches are necessary to protect the valued developments already on the barrier islands. Too many people live or vacation at Atlantic City, Miami Beach, and other barrier island towns to let their roads and buildings simply fall into the sea, these experts say.

But many environmentalists and geologists contend that roads and buildings just do not belong on the unstable, migrating islands. At a minimum, some critics say, the government should discourage further development on barrier islands by refusing to subsidize such projects as roads and waste treatment plants or property insurance for owners. Although geologists say that nothing can stop the sea's inexorable march inland, few barrier beach communities are prepared to let themselves fall into the ocean block by block. They often try to halt erosion with bulkheads or rocky seawalls that protect the structures behind them. But many scientists now say that such structures accelerate the erosion of sand on the beaches in front of them.

"Seawalls destroy beaches," says Orrin H. Pilkey, Jr., a Duke University geologist. In a recent study of scores of sites on the Atlantic Coast, Dr. Pilkey and his colleagues found that beach width on stabilized beaches was dramatically lower than on beaches with no seawalls. Many of the stabilized "beaches" had no beach at all. Although the mechanics of wave action on beaches are imperfectly understood, geologists can suggest ways in which a seawall might hasten erosion. On a gently sloping beach, a wave's force is reduced as the surge of water, or swash, rushes onshore. Much of the water's energy is gone by the time it returns down the seaward slope as backwash, so it carries off relatively little sand.

A seawall cuts this process off. Water hits it sharply and is reflected back strongly, carrying sand with it. "The wave hits the seawall and the energy doesn't dissipate," Dr. Leatherman explains. "A seawall is a last Draconian step to save property. You just kiss off the beach." Sea Bright, New Jersey, for example, once had three hundred feet of beach in front of its massive seawall, says Gilbert Nersesian, a coastal engineer for the New York District of the Army Corps of Engineers. "But the beach that existed in front of that seawall is completely gone." Maine and North Carolina have passed laws banning new hard stabilization along the shoreline. But property owners pay a heavy price for such laws. In March 1989 a series of storms off North Carolina eroded the beaches at Nags Head and Kill Devil Hills, on the Outer Banks, pitching several houses into the sea and damaging others so severely they were condemned.

"We let a lot of houses fall in," Dr. Pilkey recalls. "I'm proud of that." But houses threatened here and there are not the truest test of whether a state's politicians can withstand pressure for stabilization, he believes. Pointing to a five-story condominium on the dune line on Bogue Bank, a heavily developed barrier island in North Carolina, he asks: "Are we going to let a building like this fall into the sea?"

For centuries, people have built jetties of stone jutting out from the shore to keep sand from silting up channels. And for years they have tried to stem beach erosion by building smaller groins out from the shore to trap the sand carried along the shore by currents. Now there is wide agreement that groins can only hold sand on one beach by starving another downstream. "Most people now believe you have to be very careful with groins," says Dr. James R. Houston, director of the Army Corps of Engineers Coastal Engineering Research Center in Vicksburg, Mississippi. "They usually cause problems."

The 1938 hurricane that devastated Long Island and southern New England cut an inlet from the ocean into Shinnecock Bay, on the South Fork. The inlet and jetties built to keep it open blocked the longshore transport of sand, causing erosion at Westhampton Beach, Fire Island, and even New York City beaches, Mr. Nersesian says. Groins at Ocean City, Maryland, have had a devastating effect on Assateague Island. When the groins were built, the rate of erosion on Assateague went from two feet to thirty-six feet a year.

Many communities faced with the loss of their beaches are adopting a simple but expensive approach: replace the sand. Done right, officials at the Army Corps of Engineers say, beach replenishment can be a cost-efficient way to protect a community and provide needed recreational facilities, especially in resort towns dependent on income from tourism. Miami Beach was replenished ten years ago, in a $65 million project. "It's in magnificent condition," Dr. Houston says.

But scientists and environmentalists dispute the value of replenishment. For example, Beth Millemann, director of the Coastal Alliance, an environmental group, describes beach replenishment as "throwing dollar bills in the water." In a recent study, researchers led by Dr. Pilkey examined ninety replenished beaches on the East Coast. North of Florida, he says, none of them lasted more than five years. On the Gulf of Mexico, only 10 percent last more than five years. But Dr. Houston and Dr. Leatherman said they questioned Dr. Pilkey's findings. "Orrin and I disagree a lot on this," Dr. Houston says. "You can name a lot of beach fills that have lasted a long time."

A lot depends on how the replenishment is carried out. "If you replenish a beach with the same material in the native beach, there's nothing in the laws of physics" that would alter erosion rates, Dr. Houston says. "If the sand you put in is finer, it will go faster. If you put coarser sand, it lasts longer, but coarser sand is usually more expensive." One problem with beach replenishment programs, Dr. Leatherman notes, is that "beaches are like icebergs—only 10 percent of the active part of the beach is above water."

"The Corps of Engineers is pumping all the sand on the high part of the beach," Dr. Leatherman says. "Some of that sand has to go underwater to maintain the slope of the beach, the gentle bottom profile. So it appears that the beach is eroding like crazy." Once a beach has been replenished it must receive regular infusions of sand if it is to survive the assaults that diminished it in the first place. Dr. Houston says communities should look on this as a normal part of maintaining their infrastructure.

"It's like everything else," he says. "If you build a road you have to maintain it." In many places, property owners are being encouraged to bow to the inevitable and simply move their houses inland, a practice that has occurred on Long Island, the Outer Banks, and other high-erosion areas for decades. In areas of high erosion hazard, the National Flood Insurance Program will pay the owners of threatened buildings 40 percent of either their value or the amount of their coverage to move them inland, according to James L. Taylor, assistant administrator for the program. The money helps cover the costs of moving the house and reestablishing it on a new lot. If a property owner agrees to demolish the structure, the payment rises to 100 percent, plus 10 percent for cleanup costs. But anything later built on the property is ineligible for insurance, he said.

In Texas, the Open Beaches Act requires owners to move their houses when erosion brings the beach to their foundations. After Hurricane Alicia, in 1983, about one hundred houses in Galveston had to be moved, Dr. Leatherman said. But retreating from the shore is probably only likely in places where most structures are single-family houses and where land is abundant.

"Unfortunately," Dr. Leatherman says, "if you look at New Jersey, the checkerboard is filled." To help people who cannot or will not move their buildings, the Federal Emergency Management Agency (FEMA), working with the Corps of Engineers and the National Oceanic and Atmospheric Administration, has commissioned scientists to recommend ways that developed barrier islands might reduce storm damage. Officials from FEMA, accompanied by Dr. Pilkey, recently toured Bogue Bank, a heavily developed barrier island off the coast of Beaufort, North Carolina, to learn more about lessening property damage.

Dr. Pilkey pointed out some of the island's problems: protective dunes have been leveled for easy development; roads run straight back from the beach, providing overwash channels for floodwaters; marshes that could have absorbed storm water have been filled and built on. Even two low places on the island, where inlets formed in Hurricane Hazel in 1954, have houses on them now. And there has been no big storm to test what Dr. Pilkey calls "stack-a-shack" condominiums that line the beach in the island towns of Atlantic Beach and Emerald Isle.

The boom in coastal development over the past twenty years has occurred in a time of low storm activity. According to Max Mayfield, a hurricane specialist at the National Hurricane Center in Coral Gables, Florida, only about half as many major hurricanes have struck the United States mainland since 1970 as in the thirty years previous. To be covered by federal flood insurance, buildings must conform with codes requiring, among other things, that they be elevated and that all ground-level walls be "break-away" to allow a storm surge to pass through. But many older houses do not meet such codes and many new buildings have been modified with solidly constructed living areas on the ground level.

"They're like rotten apples in a barrel," Dr. Pilkey says. When these buildings are destroyed in a storm, surging water can crash the debris they produce into other houses. But Dr. Pilkey also points out steps property owners have taken to restore some of the protective beach landscape destroyed in development. For example, some homeowners are running snow fencing across the beach in front of their homes. Sand collects at the fences, starting the process of dune formation. Others have planted sea grass and other vegetation to hold the dunes and have built wooden walkways across them so that the plants are not damaged.

Such steps will not save structures from destruction in a major hurricane, Dr. Pilkey says, but they can reduce damage in lesser storms. What should public officials do when developed areas are threatened? In principle, says Dr. Leatherman, "I'd say, let 'em take their licks. But the reality is that Ocean City is Maryland's major resort. It's a place where the average person can afford a vacation, and it's a place the masses enjoy. I would say you can't let Ocean City fall in." For now, he believes, the best way to protect resorts like Ocean City is to nourish the beaches with sand. But as sea levels rise, such programs get more expensive. "We may have to abandon more of the beach and wall up the rest" of the island, he says.

Dr. Houston also says that while he believes undeveloped barrier islands ought to be left that way, where development exists, "Why not protect it? In California there are earthquakes. In geologic time, which is the kind of time Orrin Pilkey likes to talk about, it's probably going to be destroyed. But we don't say people can't build in California. We don't say people can't build in Kansas because of tornadoes."

But for people like Dr. Pilkey, the choice is simple: "I feel that beaches should be viewed as sacred. I can't accept the idea that any development should have any consequence as opposed to a beach." [CD]

DEAD HEAT

SPURRED by the anxiety of a public suddenly alert to the potential dangers of global warming, a small fraternity of scientists is running a high-stakes race against the environmental clock, trying to predict the precise impact of the greenhouse effect in time to take effective countermeasures. Armed with powerful computers, the scientists are using advanced mathematical models—sets of equations that express the physical workings of the atmosphere—to simulate the world's climate under varying conditions. Their findings underlie almost all current forecasts about global warming. But because the evolving art and science of global climate modeling is still so imperfect, it is often a confounding business made all the more frustrating by policymakers' demands for answers.

The scientists all say with confidence that the greenhouse effect, in which carbon dioxide and other gases combine with water vapor to trap heat inside the earth's atmosphere, is going to make the earth warmer in the decades ahead. They say climatic changes will result, with important consequences for life on earth. But they cannot prove or agree on how much the earth will warm, or how fast, or how the warming will affect individual countries or regions, or whether it has already begun.

The reason, climatologists say, is that although the climate models are steadily improving, they are still crude approximations of an atmosphere-ocean system so vast and complex that it nearly defies analysis. "They are dirty crystal balls," says Stephen H. Schneider, a climatologist at the National Center for Atmospheric Research in Boulder, Colorado. For instance, widely differing answers emerge from attempts by three of the major models to gauge the effect on the United States of a doubling of greenhouse gases in the atmosphere. Two show an increase in summer rainfall in the Southeast, but one indicates a decrease. A different two predict a drop in rainfall in the Great Plains, while the third reveals an increase. Still another combination of two shows an increase of rain in California, while the third indicates a drop.

This uncertainty intensifies the dilemma faced by scientists and the policymakers they advise. They do not have precise answers about global warming, but if they wait for the answers and if the answers are too long in coming, it will be too late to take effective action. At this stage of their evolution, climate models are as much tools for basic learning as they are practical instruments of prediction. Are they any good, then, as guides to public policy?

The scientists who use the models say yes, because though the models' estimates vary in specifics, they point in the overall direction of change. At the least, they say, the models demonstrate that weather systems everywhere are sensitive to global warming, that disruption is likely, and that a range of consequences is possible. "There are going to be sizable effects, we know that much," says Michael E. Schlesinger of Oregon State University, who heads one of the four teams in the United States that have modeled the global greenhouse effect. A team in Britain's Meteorological Office has also done so. Despite all the uncertainties, many scientists are therefore strongly convinced that it would be prudent public policy to reduce the emission of the so-called greenhouse gases, like carbon dioxide, methane, nitrous oxide, and chlorofluorocarbons, that are known to have a heat-trapping effect. This is especially so, they say, since such measures as reducing the use of fossil fuels and halting deforestation would bring major long-term environmental benefits whether there is significant global warming or not.

Perhaps no principle in atmospheric science is more firmly established than that of the greenhouse effect. In fact, the trapping of warmth has been exerting an influence since the earth first had an atmosphere. Without it, the planet would be an icy ball devoid of significant life. It is likewise well established that since the start of the industrial revolution, humans have been pouring greenhouse gases into the atmosphere at an increasing rate, virtually insuring that the earth's climate will warm significantly over the next century. But beyond these basic propositions, little is certain. How fast will the earth warm and by how much? The climate modelers' latest findings indicate that if, as projected, atmospheric carbon dioxide doubles between now and the second half of the next century, the world's average surface temperature will increase by two to nine degrees Fahrenheit.

The earth has warmed by about nine degrees since the last ice age. But the latest estimates are being questioned as scientists identify a variety of complicating climatic factors that the models do not include. Some do not reflect the impact of other greenhouse gases besides carbon dioxide. These include ozone in the lower atmosphere, methane, chlorofluorocarbons, and nitrous oxide, all of which are more effective in trapping heat than carbon dioxide. One of the five modeling groups, at NASA's Goddard Institute for Space Studies in New York, has calculated that growing concentrations of these gases could speed up the warming by as much as two decades.

Modelers think it likely that the melting of glaciers and the expansion of seawater as it is heated will cause ocean levels to rise, that warming will be greatest near the poles, that rainfall patterns will change, and that crop zones and natural ecosystems will migrate with the change in climate.

Apart from these broad considerations, what will be the local impact of global warming in specific countries or their regions? Modelers get different answers to this, and the answers are invariably qualified by "may" or "could."

At this stage of the modeling art, climatologists are mostly "learning about the mechanisms of climate," says John F. B. Mitchell of the British team. "The next stage is putting numbers on it." Some confusion has attended the greenhouse issue since dramatic testimony before Congress about global warming in the summer of 1988 coincided with that year's drought and pushed the issue into the spotlight.

James E. Hansen, who heads the Goddard team, testified that the global warming trend measured over the last century was consistent with models of the greenhouse effect, and that in his view, the pollution-induced warming was already under way. Many of his colleagues decline to go that far. They say that so far, the effects of greenhouse warming would not be large enough to be distinguished with any confidence from those produced by natural variations in the global climate. The 1988 drought, they say, was most likely a result of natural variation.

Dr. Hansen says that he has not changed his mind and that, in fact, his group's mathematical model indicates that while natural forces bring on any given drought, heat wave, or severe storm, the greenhouse effect is already acting on those natural forces to make such weather extremes more likely. Even scientists who disagree with Dr. Hansen on whether greenhouse warming has already arrived agree with him that it should become apparent in the decades just ahead, and that droughts and heat waves will become more probable. It is as if the greenhouse gases are loading nature's dice, Dr. Schneider says.

In a way, the argument over whether greenhouse warming has arrived illustrates how imprecise and ambiguous the climate models are. All the scientists look at the same data, provided by each other's models and by climate records. But some read the data more cautiously than others. "Different scientists will retain or shed their skepticism in different degrees and at different times," Dr. Schneider says.

Climate models are sets of mathematical equations that express the basic physical principles that govern the workings of the atmosphere. A computer calculates how the basic principles act on such factors as temperature and rainfall over time, under differing conditions. Scientists can insert any set of conditions they like—including any given concentration of greenhouse gases, for example—to see how this causes the scenario to change. But no computer now made or likely to be made any time soon can handle the vast number of calculations necessary to simulate the atmosphere's full complexity. Consequently, results are only approximate and can vary from model to model. This is especially true of the largest and most sophisticated of the models, called general circulation models, which cover the globe.

Scientists compromise by making calculations only at widely spaced points that form a three-dimensional grid at and above the earth's surface. A typical spacing between grid points is about three hundred miles, with the three-dimensional grid reaching to about twenty miles above the earth. The wide spacing, or coarse "resolution," as scientists call it, means that many important climatic phenomena are not included in the simulation because they are smaller than an individual grid box. Clouds are a prime example. Scientists therefore try to calculate average cloud cover for a given grid box on the basis of its humidity and the buoyant properties of the air within it.

The coarse resolution means that different models are bound to produce different pictures of climatic change for, say, regions of the United States. Climatologists are constantly trying to improve the resolution of the general circulation models. But they are hampered because the finer the model, the more computing time and capacity is required and the more expensive the exercise becomes. To cut the size of the grid squares in half would require a supercomputer to perform eight times as many calculations, says Warren Washington of the National Center for Atmospheric Research, who heads another of the five greenhouse modeling groups. "If you want to get down to the congressional district level," he says, noting that politicians have been asking him about the impact of global warming on their constituencies, "it is eight times eight times eight."

The difficulty is expected to ease as computers gain in speed and capacity. But quite apart from the resolution problem, the modelers acknowledge, their models have other shortcomings. Only one model, at Goddard, has tried to take into account the way in which greenhouse gases actually make themselves felt: gradually, over a period of years. The others mostly assume a one-time, massive doubling of the gases, which produces different results. Moreover, say those who use the models, they reflect only crudely (or not at all) such factors as soil moisture, sea ice, clouds, volcanic ash in the atmosphere, fluctuating solar radiation, evaporation, forests, and the mechanism, called convection, by which moisture is transported upward into the atmosphere. Nor, many scientists believe, do they include a number of atmospheric interactions, or "feedbacks," that might intensify, lessen, hasten, or delay global warming.

Not least of the effects that are inadequately included in the models, climatologists say, is that of the oceans. "On our planet, it is the ocean that sets the time scale for the response of the climate system" to greenhouse warming, Dr. Schlesinger says. Through vertical movement of water, the ocean takes heat from the surface down into its interior, and in so doing delays the warming. But most of the models that have so far tried to forecast the rate of global warming take account only of the upper layers of the ocean. Nor, says Dr. Schlesinger, do they adequately account for such factors as horizontal currents or salinity. Both at Oregon State and at the Geophysical Fluid Dynamics Laboratory at Princeton University, modelers have made some early attempts to include ocean actions more fully.

Preliminary findings by Dr. Schlesinger's group at Oregon State have shown that the oceans cause a delay of fifty to sixty years in warming. Earlier models had shown the delay to range anywhere from ten to one hundred years. At the Princeton laboratory, Syukuro Manabe and Kirk Bryan have demonstrated what earlier had been only suspected: that because there is more ocean in the Southern Hemisphere, it will warm up more slowly than the Northern Hemisphere. "We're confident we've got a significant result," Dr. Bryan says.

But no modelers have yet reliably coupled two equally sophisticated general circulation models, one for the atmosphere and one for the ocean, in a synchronized fashion, and used the coupled models to assess the greenhouse effect. "It's likely to take ten or twenty years before we do that well," Dr. Hansen says. And until that happens, modelers say, reliable predictions about the regional effects of global warming are going to be extremely difficult.

Some climatologists are beginning to talk about merging their efforts to construct one comprehensive model that they hope would eliminate most of the big uncertainties. Until such a comprehensive model emerges, they say, greenhouse warming is likely to cause many surprises. "We can promise change," Dr. Schneider says, "but we can't tell you what in terms of Iowa versus India." [WKS]

A HOTTER AMAZON?

Converting all of the Amazon's tropical forests to pastureland would reduce rainfall there by 20 percent and make the region substantially hotter, a British study has found. In a computer simulation, the scientists examined what would happen if forests and savannas were converted to pasture over the whole Amazon basin. Farmers are doing just that in parts of the region, which contains half of the world's tropical forests. The three-year simulation study, published in the journal *Nature* in 1989, was done by scientists in England's Meteorological Office and Department of the Environment.

Half the rainfall in the Amazon basin is believed to consist of water that evaporates from the forests. Evaporation is lessened when the trees are cut. This not only reduces rainfall, but because evaporation draws heat from the land, cooling the earth, it also makes the surface warmer. Previous studies had suggested that deforestation reduced evaporation and made the surface warmer, but did not establish any clear indication of reduced rainfall. The study's authors say that the conclusions are tentative and that more extensive ground-based observations in Amazonia would make the simulation more realistic. But, they wrote in *Nature*, the results "indicate that deforestation can cause significant local climatic perturbations."

BY ANY MEANS NECESSARY

A S scientists and public officials urge painful, expensive measures to prevent climate change—reducing energy use, developing alternative sources of power, curbing production of destructive pollutants—a breed of visionaries is dreaming of more direct, if seemingly fantastic, countermeasures. These experts are exploring extraordinary ways to combat pollution in the earth's atmosphere and the threat of climatic upheaval. They envision environmental wars fought with lasers that blast apart harmful chemicals, satellites that beam safe energy to earth, microorganisms that soak up pollutants, and chemical-releasing airships that replace critical elements being removed from the earth's atmosphere.

One aim is to counter the gases that produce the greenhouse effect, which is thought to be gradually warming the earth, threatening to raise oceans and damage agricultural production. Another is to halt the breakdown of the ozone layer, which blocks the sun's ultraviolet rays. Work on futuristic cures for these pollution problems started in the 1970s and accelerated in the 1980s as concern has grown about potential damage to the atmosphere. Experts involved in the research stress that the ideas are often untested and in some cases could be risky. Most experts agree that for now, preventing the release of harmful gases into the sky deserves far higher priority than efforts to fix a damaged atmosphere. Yet the surge of creative thinking, they add, could ultimately play an important role in the battle to stabilize the earth's changing weather, especially if conditions take a dramatic turn for the worse.

"Some of this is mighty speculative," says Thomas H. Stix, a Princeton University physicist. Key questions are whether the techniques are safe and feasible and, if so, whether the potentially huge cost of carrying them out would be economically justified. In addition, some experts warn that large-scale intervention in the earth's delicate and poorly understood cycles of climatic chemistry could trigger unwanted side effects.

Backers of the research say its value often lies not so much in sparking concrete plans as in inspiring thought about how to ultimately deal with climate problems. "This area has been given little thought," says Wallace S. Broecker, professor of geochemistry at Columbia University. "At a minimum, a rational society needs some sort of insurance policy on how to maintain a habitable planet."

One futuristic idea is to use giant lasers atop mountains to scrub harmful chemicals from the earth's atmosphere, a concept pioneered by Dr. Stix at Princeton. He calls it "atmospheric processing." His lasers would be aimed at industrial chemicals known as chlorofluorocarbons, or CFCs, which are used mainly in refrigerants and in the manufacture of plastic foams. Once released in the air, they rise high into the stratosphere, where they destroy protective ozone. Stratospheric ozone helps block ultraviolet light, which can cause skin cancer and eye damage and harm natural systems.

The laser system would break apart CFCs in the lower atmosphere before they had a chance to damage the ozone layer. The concentrated light would be tuned to a frequency most easily absorbed by CFCs, which is in the infrared part of the electromagnetic spectrum. Dr. Stix calculates that an array of infrared lasers around the world could blast apart as much as one million tons of CFCs a year, equal to the current annual flow into the atmosphere. In terms of feasibility, Dr. Stix says: "Some of the answers are known. Many aren't. A major question is whether you can get the laser's energy absorbed by CFCs and not other molecules, such as water vapor or carbon dioxide." Another issue is to what extent the general atmosphere would absorb laser energy, limiting propagation of the beam.

A less exotic cure for ozone depletion would be to simply replace it. Experts have proposed that bulk ozone be produced on earth and lofted into the stratosphere in rockets, aircraft, or balloons. Other ideas include firing aloft "bullets" of frozen ozone or placing solarpowered ozone generators in high-altitude balloons. Since ozone consists of three oxygen atoms and atmospheric oxygen has two such atoms, raw material would be plentiful. Leon Y. Sadler, a chemical engineer at the University of Alabama, writing in *Chemical and Engineering News*, has proposed using a fleet of jets to dispense ozone. His calculations showed that the number of planes needed to replenish the ozone would be equivalent to less than 2 percent of the aircraft that haul freight in the United States. But other scientists have calculated that the job would be many times larger. "Ozone replacement would be a really massive, massive thing to do," says Michael C. MacCracken, head of atmospheric sciences at the Lawrence Livermore

National Laboratory in California. For now, all scientists stress the importance of reducing emission of ozone-destroying chemicals, as is now called for by international treaty.

On another front in the pollution war, scientists are proposing to counter the effects of carbon dioxide and other trace gases that are rapidly accumulating in the atmosphere. These chemicals, the by-products of fossil fuel combustion, other industrial activities, and deforestation, trap heat from the sun that would otherwise be radiated back into space, acting like a greenhouse. One way to thwart such gases would be to increase the reflectivity of the earth's atmosphere so that more sunlight is reflected back into space, says Dr. Broecker, the geochemist at Columbia University. This happens naturally when volcanoes spew sulfur dioxide into the atmosphere. In 1982, for instance, El Chichón, a volcano in Mexico, pumped about 8 million tons of sulfur dioxide into the atmosphere, causing a slight planetary cooling. Dr. Broecker says about 35 million tons of sulfur dioxide would have to be transported to the stratosphere each year to counter the global warming produced by a doubling of the carbon dioxide in the earth's atmosphere, which is expected in the next century. The job would require a fleet of several hundred jumbo jets.

"This is not a big expense compared to totally changing our reliance on fossil fuels," Dr. Broecker says. But the method would have major drawbacks: it would increase acid rain and give the blue sky a whitish cast. "The point is not that the strategy is necessarily a wise one, but, rather, that purposeful global climate modification lies within our grasp," Dr. Broecker wrote in his book, *How to Build a Habitable Planet*. He added: "One hundred years from now the temptation to take such action may be high."

Other proposed ways to increase the earth's reflectivity are equally drastic. They include covering much of the world's oceans with white Styrofoam chips, which would reflect more sunlight back into space than regular ocean water, and painting the roofs of all houses white. It is not known to what extent such actions might offset global warming. Some dream of blocking sunlight before it ever reaches the earth. Giant orbiting satellites made of thin films could cast shadows on the earth, counteracting global warming. Scientists have calculated that a series of satellites with areas equivalent to 2 percent of the earth's surface could compensate for a doubling in carbon dioxide. Some space scientists have contemplated using such shields to make Venus less hot; the costs and benefits on earth have not been determined.

Still another way of coping with greenhouse gases would be to try to remove them from the atmosphere. Measures to conserve and plant new forests, which absorb carbon dioxide, are already moving ahead. More radical thinkers imagine encouraging the growth on a vast scale of tiny ocean organisms that soak up carbon dioxide. Already, the oceans are believed to be dissolving much of the extra carbon dioxide in the atmosphere. The bodies of microscopic creatures, for instance, incorporate carbon dioxide and, when they die, they sink to the bottom of the sea and turn into limestone. To increase this effect, scientists have proposed fertilizing the oceans to spur the growth of phytoplankton, microscopic plants that are a key element in the ocean's food chain. But this method has serious drawbacks. It would alter the marine food chain, and, if pursued too vigorously, the cascade of carbon dioxide into the deep ocean would eventually eliminate oxygen there, killing most life.

One group of experts holds that proposals to counteract climatic damage are misguided, and that advanced technologies should be used to prevent such problems in the first place, mainly by eliminating dependence on fossil fuels. Dr. Peter E. Glazer, a vice president at Arthur D. Little, Inc., a consulting and research firm in Cambridge, Massachusetts, has long advocated building a fleet of solar-powered satellites in space. Illuminated by the sun twenty-four hours a day, these spacecraft would turn sunlight into electric power, which would be beamed to earth in the form of either microwave or laser beams, and then turned back into electricity. First proposed in the 1970s, the idea has aroused renewed interest because of its environmental allure.

A more down-to-earth way to generate power without producing carbon dioxide is to harness nuclear fusion, the process in which hydrogen isostopes fuse together to release energy in the kind of process that powers the sun, the stars, and hydrogen bombs. While conventional nuclear power plants also supply energy without carbon dioxide pollution, they are mired in political, environmental, and economic disputes. The problem is achieving the fusion technique. "Fusion is still a matter for physicists to grapple with," Dr. Glazer says. "On the other hand, solar power satellites are just a matter for aerospace engineers to design and build."

While futurists clash over the best way to save the planet with exotic technologies, other experts warn that the debate should not deflect attention from the more pragmatic approaches to dealing with climate ills. "It's reasonable to ask if there's anything to do to repair the damage," says Michael Oppenheimer, an atmospheric physicist at the Environmental Defense Fund, based in New York. "But the effort and scale of some of these things is huge. It's probably cheaper in the long run to rely on prevention rather than unusual cures." [WJB]

WHEN A COW BELCHES

The average cow belches up to four hundred liters of methane a day, a situation that worsens the global greenhouse effect and ought to be curbed, a Colorado State University scientist says. The greenhouse warming of the earth's atmosphere is caused mainly by carbon dioxide in the air, much of it generated by the burning of fossil fuels. Carbon dioxide traps solar energy and raises global temperatures.

But scientists have recently discovered that methane, commonly produced by fermenting vegetation, also contributes strongly to climatic warming. A large share of the methane released into the atmosphere comes from the chemical degradation of cellulose in the guts of termites, rodents, ungulates, and other eaters of wood and straw. The methane molecule, consisting of one atom of carbon and four of hydrogen, is produced in environments where no oxygen is available, as in the rumens of cattle. The rumen is a separate compartment of a cow's stomach, where cellulose is broken down by symbiotic bacteria into digestible cud and methane gas.

The production of methane in the rumen can be greatly reduced by feeding cattle substances called ionophores, which Dr. Johnson describes as "weak, selective antibiotics." These substances, already widely used in the feeding of beef cattle, selectively inhibit the reproduction of methane-producing bacteria in the gut, leaving room for other bacteria that help digestion more efficiently. Ionophores are not ordinarily used with dairy and other cattle, however. Dr. Johnson believes that other possibilities for reducing methane production should be developed.

NUCLEAR GARBAGE

WHEN the Department of Energy declared that it had lost confidence in its latest effort to find a safe place to bury nuclear waste, a gnawing question resurfaced: is the task possible? Some of those closest to the issue say that even now, more than forty years into the nuclear age, the answer is not clear.

The problem is formidable because some radioactive materials created in nuclear reactors will last so long that from the human standpoint they might as well last forever. And these long-lived elements, especially plutonium, can be poisonous in minute quantities. They must be isolated from contact not only with people, but also with anything that might come into contact with people, like water supplies or the food chain. Since the nuclear age began, engineers have considered everything from shooting the wastes into outer space to putting them under the polar ice caps. For now, most of the wastes are sitting in pools of filtered, cooled water next to the reactors that created them, awaiting more permanent disposal. To that end, scientists and engineers have adopted three major overlapping strategies.

Official policy in the United States is that the safest method is burial in a "deep geologic repository," someplace dry, stable, and desolate. But finding the spot is proving tough. A second approach, used mainly in France but also in West Germany and Japan, is chemical treatment of the waste to remove the plutonium for use in a reactor. There its atoms are split, giving off heat to make electricity and turning the plutonium into elements whose radioactivity is shorter-lived. But civilian reprocessing has been rejected in this country due to its high cost and the danger that commerce in plutonium could lead to the spread of nuclear weapons. Moreover, the operation can cause nuclear pollution, and finding a place for the reprocessed wastes is still a problem. Most military wastes have been reprocessed, however.

A third option, chosen by many smaller nations, is to store the wastes above ground, in steel or concrete casks thick enough to block the radiation or in thinner containers housed in solid, earthquake-resistant buildings. Unlike storage in pools next to reactors, this does not require pumps and filters, which can break down. The casks and buildings will not last forever, but the expectation is that they will keep well into the next century, when scientists will presumably have found a solution to the problems of underground disposal.

Most of the "high level" waste that the government hopes to bury is the spent fuel of nuclear reactors. Reactors run on uranium, which is only slightly radioactive. But the reactors create radiation. To make heat, they split the uranium atoms, leaving fragments that are inherently unstable. After their creation, these fragments regain their stability by throwing off subatomic particles or energy waves called radiation. The rate at which the fragments give off their radiation and return to stability varies from seconds to millennia, depending on the material involved. Iodine 136, for example, has a half-life of eighty-three seconds, meaning that its radiation level falls by half in that amount of time. Cesium 137 has a half-life of thirty years.

The rule of thumb is that after ten half-lives, the material has reached the background radiation level, meaning the level that exists in nature. For most fission products, ten half-lives is a period measured in hundreds of years. But reactors create a second category of radioactive materials, many of which are longer-lived. These "activation products" have become unstable by absorbing a subatomic particle called a neutron. The most important of these is plutonium 239, with a half-life of 24,390 years. Plutonium is not found in nature. Plutonium can itself be used to run reactors or to fuel bombs. The Energy Department and its predecessor agencies have run reactors simply to make plutonium, processing the spent fuel by chemical and mechanical means to recover that material for weapons. France, Great Britain, and Japan do the same, to varying extents, to gain plutonium for reactor fuel. France has an active program to commercialize breeder reactors, which make more plutonium than they consume.

Those reprocessed wastes are in liquid form, but France solidifies them in glass. The Energy Department is building plants to do the same with its liquids in preparation for burial. The reprocessing and solidification increases the volume of wastes, however, and even after reprocessing, some very long-lived material remains. Iodine 129, for example, has a half-life of 17 million years.

In this country progress on geologic disposal has been slow, and today even the definition of the job is sketchy. Since its old performance standards for underground repositories were thrown out by a federal judge in 1987, the Environmental Protection Agency (EPA) has been drafting new ones, specifying the maximum permissible leaks after one thousand and ten thousand years. The success in containing the wastes will depend on the integrity of the packaging materials and the chemistry of the surrounding soil. Government scientists say that the soil at their leading site, Yucca Mountain, on the edge of the Nevada Test Site one hundred miles northwest of Las Vegas, is favorable, tending to bind up radioactive material rather than allow it to travel with underground water. They are still developing the packaging, but say that it should be possible to build a container that will hold the waste for many centuries. But predicting leaks for ten thousand years is a new challenge. Though plutonium remains dangerous much longer than that, for 240,000 years, the EPA believes that so much geological change will take place over 10,000 years that it is pointless to try to plan for longer.

"I don't know of any estimation model on the face of the earth that could even talk about a thousand-year projection," says Dennis O'Connor, a special assistant in the EPA's Office of Radiation Programs who is working on the project. "It's darn difficult to talk about a health risk ten thousand years in the future." Others are far more optimistic, saying that with adequate skill and resources, a place can be found to bury the wastes that will keep them isolated until they are harmless. "Geologic processes are very slow," explains J. Carl Stepp, a seismologist at the Electric Power Research Institute, a utility group in Palo Alto, California. "It's quite within the state of our scientific ability to understand those processes and predict their rates, with some uncertainty, of course."

Carl P. Gertz, the manager of the Department of Energy's waste project, points out that many natural features, from caves to rats' nests, have endured more than ten thousand years. So government scientists and engineers have set out to find a suitable geologic structure. Early in the process they focused on salt formations because they are stable, their existence indicates that very little water has been present to dissolve them, and over time they tend to seal up all openings dug in them. But salt presented two problems. First, although many salt formations are low in moisture, critics said a hot object in salt would attract water. The wastes generate heat as they give off radiation, and some scientists say water concentrating around them could form a highly corrosive brine, eating through the waste containers. Second, the salt formations are in Texas, Louisiana, and Mississippi, which had enough political muscle to frustrate the plans for a repository.

Another geologic structure that raised the hopes of planners was granite, because it is solid. Much of the granite is found in wet areas, however, and water could disperse the wastes. But the investigation of granite was halted for a different reason: the political storm that greeted the Energy Department in potential locations like Michigan, New Hampshire, and Maine. The current candidate for the wastes, Yucca Mountain, a volcanic ash structure, was promoted to front-runner by Congress in a 1987 amendment to the Nuclear Waste Policy Act. The amendment instructed the Energy Department, which is to build the repository to meet the EPA specifications, to evaluate Yucca's suitability and then build there unless an insurmountable problem is found.

Volcanism may be such a problem. A volcanic cone about five miles away was once believed to be more than three hundred thousand years old and thus probably dormant. But it is now believed to result from an eruption less than twenty thousand years ago, raising the possibility of another eruption, which might breach the repository. The chief official in Nevada's effort to block the repository stresses this possibility; the head of the project, however, suggests that not only is an eruption nearby unlikely, but it would not necessarily affect the repository.

At the Nuclear Regulatory Commission, which is to use the standards developed by the EPA to evaluate the Energy Department's license application for the repository, John Trapp, the senior geologist, told his superiors in a 1989 memorandum that the "Yucca Mountain site would have a very hard time passing a licensing hearing, strictly on the volcanics issue." He suggested that the site be dropped.

The initial attraction at Yucca was its desert climate. With very little rainfall, the area has no surface water and a very low water table. But the water table, eight hundred to twelve hundred feet below the proposed site, is another scientific question mark. Curiously, it rises in one corner. Scientists do not know why, or whether the water level could rise further. Nor is it certain that the area will always be a desert. Historically, rates of rainfall change very slowly, but there is growing consensus that the earth is entering a period of climatic change induced by human activity, in which the past may not be a good guide to the future.

Scientists are also discussing how to mark the repository and its lethal contents for future generations, who may not speak any language now spoken. For guidance, the government in 1982 commissioned an archaeological survey of the oldest man-made objects to determine which were still comprehensible and why. Among other complications, the survey noted that acid rain is fast obliterating many ancient monuments and that any monument made of a useful material, from metal marker to stone block, might end up being recycled in some future generation. Among the long-term possibilities for marking a site: creating a magnetic anomaly, presumably by burying large magnets, or changing the local vegetation.

The alternative to burial is above-ground storage, either in a central location or at the reactors where the material is produced, an idea that few people like but that some dislike less than burial. "The best way to handle that material is to require it to be stored where it is generated," says Hugh J. Kaufman, assistant to the director of the Hazardous Site Control Division at the EPA. He emphasizes that this is his own opinion, which runs counter to the path the government has been pursuing.

The grounds around a nuclear power plant are a better place to store highly radioactive waste than a nuclear dump would be, he explains, adding: "The people who work at nuclear power plants have Ph.D.'s in chemistry and physics. They know how to handle it. And handling it is much easier than running the plant." In contrast, the people who operate a repository are likely to be much less well trained.

Another alternative is above-ground buildings that would hold the wastes well into the twenty-first century, although finding the host for such a complex would be difficult if it appeared permanent. Lawrence T. Lakey, a nuclear consultant who formerly managed the International Programs Support Office of Pacific Northwest Laboratories, which tracks disposal efforts around the world, says he once favored underground disposal. But now there is so much public anxiety that "I don't see how they're going to get the okays to do it. If I had my druthers, I'd let it sit there and let the whole hullabaloo die down. That's what's happening in other countries."

But some environmentalists disagree. Dan W. Reicher, the lawyer at the Natural Resources Defense Council who successfully argued the case against the EPA's original standards, believes that a geologic repository is the only solution. "It's almost philosophical. We simply can't rely on distant generations to have the same sort of institutional controls over this very dangerous material that current society does. We need to isolate this from the environment to the greatest extent possible, so we don't have to rely on distant generations to make sure that it's kept safely." [MLW]

THE GOOD NEWS ABOUT SMOG

Environmental scientists find themselves facing a seeming paradox: one group of researchers determined that the ozone shield, which screens out harmful ultraviolet radiation, was slowly being depleted over North American latitudes. But contrary to what might have been expected, other researchers found no increase in the amount of ultraviolet radiation reaching the ground in eight American cities. Some scientists are suggesting a solution to the apparent contradiction. They say people in metropolitan areas appear to have a built-in protection against increases in ultraviolet rays that might penetrate the high-altitude ozone layer.

But no one is cheering. The protection, the scientists believe, comes largely from low-altitude air pollutants, including ozone, that screen out ultraviolet radiation before it reaches the earth's surface. The price is obvious. Urban dwellers may be shielded from solar rays that can cause cancer, eye problems, and severe sunburn, at least for now. But they are breathing in more smog. "It turns out that around urban centers, we really don't have to worry about increased UV in terms of its effects on people," says Mario Molina of the Jet Propulsion Laboratory at the California Institute of Technology, an expert on the ozone problem. "Instead, you worry about breathing ozone."

No one has demonstrated this definitively, but a number of scientists have been gravitating toward Dr. Molina's view ever since the seeming paradox arose.

WASTE NOT, WANT NOT

FORCED into action by the cost of its worsening waste-disposal crisis, the United States is creating an elaborate new recycling system that promises to put the nation on a use-it-again rather than a throw-it-out footing. The effort goes far beyond the grass-roots recycling movement of the early 1970s, when environmentalists, appealing to the nation's conscience, tried but failed to get recycling to catch on as a general way of life. And it involves much more than the curbside collections of recyclable newspapers, bottles, and cans that are becoming a familiar feature of life in much of urban and suburban America. While necessary and critical, that is only a first step, one that becomes futile unless the materials can also be reprocessed, sold, and recast into new products.

To meet that need, there is gradually emerging a broader recycling structure founded on government initiative, industrial enterprise, and new technology. Laboratories are designing new systems for separating the many varieties of recyclable waste and converting them back into high-quality raw materials for industrial use. States, localities, and private companies are starting to build or contract for the use of such systems. When Rhode Island opened an elaborate, highly automated new processing plant it was designed to be a key element in that state's pioneering recycling system.

New techniques for recycling plastics are creating expectations that the plastics industry will soon follow the recycling lead of the paper, glass, and metal industries. The Du Pont Company and Waste Management, Inc., of Oak Brook, Illinois, increased those expectations by announcing that they would jointly build several plastics recycling and reprocessing centers around the country. Waste Management, the country's largest waste handler and major operator of landfills, is now moving heavily into recycling.

At least ten states, including New York, New Jersey, and Connecticut, now have mandatory recycling laws. More than five hundred cities around the country have regular curbside collection of recyclable materials. They are concentrated in the Northeast, on the Pacific Coast, and in the upper Midwest, but they are found in all regions. The Environmental Protection Agency has set a goal of eliminating 25 percent of the country's trash through a combination of recycling and waste reduction by 1992. About 10 percent of the country's waste is now recycled, a proportion that has remained essentially the same throughout the 1980s.

This includes 33 percent of the nation's newsprint, 10 percent of its glass, 8 to 10 percent of its metals (half its aluminum), and 1 percent of its plastics. But many localities have set targets well beyond the federal goal and are already achieving them, and some experts believe that half the "waste stream" is easily recyclable and that 80 percent is potentially so. This broad effort is still in its early stages, and the system is developing unevenly. A number of experts fear that the market for recyclable materials is not expanding fast enough to handle the growing flood of curbside collections. The market for used newspapers in the Northeast has already become so flooded that some communities are paying brokers to take them away. Not long ago, the brokers paid the communities.

Some authorities say the whole movement could be aborted in disillusionment if the market fails to expand. Others say that if a steady, stable supply of high-quality recyclable materials is achieved, the market will respond and industry's investment in recycling will grow. If it does, experts say, environmental pollution would be reduced and the country would reap great savings of natural resources, energy, and money. Many other nations have long since discovered and acted on this. But despite appeals by environmentalists in the 1970s for the nation to come to grips with the debris of affluence, the country failed to develop an integrated structure for recycling.

Now the country's solid-waste management system "is in a state of transformation," says John F. Ruston, the chief specialist on recycling of the Environmental Defense Fund, a nonprofit advocacy organization. The impetus is the growing crisis that is forcing local governments to close landfills as they fill up. More than a third of the landfills will be full in two to three years, according to the Environmental Protection Agency (EPA). Local governments are having to bear soaring costs, now up to one hundred dollars a ton and rising, to truck waste elsewhere, often hundreds of miles away, to dump or incinerate it. The result is that economics is forcing the country to face up to recycling in a way idealism could not.

A broad consensus among government, industry, and environmentalists now places recycling at the top of a hierarchy of measures that taken together make up what is widely referred to as a developing new strategy of "integrated solid-waste management." In this strategy, which has been officially adopted by the EPA as a national approach, recycling is the preferred method of dealing with trash, along with an overall reduction in the amount of waste through such measures as reuse of products, selective buying habits, and redesign of packaging. For waste that cannot be recycled, incineration in modern "combustors" that minimize pollution is the preferred option. Landfills, the keystone of waste disposal in this country for decades, are now considered a last resort.

What happens immediately after recyclable materials leave the curbside is crucial to producing the high-quality materials that industry will accept. When the materials are picked up, they typically have been separated by householders into groups of three or so: perhaps newspapers in one group, other waste paper in a second, and glass, metal, and plastic containers in a third. As municipal composting programs gain in popularity, some communities are requiring that grass cuttings and other yard waste also be separated. To require householders to do more, experts are concluding, risks losing their cooperation.

Among the technologies now emerging to handle the next step in the process is the one introduced recently in Rhode Island. Designed and operated by a Massachusetts company, New England CRInc., in conjunction with Maschinenfabrik Bezner, a West German affiliate, the system processes more than eighty tons a day of mixed recyclables, using six workers. This and other kinds of processing centers are being built, or will be, in a number of states.

The technology of plastics recycling has lagged behind that of other recyclable materials, but strides are being made. The Center for Plastics Research at Rutgers University, for instance, has developed an automated system that converts plastic soft-drink bottles and milk jugs back to raw material from which products ranging from park benches and fence posts to paintbrush bristles and carpet backing are made. The system is being licensed for use by communities in Ohio and New Jersey and by a number of localities abroad. A few cities, like Seattle, have come close to putting all the pieces of the emerging new system together. Seattle's program begins with a powerful economic incentive: the more waste Seattle residents put out for regular trash collection, the more the city charges them. But there is no such penalty charge for collecting recyclable materials, including yard waste, in separate containers. The yard waste is composted, and other recyclables are processed for reuse at one of two plants operated in the city by private contractors, one of them Waste Management. The city sells tin cans and glass to local mills, newsprint and cardboard to other mills in the region, and mixed paper to Pacific Rim countries for reprocessing into a variety of paper products.

But as more and more states and local governments gear up for recycling, the market may not be able to bear the load, a number of authorities fear. The metal, glass, and plastics industries have not yet experienced such a crunch. But the paper industry has, particularly in the Northeast. "If they don't figure out ways to develop markets, either the whole system is going to collapse or they will end up land-filling or burning" recyclable paper again, said William L. Kovacs, who served as chief counsel to the congressional subcommittee that drafted the Resource Conservation Recovery Act, the federal law governing solid waste. Mr. Kovacs, now in private practice, has represented the paper recycling industry.

The industry has been running at capacity in trying to recycle the growing flood of waste paper but is nevertheless building up unused inventories, according to William E. Hancock, a recycling expert at the American Paper Institute, the industry trade organization. In the Northeast, he says, "a kind of bidding contest" has developed in which communities compete to pay brokers to take their paper. Although this is often less expensive than disposing of the trash conventionally, Mr. Hancock says that many communities might give up in disgust. That would defeat the entire enterprise, he says, because the mandatory recycling programs are essential in that they "demonstrate to future investors in recycling that there's a collection mechanism that will deliver."

Lorie Parker, the manager of Seattle's recycling project, believes that "the people who market recyclable materials say there will be a short-term problem, but that markets will respond when they do see a steady flow. You have to show that steady flow before they make the investment. Until they do, we might be in a tight place." [wks]

WIRED

IN the century since electric power revolutionized human existence, most people have scarcely thought, if at all, about whether it is safe to live with the electromagnetic fields radiated by the cables, wires, fixtures, and appliances all around them. When the question did come up, scientists generally assured the public that there was no health danger. They are no longer so certain. While virtually all experts still say no proof yet exists that electromagnetic fields pose any health threat, accumulating scientific evidence has convinced many that there is cause for concern.

Laboratory studies on animal cells have shown that electrical current alternating at 60 cycles per second, or 60 hertz, the kind that comes into almost every American home, emits radiation that can cause biochemical changes. Some of the changes might conceivably cause adverse health effects if the cells in the human body are similarly affected. And three epidemiological studies have demonstrated a statistical association between exposure to power distribution lines and cancer in children, although two other studies have not.

The rising sense of concern—and the uncertainty engendered by ambiguous and often contradictory data—was brought into sharp focus in a comprehensive background paper issued by the Congressional Office of Technology Assessment. "The emerging evidence no longer allows one to categorically assert that there are no risks," said the report, prepared by a team at Carnegie-Mellon University. "But it does not provide a basis for asserting that there is a significant risk. It is now clear that 60-hertz and other low-frequency electromagnetic fields can interact with individual cells and organs to produce biological changes," the report concluded. While the nature of these interactions is "subtle and complex" and their implications for public health remain unclear, the study said, there are "legitimate reasons for concern."

The concerns could evaporate in the face of further research. All of the findings are still considered preliminary at best, grist for hypotheses to be rigorously tested through the scientific method. Most of the laboratory studies that have found biochemical changes have not yet been successfully repeated, and even if their findings are borne out, no one knows how much of a health risk, if any, that would mean. Other studies have found no effects at all. Virtually no one who has studied the problem believes that whatever risk might be posed by 60-hertz fields is anywhere near the risk posed by cigarettes, asbestos, automobile accidents, or a whole range of other familiar hazards. And no one is yet advocating the rewiring of America.

Nevertheless, the accumulating evidence has moved the issue squarely onto the public agenda. Congress has held hearings on the question. Eight states, including New York and New Jersey, have regulated the intensity of the electrical field transmission lines can generate. The press has focused attention on the issue. (In 1990, *The New Yorker* ran a three-part series.) And, after years of flat or decreasing research expenditures, a worldwide research effort aimed at clarifying the question is gathering momentum.

"The whole thing is very worrisome," says Dr. David O. Carpenter, the dean of the School of Public Health operated jointly by the New York State Department of Public Health and the State University of New York at Albany. "We see the tips of the iceberg, but we have no idea how big the iceberg is. It ought to concern us all." Dr. Carpenter was the executive secretary of the New York State Power Lines Project, which carried out a major study on the health effects of 60-hertz electromagnetic fields; its major findings were reported in 1987. The power lines study identified "several areas of potential concern for public health," particularly the possibility that magnetic fields near homes had been linked by epidemiological studies to cancer, but said final conclusions must await more research.

Even some experts associated with the electric power industry, which has long asserted that electromagnetic fields pose no risk, concede that the results of the research raise serious questions that must be answered. "Until a couple of years ago it looked like there was nothing here at all," says Dr. Leonard Sagan, who directs radiation studies for the Electric Power Research Institute of Palo Alto, California, which is supported by the power industry. Now, he said, "all these things coming together can't be ignored." While stressing that "I don't think there's evidence of a risk to human health that's been demonstrated," he said, "I don't mean to exclude the possibility of such a risk."

When Dr. Samuel Milham, head of chronic disease epidemiology for the Washington State Department of Health, first looked into the matter, he thought it was nonsense. "I said it's got to be voodoo," recalls Dr. Milham, who conducted several occupational studies that find higher cancer rates in people who work around electromagnetic fields, like power company employees and electricians. "Now I believe something is going on. What its extent is, I don't know."

High-voltage transmission lines have for some years been the focus of protest by citizens living near rights-of-way. But if electromagnetic fields do turn out to pose a health risk, transmission lines may be the least of the trouble. Distribution lines, home wiring circuits, appliances, lighting fixtures, even electric blankets also create 60-hertz fields, and the study for the Office of Technology Assessment concluded that they "could play a far greater role than transmission lines in any public health problem." The office's report was prepared by Indira Nair and M. Granger Morgan, both physicists at Carnegie-Mellon University's Department of Engineering and Public Policy, and H. Keith Florig, a research fellow in the department.

The research effort aimed at understanding the relationship between electromagnetic radiation and health is an extensive one, and, according to the Office of Technology Assessment report, most of the work is of "very high quality." But so far it has focused on identifying some possible effects of such radiation on living tissues and organisms, and it has raised more questions than it has answered.

Any electrically charged conductor generates two kinds of invisible fields, electric and magnetic. Taken together, they are called electromagnetic fields. Sixty-hertz fields fall into what is called the extremely low frequency (ELF) range of the electromagnetic spectrum. The electrical field can be blocked with shielding, but the associated magnetic field cannot easily be blocked. Electromagnetic fields at these low frequencies do not produce the same dramatic effects on the body as are produced by two other commonly used forms of electromagnetic radiation, both at much higher frequencies: X rays, which have enough energy to break apart DNA, the carrier of genetic information; and microwaves, which cause damage from heat. For years scientists asserted that because 60-hertz fields can do neither, they could not possibly cause significant changes in the body.

That assertion is now being challenged by laboratory experiments on living cells and animals and by epidemiological studies that have shown a statistical association—but no cause-and-effect relationship—between cancer and exposure to electromagnetic fields from wires that carry electricity through neighborhoods and into homes. Findings from what is still a young and growing body of laboratory experiments with human cells and animals suggest that electromagnetic fields can interfere with the functioning of DNA and RNA, the controllers of cell reproduction; that they can cause reproductive disorders and birth defects in chicks; that they stimulate activity in biochemicals linked to the growth of cancer; and that they affect other substances that are critically involved in the workings of the central nervous system.

At the cellular level, the evidence so far identifies the membrane that envelops the cell as the main site of interaction with electromagnetic fields. The membrane governs some of the cell's most critical functions, like the flow of material, energy, and information from the outside to the interior. If its function were severely disrupted, cell-to-cell communication and immune response could also be disrupted, scientists say. Some experiments have found that exposure to electrical power fields alters the flow of calcium across the cell membrane, although the health significance is not clear. Calcium governs cell division and egg fertilization. One experiment has shown that electromagnetic fields change the rate of synthesis in DNA and another that they interfere with the functioning of RNA in converting the instructions issued by DNA into production of proteins. Still another has shown that the radiation spurs the action of an enzyme associated with the growth of cancerous tumors.

One cellular experiment has shown that 60-hertz electromagnetic fields alter the action of hormones called neurotransmitters, which send signals between nerves, and of hormones that control the biological clock. If the same effect occurs in humans, the Office of Technology Assessment study said, this might play a role in a number of disorders, including altered sensitivity to drugs and toxins, disruption of the biological clock, mood and sleep disorders, and chronic depression. In what has been called the "henhouse project," six independent laboratories in the United States, Canada, and Europe exposed fertilized chicken eggs to pulses of 60-hertz radiation like that emanating from video display terminals and television sets. Taken together, the experiments showed an increase in the proportion of abnormal embryos in eggs that had been radiated. The same effects have not been observed when eggs are exposed to steady, rather than pulsed, radiation of the sort generated by electrical appliances and power lines.

The studies that have gained perhaps the most attention, however, have sought to assess the association of electromagnetic fields encountered in everyday living with adverse health effects in humans. In one, researchers at the Kaiser-Permanente Medical Care Program in Oakland, California, reported that women who used video display terminals for more than twenty hours each week in the first three months of pregnancy suffered almost twice as many miscarriages as women doing other types of office work. The study's authors were careful to say that the findings did not necessarily mean that the terminals themselves had caused the miscarriages and that other factors, like stress, could have been responsible.

Other epidemiological studies have tried to find out whether there is an association between exposure to 60-hertz fields and childhood cancer. The first of these, by two Colorado epidemiologists, Nancy Wertheimer and Ed Leeper, compared children who died of cancer in Denver between 1950 and 1973 with a control group of other children. The study found that children who lived near electrical distribution lines were twice as likely to develop cancer as those who did not. The study was widely debated and criticized for what were seen as biases and failure to account for all variables. A subsequent study was commissioned by the New York State Power Lines Project to test the findings. The new study, headed by David A. Savitz, an epidemiologist now at the University of North Carolina, attempted to repeat the earlier investigation while eliminating what critics saw as its flaws. It was also carried out in Denver but involved children diagnosed from 1976 to 1983 as having cancer. To the surprise of the critics, the Savitz study bore out the earlier findings.

The scientists who designed the Savitz study "were very skeptical," says Dr. Carpenter, the Power Lines Project's executive secretary. "None of us expected the Savitz study to replicate the Wertheimer results. But the study did in fact replicate the results in almost every degree." As much as anything, the study is widely credited with persuading many skeptics to take the issue seriously. Dr. Savitz himself is the first to warn against drawing unwarranted conclusions from the study. At best, he said in a formal statement after the study was made public, "there is a suggestion of a possible hazard which has yet to be resolved." Even if the study provided conclusive proof, he said, the risk of cancer would be 1.5 or 2 cancers in 10,000 children a year. "This would be very important," he said in the statement, "but minor relative to childhood injuries or risks from known cancer hazards to adults such as cigarette smoking or asbestos exposure."

In an interview, he summed up what he and many others consider the mainstream scientific opinion about any cancer threat posed by electromagnetic fields: "Is there reason for concern? An honest, objective review would have to say yes. Is there persuasive evidence that ELF radiation causes cancer? The answer is a clear no." Scientists are nowhere near establishing any risk standards for electromagnetic exposure. Even if that should turn out to be called for, the task is expected to be more complicated than in the case of other environmental risks to health, such as those posed by chemicals. Partly, this is because electromagnetic fields often appear to affect living tissue only at certain thresholds or "windows" of intensity. Biological effects may be triggered at lower intensities rather than higher ones: more may not be worse.

Present efforts in several states to set "safe" upper limits on the strength of radiation from transmission lines, the Office of Technology Assessment report said, are therefore scientifically insupportable. And in fact, reducing the intensity of fields to which people are exposed might even make things worse, said Mr. Florig of Carnegie-Mellon. Experts at this point generally do not urge that 60-hertz fields be aggressively regulated. The Carnegie-Mellon team, however, has advanced what it calls a policy of "prudent avoidance" until more definitive knowledge develops. They advise taking relatively low-cost, low-effort steps like these to limit exposure to electromagnetic fields:

- Try to route new transmission lines so they avoid people and widen transmission rights-of-way, but do not tear out and rebuild old lines.
- Develop new approaches to house wiring that minimize electromagnetic fields.
- Redesign new appliances to minimize or eliminate fields, but do not throw out all old appliances before they wear out.
- Use electric blankets only to preheat the bed, or eliminate them.
- Move an electric alarm clock as far from the bed as is practical.

Dr. Savitz said that while concern may be justified, "our study is not sufficiently convincing to warrant drastic action by homeowners." But in Canada, an arbitration board in a labor-management case adopted a preventive precedent in ordering that an employer cooperate in the development of a prototype shielding mechanism for video display terminals. The board reasoned that although the evidence of harm from video display terminals was merely suggestive and inconclusive, "there are simply too many incidents where the environment has been invaded by unknown factors which have come to light only after the harm has been done."

Virtually everybody who has looked at the problem advocates a major research effort. "There is enough concern that some key questions have to be answered," says Richard Phillips of the Environmental Protection Agency's Health Research Laboratory at Research Triangle Park, North Carolina. Dr. Phillips, a recognized authority in the field, edits the journal *Bioelectromagnetics*, in which much of the research on the subject has been published. "The big question," he said, "is the cancer issue." Despite the proliferation of studies, the Office of Technology Assessment report found, there has been a marked decrease in the level of federal research support in recent years. Congress allocated $3 million for that purpose to the Department of Energy in the fiscal year 1989. Because of budgetary restraints, most of the environmental agency's projects dealing with ELF fields were shut down in 1986.

A number of states have undertaken modest research programs, and the electric utility industry has backed others. The major utility researcher, the Electric Power Research Institute, spent $1.7 million in 1986, but in a measure of increasing concern spent $5.5 million in 1989. The institute's is the largest single program in the world. Scientists in other countries are also pursuing the subject, and Dr. Sagan of the power institute says, "One sees a worldwide research effort developing." Moreover, American scientists are showing greater interest. "Until very recently," he says, "it was very difficult to get the attention of the scientific community." That is no longer true. "My hope is that we will have a definite answer one way or another in four or five years," he says. "The bad news would be if it continues to be ambiguous." [WKS]

SAVED BY THE PHILODENDRON

The lilies of the field may toil not, but those plants in the living room and office, philodendron and ficus, daisies and mums, are much more than decorative. Scientists are finding them to be surprisingly useful in absorbing potentially harmful gases and cleansing air inside tightly sealed modern buildings. In studies of biological methods for purifying air in space stations, environmental specialists at the National Aeronautics and Space Administration (NASA) have determined that several common houseplants reduce the amounts of formaldehyde, benzene, carbon monoxide, and nitrous oxide in indoor air.

Scientists say polluted rooms cannot be cleaned up by plants alone. No amount of greenery, for example, is likely to reduce the smoke and dust in a room. Still, "It's a whole world of environment studies we're just beginning to touch," says Dr. B. C. Wolverton, the NASA investigator in charge of the plant research. "If you put plants in buildings, will it help improve air quality? We say, from our tests, yes indeed. The more foliage, the healthier the environment is going to be."

According to Dr. Wolverton, the absorption of polluting chemicals by plants is part of the photosynthetic process in which plants, to live and grow, must continuously exchange gaseous substances with the surrounding atmosphere. The plant leaves take in carbon dioxide and exchange it for oxygen and water vapor, which is returned to the atmosphere. It now appears that plants can also take in other, more dangerous, gases through the tiny openings, or stomates, on the leaves.

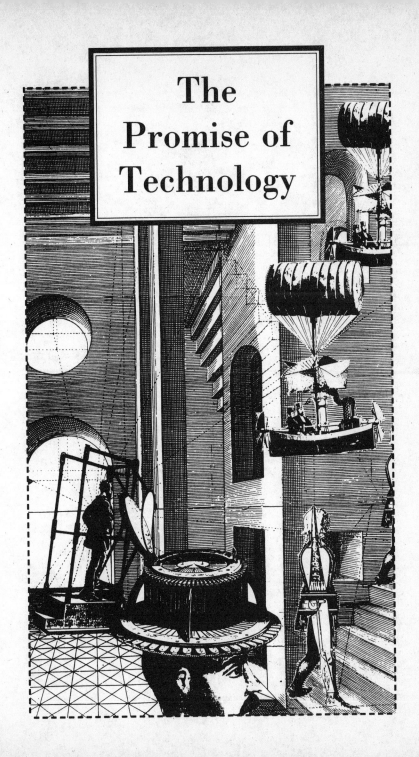

The Promise of Technology

FAKING IT

NO longer content with dissecting tissues, analyzing proteins, and breeding fruit flies, an increasingly diverse group of scientists has decided that the best way to study life is to make some of their own. They are creating a field called artificial life, mixing the impulses of biology with the tools of computation. By looking beyond the usual materials of life—beyond the familiar biochemistry of earthly animals and plants—they hope to capture its spirit: the animated, the energetic, the replicating, the evolved.

Most of the would-be organisms of artificial life exist solely in the electronic environment of the computer, where they are in little danger of being confused with the real thing. The first conference on artificial life, held in 1987 at Los Alamos National Laboratory, offered models of processes from protein formation to plant growth to animal predation—processes meant to be, if not life, then at least lifelike.

The simulations of biology address some of the most troubling questions of the life sciences: how the primitive precursors of DNA gained the ability to store information and copy it, how the senseless force of natural selection created structures of such extraordinary complexity and beauty, and how the laws of ecosystems arise from the whims of individual animals.

They also reflect an expanding sense within science of what life is. Artificial life seeks "the ghost in the machine," as the conference organizer, Christopher Langton of Los Alamos, put it—an essence arising out of matter but independent of it. For the first time in generations, some researchers believe, science has a legitimate way of talking about life's soul.

"It lies in the complexity of organization," said Richard Dawkins, an evolutionary biologist at Oxford University. "It's not a substance; there's no living material. It's just an incidental fact that in real living things the entities that happen to be organized happen to be made of organic, soft, squishy stuff, whereas in a computer they're made of hard, nonmoving chips."

The creatures of artificial life already make up a strange menagerie. There are flocking birds and schooling fish just a few generations removed from the cartoons of Walt Disney. Invisible bugs breed and die out as they leave trails through a mound of electronic food. Computer flowers bud and unfold, their timing controlled by computer chemicals running up and down computer stems. Stick-figure shapes evolve in a few dozen generations into startling butterflies and shellfish.

Some simply imitate real organisms. Most, however, depart from reality to capture some abstract quality of living things, preferably a quality that arises not from the designer's intent but from unplanned processes.

"What keeps me awake at night is not correspondence to reality," said Steen Rasmussen of the Technical University of Denmark. "I want to know what is the soul in this that creates order—what is the engine."

Stripped of bone and sinew, leaf and petal, ribosome and chromosome, life still has a unique logic that can be abstracted in a computer—that, at any rate, is the belief driving the new discipline. Nor is the computer essential. Some scientists are trying to create microscopic carriers of information in fragmentary protein strands or pieces of clay crystal.

"Surely there must be a more general sort of biology," said Graham Cairns-Smith of the University of Glasgow, author of *Seven Clues to the Origin of Life*. "This is the aeroplanes-don't-have-feathers principle. Yes, birds have feathers and fly beautifully, but we have different requirements."

Those explicitly seeking to create life within a computer or a test-tube biochemical system form a group that now brings together microbiologists, evolutionary theorists, physicists, chemists, and computer scientists. At Los Alamos, they spawned rooms full of computer demonstrations, wandered from place to place wearing buttons asking "What is a genetic algorithm?" and showed videotapes of robots taking five hours to weave across a room.

They face a problem of definition. Most modern biologists think of an organism's abilities to process matter and energy, to replicate itself, and to evolve as the essential, defining qualities of life. Some computer models already have those abilities, in more or less trivial ways.

So scientists debate the question of how they would recognize a genuine artificial creature if they had one. After one particularly testy exchange, a scientist proposed that a key criterion should be "irritability." Others recommended purposefulness and unpredictability as qualities any good organism should have.

Gerald F. Joyce of the Salk Institute in San Diego suggested the biologist test: put the artificial organism into a room with a biologist. If the biologist comes out and says it's alive, that would be encouraging.

"And if your organism comes out and says it's alive, then you're on the right track," Dr. Joyce said.

Many of their colleagues will accuse those promoting artificial life of overoptimism and exaggerated claims, particularly about the capabilities of computers. They remember the overselling of artificial intelligence in the 1970s, and the Los Alamos conference, too, drew a healthy contingent from what one biochemist, Hyman Hartman of the University of California at Berkeley, called the "computers are the next form of life, so let's get on with it" school.

Dr. Hartman warned against relying too blindly on computer models. As his own model undulated and sparkled hypnotically on the giant screen behind him, he told the audience, "One of the great dangers of artificial life is that you can be very, very clever and invent beautiful machines that do beautiful things, but you've gotten very, very far away from what you're trying to understand."

His simulation, a checkerboard of sixty-five thousand cells that changed color according to simple rules, was meant to show how simple processes on the surface of a clay crystal might generate complexity. Indeed, as in several other demonstrations, strikingly rich and irregular patterns arose—large-scale structure emerging from the interplay of small-scale rules.

The recognition in recent years that complexity can arise spontaneously from simple systems gives the field of artificial life its strongest motivation. The scientists agreed that the most promising demonstrations were those whose lifelike qualities emerge unbidden, surprising even their programmers.

They want to play god, but not a god who directs the motion of every sparrow, as Dr. Langton said. "We don't want to do that—no global controller," he said. "And no miracles—miracles aren't allowed except at the very beginning."

A computer graphics expert trying to create a flock of birds that will fly convincingly around obstacles, for example, must create a free-floating yet tightly coordinated pattern of motion. Instead of programming a flock from the top down, Craig W. Reynolds of Symbolics, Inc., let each of hundreds of imaginary birds follow a set of rules for avoiding their neighbors.

A natural-looking flock took shape, sweeping gracefully but not rigidly around blocks and cylinders. And unexpected behavior emerged as well—one bird crashed into an obstacle, fluttered in a momentary daze, and then staggered onward.

The spontaneous emergence of organization is a central problem of life at all scales. Those studying the origin of life are acutely aware that, without some self-organizing principle, it would take many times the age of the universe before chance would bring amino acids together in just the right combinations necessary to form the elaborate machinery of DNA.

Self-organization must also guide the combination of embryology—the unfolding of individual creatures according to the rules of development built into their genes—and evolution. These remain deep mysteries, and computer models are intended to show not how they do occur, but how they might plausibly occur.

Scientists have discovered in recent years that some seemingly complicated patterns, like the branching, jagged structures of plants, have simple descriptions in the language of fractal geometry, in which patterns are built up from rules repeated on different scales. No one knows just how such rules are encoded in the genes of real plants; nevertheless, several demonstrations at Los Alamos created lifelike ferns, trees, and even flowers from relatively modest fractal instructions.

One program, by Przemyslaw Prusinkiewicz of the University of Regina in Canada—the winner of an "Artificial 4-H Contest" for most lifelike organism at Los Alamos—mimicked the growth of a variety of flower species. It combined geometric instructions with a set of timing signals, like the chemical signals that real plants use to control branching and budding. The results were vivid images of plant growth.

Such models illustrated rich development with no possibility for evolution. By contrast, Dr. Dawkins, the Oxford zoologist and author of *The Blind Watchmaker,* offered a stick-figure version of embryology with surprising evolutionary power.

Through random mutation and a somewhat arbitrary version of natural selection, the program manages to evolve into shapes with surprising complexity and often a surprising resemblance to earthly creatures. Each experience with the model brings new evolutionary paths, none of which could have been predicted.

The results are just drawings on a computer screen, with neither the attributes nor the potential of real life, as Dr. Dawkins himself noted. In the long term, he said, electronic versions of evolution could produce something more.

In general, by creating a variety of computer environments, universes with their own sets of rules, scientists intend to provide ways of thinking about universal principles of life—principles more general, perhaps, than those observed in nature. Computer scientists since John von Neumann, one of the fathers of computing in the 1940s and 1950s, have known that such artificial environments can create "self-replicating automata," organized structures that reproduce themselves.

"If they don't have the whole enchilada, at least they have a few pieces of lettuce," said A. K. Dewdney of the University of Western Ontario, *Scientific American*'s computer columnist. "If we have a system that can organize itself or can evolve in some sense, then the hope is that down the road the system, without any thought or care on our part, will become intelligent."

For artificial life to become a successful approach, Dr. Dewdney and other scientists said, models will have to become much richer than the first efforts. They will have to combine processes of growth, competition, and evolution, only pieces of which have been seen so far.

Still, many of them are optimistic—willing descendants of Dr. Frankenstein, whose creation remains "the bugaboo metaphor for artificial life," in the words of J. Doyne Farmer of Los Alamos, an expert on chaotic dynamics who is modeling the body's immune system. He echoed some other scientists in calling the prospects frightening, perhaps not so much because of what might be created as because of what it might tell us about people.   [JG]

DREAMS OF LEVITATION

A PHYSICIST imagines that he is momentarily annoyed by the big conference table occupying the middle of his office. He gives it a shove with one hand and, in his imagination, it floats away, drifting lazily toward the corner, until finally it stops with a bump against the wall.

So it might, in a speculative future. For now, the physicist, Praveen Chaudhari, vice president for science at the International Business Machines Corporation, is engaging in a reverie about superconductivity and its most bizarre by-product: the phenomenon of levitation, science's answer to the flying carpet. "You don't even have to make cars," he says. "You could make little gizmos, you could put on a pair of special shoes and make little tracks along which you as a human being could push yourself and keep going. Nothing to stop you, right? I see the whole transportation system being very different. At airports, instead of these long conveyor belts we have, you could get onto one of these platforms that are levitating and just stay on it while it takes you around."

Levitation is not just the strangest but also one of the most practical prospects raised by the recent boom in superconducting materials. Floating trains, floating furniture, floating toys, floating people—otherwise sober scientists are talking about applications that used to belong to the realm of science fiction. They are amused, but they are serious. If the new materials fulfill their early promise, and especially if a room-temperature version can be made practical, the ability to lift objects off the ground and free them from mechanical friction could bring surprising rewards.

Levitation comes from a property of superconductors only indirectly related to the property that gives them their name: the ability to carry electric current without any loss due to resistance. With or without electric current, a chunk of superconductor placed above a magnet settles calmly in midair. For that matter, a chunk of magnet placed above a superconductor also hangs in the air—levitation works either way. The superconductor has the peculiar property of pushing out any external magnetic field, so the magnet cannot approach. It just hovers, in the soft grip of an invisible hand.

The phenomenon has a certain built-in measure of stability. "Levitation means that your piece of metal sits on a magnetic pillow," says Vladimir Z. Kresin of the Lawrence Berkeley Laboratory in Berkeley, California. "You can move right, left, forward, back—because of the configuration of the magnetic field, you have a real equilibrium position in the center. It's a system trying to keep everything in the middle." Any mechanical device that requires bearings—any device, for example, in which something must rotate at high speed, like a generating turbine or a gyroscope—could use levitation to eliminate friction. Friction typically sets the limit on the speed of rotation, and it also produces waste heat that must be drawn away.

Scientists have let themselves fantasize about levitation for decades. Twenty-two years have passed since a Stanford physicist, William A. Little, writing in *Scientific American,* proposed not only superconducting hovercraft but also a sort of physicist's theme park, with people "riding on magnetic skis down superconducting slopes and ski jumps." To date, the single real-world anchor for such whimsy is the levitating train, a transportation system whose feasibility has been demonstrated by a Japanese National Railways prototype. The experimental trains carry powerful superconducting magnets, cooled, expensively, by liquid helium.

Smaller-scale, less expensive technologies await superconductors that require less cooling, and such superconductors have been found in a series of recent breakthroughs that have brought superconductivity out of the shadows of scientific esoterica. A new class of materials, easy to duplicate and inexpensive to produce, makes the sudden transition to superconductivity at record high temperatures, though still several hundred degrees below zero. Those materials, requiring cooling by liquid nitrogen, are enough to make possible such non-floating applications as highly efficient long-distance transmission lines and fast, small supercomputers, applications with vast commercial promise.

But another class of applications—the kind that would transform a host of ordinary, visible aspects of everyday life—demands a superconductor that would require no refrigeration at all. Physicists have been reporting signs of this elusive new room-temperature superconductor, seemingly making itself felt in several different laboratories, though still impossible to isolate and stabilize. So scientists are allowing themselves to hope that the room-temperature superconductor will become a reality, and they are thinking more seriously than ever about a world in which objects could be made to float. "If it's going to come into society, you could think of assembly lines, guiding materials around in this innovative way," says Theodore Geballe of Stanford University.

Personal transportation could be freed from two dimensions, especially in cities, "where you get gridlock," he said. "You could just go into three dimensions with small, guided transportation, not the big levitated trains." Metal tracks would have to be built at different levels, floating people along. The capital cost would be considerable, and the temptation to slide through the air at unsafe speeds could be a serious concern.

Propulsion might involve magnets, air jets, or muscle power—in any case, starting and stopping would be nontrivial engineering problems, as would the question of air-traffic control. The consequences of making powerful magnets a ubiquitous feature of everyday life are far from obvious. In terms of pure science, however, the fundamental principles are well understood.

Levitation begins with the fact that a magnetic field creates a current in any conducting material—the principle at work in electrical generators. A current creates its own magnetic field—the principle at work in electric motors. But superconductors are special. If a magnetic field penetrated a superconductor, it would create a current that would set up exactly the opposite field. Wherever the magnet moved and however it was oriented, it would see its ghostly counterpart below, repelling it. So the external field cannot penetrate—it is expelled, an effect known as the Meissner effect.

Even on the small scale of the laboratory, the results are uncanny. A piece of flat iron magnet sits on a table. A chunk of the new superconducting material, a dull gray ceramic, is dipped into a Styrofoam cup full of liquid nitrogen to cool it. Then the superconductor is put above the magnet, where it floats. It can be poked, spun, and nudged from place to place, but it remains suspended until it warms up. Then, making the transition from superconductor to ordinary ceramic, it settles to the ground.

The technology designed for high-speed trains uses a variation of the physics of levitation, set in motion. The train is equipped with superconducting magnets, coils of wire that become magnetic when a current is passed through them. Because there is no resistance to electricity, the current does not need to be maintained with a continuous power supply. Once it is started, it continues forever. The train sits on a track of ordinary metal, such as aluminum. As long as it is motionless, it just sits, but when it begins moving forward, the magnets induce a current in the aluminum, setting up another repulsive magnetic field. The effect is instantaneous and short-lived.

"The magnet in the vehicle has to think it sees an equal and opposite magnet down below," Dr. Geballe says. "A few milliseconds, that's enough time. Then you move on to a virgin piece of aluminum." The train starts on wheels and then, at about fifteen miles an hour, lifts off the ground. For forward propulsion, the train relies on separate magnets embedded in the track. In case of a complete power failure, the train would simply settle gradually back down onto its wheels.

The Department of Transportation investigated levitating trains, among other futuristic transportation ideas, a decade ago, but interest waned, in part because the United States, with its spread of urban areas and its love of private automobiles, seems less than ideally suited for large-scale rail transport. Some scientists complain that the federal government has long been too reluctant to support research into innovative technologies of transport. The Japanese, however, went ahead with a program of trains using superconducting magnets, while West Germany sponsored an experimental train using a different magnetic technology. As a result, much of the engineering has already been done. "It's entirely feasible—the Japanese and the Germans could implement it right away," says Francis C. Moon, chairman of the department of theoretical and applied mechanics at Cornell University. Dr. Moon conducted research on the stability of levitating trains and observed tests of the Japanese prototype, flying six to eight inches above its track at speeds of two hundred to three hundred miles per hour. Track wear and tear is not a problem; nor is noise. "The train goes by in a whisper," he says. "It's weird to see."

Years of imagination and hard engineering have gone into the levitating train. The next generation of levitating objects can only be guessed at. But design work and engineering calculations have also been applied to the problem of replacing bearings with superconducting magnets. In an engine or turbine where one ring rotates inside another, the principle would be the same as in a levitating train.

Less industrially, Dr. Chaudhari's floating furniture would be guided by wires embedded in the floor, he suggested. To allow a table or chair to rise and sink, designers could use small electromagnets that could be controlled with a handy dial. "It just pops up," he said.

"The strength will determine how high or low it goes." Others, when they witness levitation, cannot help but think of toys. "It's funny by itself," said Dr. Kresin of Lawrence Berkeley Laboratory. "Maybe one application will be for the toy industry. You can apply huge fantasy using these principles." [JG]

HARNESSING THE SUN

A CENTURY and a half after the French physicist Edmund Becquerel discovered that light can be transformed directly into electrical energy, the photovoltaic exploitation of sunlight finally seems on the verge of broad commercial use. So far, solar electric cells have largely been sold for specialized purposes where cost was less important (as in spacecraft) and to power isolated communications networks, navigation aids, recreational vehicles, and homes remote from power lines. Now, because of technological advances, reduced manufacturing costs, and other factors, a few solar plants are being built to sell electricity to utilities. And within a decade or two, experts say, solar plants could make a significant contribution to the nation's electricity supply, especially in periods of peak demand when excessive loads on conventional generating capacity can cause "brown-outs" or even blackouts.

Significant progress with solar cells would help speed the transition from fossil fuels that experts say is needed to combat the greenhouse effect, climate changes induced by the rise in carbon dioxide in the atmosphere. A major technological advance has transformed thinking about the ultimate prospects for solar electricity. A research program sponsored by the Department of Energy reached a milestone that experts had compared with the running of a mile in less than four minutes, a goal that was regarded as impossible until it was achieved. Sandia National Laboratories in Albuquerque, New Mexico, built an experimental device called a mechanically stacked multijunction solar cell that converts a phenomenal 31 percent of the light striking it into electricity. Previously, many experts had considered

30 percent efficiency unattainable for a photovoltaic cell. (By comparison, the solar cells in pocket calculators have an efficiency of less than 3 percent.) Scientists at Sandia believe that, with the device's capacity to use cells in tandem, efficiencies of up to 35 percent will be achieved in the next few years. For the first time, therefore, the conversion of solar energy to electrical power could become comparable in efficiency to conventional power generation; coal-fired and oil-fired electric plants have an average efficiency of 34 percent.

High efficiency is vital to making economical use of light from the sun, because an inefficient cell costs just as much as an efficient one to install in the necessary supporting machinery. Even before the technological breakthrough, which will not find its way into commercial plants for several years, the cost of manufacturing and operating efficient photovoltaic cells had begun to fall. The trend was dramatically underscored in the fall of 1988 with an announcement by the Chronar Corporation of Princeton, New Jersey, that it intends to build a fifty-megawatt photovoltaic electricity plant near Los Angeles. The plant, which will use an older technology to supply power to the Southern California Edison Company, will not be the first of its kind for the region. But it will have a capacity seven times as great as the next-largest solar plant.

The plant will have a conversion efficiency of only about 6 percent, but in the next few years plants are likely to be more than twice as efficient. Further down the line, the new tandem cells and other new technologies could bring efficiencies to 38 percent. Production costs could be substantially reduced with a new generation of "thin-film" photovoltaic cells produced in endless sheets, in somewhat the way photographic film is manufactured.

Solar energy enthusiasts have long dreamed of replacing fossil fuels and nuclear power, plagued by environmental and political concerns, with energy extracted from sunlight. The development of solar power has been sporadically accelerated in the past three decades by oil shortages, which led to government subsidies for solar development, and by the need for virtually inexhaustible power supplies aboard space vehicles, which spurred technological innovation. But formidable obstacles have snagged progress. Falling oil prices in the mid-1980s reduced the incentive to develop alternatives to fossil fuels. At the same time the United States government reduced its support to the leading photovoltaic program from a peak of $150 million a year in the oil shortage in 1981 to $35 million. Basic research in general has failed to realize the optimistic forecasts made in the 1970s for solar energy.

Another problem is that the efficient exploitation of solar power requires lots of space and abundant sunlight. A typical solar electricity plant that now supplies 1 million watts of power per hour to the Southern California Edison commercial grid sprawls over twenty acres of land near Hesperia, California, east of Los Angeles. A coal-fired plant producing ten times as much power would probably occupy only one tenth the space—though it uses more space elsewhere in mining and transportation. But improved understanding of the physics of materials and new engineering techniques seem to herald a new period of rapid progress. Dr. Dan E. Arvizu, director of the photovoltaics program in Sandia, believes that despite the disappointments of the past decade, the recent progress indicates that solar photovoltaic energy may provide as much as 1 percent of America's electricity in the early years of the twenty-first century.

"That may not sound like much, but it would be an enormous amount of power available during times of peak consumption," he says. The present United States power generating capacity is about 600 billion watts. One percent of this would be 6 billion watts, which would supply the needs of more than 3 million people. When the cost of solar power is brought down to about twelve cents per kilowatt-hour, Dr. Arvizu says, solar power will become attractive for use in peak periods, and this goal is in sight.

Although many systems for exploiting solar energy have been devised, experts think that photovoltaic cells offer the best prospect of efficient commercial electricity generation. Such cells contain crystalline material, in which each atom is linked to neighboring atoms in a rigid lattice structure. The links serving as bonds in the structure are electrons. When light passes through a photovoltaic material, some of the photons in the light strike electron bonds between atoms and knock the electrons loose. Once freed from its lattice, a negatively charged electron begins to wander through the photovoltaic material. The empty, positively charged space it leaves behind in the crystalline lattice is called a hole, and since a neighboring electron is likely to jump into a fresh hole and thereby leave another hole, holes, in common with loose electrons, are said to wander.

The trick in making a photovoltaic cell is to continuously sweep the loose, wandering electrons out of the cell as fast as they are produced, and use them as electric power before they have time to affix to wandering holes. If this happens, the solar energy that was imparted to the electron in knocking it loose is lost as heat. One way to draw off the electrons is to introduce a few foreign atoms (typically, boron and phosphorus) into the lattice in such a way as to create permanent positively charged and negatively charged regions within the material, separated by a thin zone known as a junction. The resulting electrical field across the junction forces wandering, negatively charged electrons to migrate to one region and positively charged holes to the other. An external electrical connection between these regions allows the electrons to flow as current, which can be used to light lamps, drive motors, and so forth.

Early photovoltaic devices, used in light meters and electric eyes, were generally based on a metal-like element called selenium. But in 1954, Bell Laboratories discovered a way to use silicon crystals as photovoltaic power sources, and most photovoltaic cells developed since then have incorporated silicon. Part of the new Sandia mechanically stacked multijunction solar cell is made of silicon, but another part is made of a different photovoltaic material, gallium arsenide. The cell, called a tandem device, is actually two cells in one, manufactured separately and glued together with a transparent adhesive.

The top cell, produced for Sandia by Varian Associates, Inc., of Palo Alto, California, consists mainly of gallium arsenide, and it absorbs sunlight from the blue end of the spectrum to make electricity. The upper cell is nearly transparent to other wavelengths of light that pass through it to the lower cell. To make the gallium arsenide cell as transparent as possible, the necessary metal connections on its top and bottom are formed of ultrathin wires arrayed in spiderweb patterns. The lower cell, made for Sandia by Stanford University and consisting mostly of silicon, generates power using the residual reddish light that passes through the upper cell.

A complex system of microscopic interconnections links the upper gallium arsenide part with the lower silicon part so that their electrical output can be combined smoothly. Dr. Arvizu says that, in principle, many different cells might be stacked, each one extracting energy from a narrow band of wavelengths of sunlight. For the most efficient conversion of sunlight, however, three layers is likely to be the practical limit. To reduce the cost of generating power with expensive high-efficiency cells, sunlight must be concentrated rather than allowed to fall directly on the cell. "It's much cheaper to make lenses than semiconducting material," Dr. Arvizu says, "so we like to use a large lens focusing light five hundred times as strong as ordinary sunlight on a small cell."

Meanwhile, important progress has been reported in the development of efficient thin-film photovoltaic cells, which do not require lenses and are much cheaper to make than layered crystalline silicon or gallium arsenide devices. Thin-film cells based on noncrystalline (glassy) silicon are produced in great numbers (mostly by Japan) to power pocket calculators and similar devices. Although very cheap, these cells offer efficiencies of less than 3 percent and wear out rapidly. An advantage of such cells is that they continue to produce power in limited amounts even when the sun is not directly focused on them or on cloudy days. By contrast, high-efficiency cells of the type Sandia is developing work properly only in bright sunlight and they must be oriented to face the sun directly.

One of the latest thin-film cells, said to have reached efficiencies up to 14 percent, does not degrade in long use. It is based on a substance called copper indium diselenide, or CIS, which the Boeing Aerospace Company showed in 1981 could be made into very efficient thin-film cells. The main commercial developer of these cells is Arco Solar, Inc., of Chatsworth, California, a division of Atlantic Richfield Corporation. In the photovoltaic cell developed by Arco Solar, a thin layer of CIS is the electrically positive part and layers of cadmium sulfide and zinc oxide make up the negative part. The combined thickness of all the active layers in this cell is less than one thirtieth the thickness of a human hair. When light shines on the cell, electrons are forced into any external circuit connected to it, creating a current. Arco Solar's manufacturing innovations, recently hailed by the Electric Power Research Institute as revolutionary, have made the production of such cells profitable.

According to Arco Solar's senior vice president, Dr. Charles F. Gay, the company is making "table-sized" CIS cells by depositing five very thin layers of material on a backing of glass. Eventually, however, the company hopes to produce cell material continuously on a flexible plastic backing that can be wound from one spool to another. Dr. Gay says worldwide sales of photovoltaic cells amount to about $100 million a year, of which 25 percent to 35 percent go to Arco Solar. "The business is big enough to support further research and development," he says, "and as the price of photovoltaic cells comes down, new markets are emerging. The next big plateau is to start replacing diesel-electric generators in isolated places like Hawaii with solar generators. Many different kinds of power will have to be developed in the future, but it looks as if solar photovoltaics will have an assured role." [MWB]

THE SEARCH FOR SECURITY

JUST how difficult would it be to steal the materials needed to make a nuclear bomb or to penetrate the barriers that safeguard bank transactions and vital corporate secrets? In the last few years, terrorists, spies, and other potential intruders into sensitive facilities have become more adept and better equipped than ever. Experts note, for example, that terrorists from the Middle East, Japan, and Europe are increasingly equipped with advanced weapons and bullet-resistant jackets. Even more disquieting is the fact that terrorists are becoming skilled in circumventing technologically advanced security systems, including those controlled by computer programs.

But in the continuing battle of wits, officials say that technology and tactics designed to keep intruders out of buildings or to protect sensitive materials and documents are keeping pace. Security devices and systems that a few years ago existed only in science fiction or as prohibitively expensive prototypes are becoming available even to ordinary businesses. Jails, banks, and modest-sized companies are already using some of the new technologies.

The toughest security measures are reserved for facilities that make or store plutonium and uranium 235 as well as the equipment a terrorist could use in assembling an atomic bomb. The would-be thief, spy, or terrorist now faces hedges that are really walls, invisible sensors made of tamper-proof radiation beams, and foam that can mire the intruder like a fly in meringue. Traditional identification systems based on photographs and handprints are being replaced by advanced systems that can speedily distinguish one person from all others by biological measurement. In one system coming into widespread use, a scan of the pattern of blood vessels at the back of a person's eyeball is used to determine whether the person is authorized to enter a restricted area. Supervising the operation of all this is an

308 • THE PROMISE OF TECHNOLOGY

army of 6,800 Department of Energy "inspectors," crack fighting men and women as skilled in hand-to-hand combat as in operating computers.

There was a dramatic reminder in the fall of 1987 that defense laboratories face a continuing threat. A man planted a powerful bomb beneath a car parked across the street from the sprawling Lawrence Livermore Laboratory in California. Although the bomb caused neither casualties nor serious damage, it destroyed the car and alarmed officials at Livermore, one of America's main designers of nuclear warheads and anti-missile weapons. The Livermore incident and other developments have spurred a reexamination of security at many top-secret defense facilities. In an unusual public disclosure of some of the steps their own security experts were taking, Livermore officials recently described how the compound was being compartmentalized, how risks were being identified, and what kinds of systems were under development to detect penetration or the theft of nuclear explosives and secrets.

The laboratory's powerful computers were programmed to predict the possible consequences if any given Livermore employee were to become a spy and exploit the secrets and materials to which he had access. The computers were also assigned to evaluate tactics and priorities for minimizing the risks. Officials say the results are significantly improving security. "Over the years security threats have become more technically sophisticated," a Livermore report said. "New security risks have also been created by the increase in the number of terrorist organizations and forces. Although most adversaries have, in the past, adopted stealth, deception, and other nonviolent approaches, we must also be prepared for attempts by armed force."

The Department of Energy administers the laboratories and plants that make nuclear explosives and reactor fuel, as well as some of the most advanced weapon systems under development. In 1984, to meet the perceived danger facing its installations, the department created a school for its paramilitary forces, the Department of Energy Central Training Academy at Kirtland Air Force Base, Albuquerque, New Mexico. The academy trains Department of Energy troops at a rate of thirty-eight hundred a year. Besides protecting Livermore, they guard Los Alamos, Argonne, Sandia, and Brookhaven National Laboratories, as well as all the facilities making or testing nuclear explosives, including the Nevada Nuclear Test Site and the Hanford, Washington, Savannah River, South Carolina, and Oak Ridge, Tennessee, processing plants. Dennis C. Wilson, acting director of the academy, says that his organization also evaluates new anti-terrorist inventions and systems and that although the great majority prove to be mere gimmicks, some have significantly improved security.

Among the latest innovations is a pulsed infrared beam used around the perimeter of some facilities. In old-fashioned electric eye systems, a beam of light was aimed at a photoelectric cell, and if the beam became obstructed, an alarm would sound. Later versions of the system used invisible infrared radiation. "Nevertheless," Mr. Wilson says, "potential intruders found that you could place mirrors in the beam path to divert it out of the way but still allow it to reach the light sensor. That way, a space was opened that an intruder could step through without triggering an alarm. But in a new version, the system measures the time the infrared light takes to reach the sensor, and if it takes too long because of having to travel along a lengthened path, an alarm goes off."

The traditional first line of defense against terrorists was the chain-link fence, but nuclear facilities and military installations have begun replacing these with hedges or "living fences" that are said to be more effective. A type of tree imported from China, trifoliate orange, was chosen because it develops strong, densely intertwined branch systems studded with dagger-sharp four-inch thorns. Barrier Concepts, Inc., of Oak Ridge, Tennessee, the company that developed and grows "living fences" for government installations, contends that a fully grown hedge about four feet thick can stop a light truck. The hedge, which is initially grown at a nursery, is transplanted while still immature and reaches its full size in about a year. James Passmore, an executive of the company, said that even if an intruder were to use a chain saw or bolt cutter against the barrier, the intertwined branches would make them almost impossible to remove.

Fences and barriers must have entry gates, however, and a gate might be vulnerable to a terrorist driving an explosive-laden truck at high speed. Several companies have developed fast-acting emergency road barriers to meet the threat. A device marketed by Barrier Concepts uses a hydraulic ram to raise a heavy steel blade that can block a roadway in less than one second after a sensor recognizes a threat. The company says the barrier, already in use at various defense plants and under development for some embassies abroad, can stop a 15,000-pound truck hitting it at fifty miles an hour. But security officials believe that spies posing as legitimate employees represent a greater potential threat than outsiders. The thrust of many new measures is toward sealing off sensitive areas from everyone without good reason to enter. When entering or leaving a high-security office, workshop, storage bunker, or plant, an employee must pass through a chamber reminiscent of the air locks used by astronauts to enter or leave orbiting spacecraft.

The employee or visitor is locked in the chamber until cleared to enter or leave the security zone. Standing on an automatic scale to be weighed, the visitor first types an identification number on a key pad and inserts an identity badge into a slot. Reading the card with a magnetic sensor or laser, a computer extracts such personal information as the visitor's weight and security clearance level. The computer checks to see that the person's measured weight matches the weight registered on the badge and that the person is authorized to enter at this time on this day. Meanwhile, a television camera sends an image of the person's face to a central control room. If the person is leaving an area where nuclear materials are stored, the exact weight at the time of leaving will be compared with the weight measured at the time of entry, and a significant discrepancy will sound an alarm. "A person isn't going to eat a sandwich or drink anything in that kind of environment, so his weight shouldn't change," a security official says.

An important step in the access procedure has recently been added: the scan of the blood vessels in the eye. A decade ago, Robert B. Hill of Portland, Oregon, the son of an ophthalmologist, began work on an instrument that could rapidly scan and record the pattern of blood vessels in the fovea, a small spot near the center of the retina at the back of the eye. The device is now marketed by EyeDentify, Inc., the company Mr. Hill founded. A low-power infrared light source directs a beam into the eye, and the image reflected by the fovea goes back through the lens of the eye to a video camera. The device scans the resulting image electronically. The intensity of light at each point on the image corresponds to the pattern of blood vessels surrounded by the brighter reflection of retinal tissue. The pattern of light and dark in the image is as distinctive as a fingerprint, but is much easier and faster than a fingerprint or handprint to encode in digital form for storage in a small microcomputer.

Once a pattern is registered in the computer's memory, the computer can recognize the pattern's owner in less than two seconds. The person merely looks into an eyepiece, focuses the eye on a target, and presses a button. Developers say that when only one eye is scanned, the chance of a mistake is one in a million. If the system is used to scan both eyes, the chance of admitting the wrong person is reduced to one in one trillion. Early models of the retina scanners cost up to $60,000 in 1984. Since then, EyeDentify's retina scanners have rapidly improved, and some were selling in the late 1980s for less than $7,000. The Departments of Energy and Defense, the Central Intelligence Agency, and other government organizations have bought them. Of more than five hundred devices the company has sold in the United States and abroad, most have been purchased by governments, but private institutions are also acquiring them. Retinal scanners now restrict access to sensitive computer centers of American Airlines and the Boeing Company, to the Utah State Prison, to many research laboratories, and to several banks.

A spokesman for EyeDentify predicted that retinal scanners would one day replace the plastic cards used by bank customers to operate automatic cash machines. The company is already selling a retinal equivalent of the traditional industrial time clock; instead of clocking in or out by inserting a paper card in a machine, a worker merely looks into an eyepiece. EyeDentify contends that the average American company could save $17,350 a year in wages overpaid to dishonest employees.

Despite the new safeguards, spies and thieves may still find ways to enter high-security areas, so human and robotic sentries have been reinforced by an arsenal of new sensors capable of sniffing out trouble. One instrument containing a gamma-ray spectrometer automatically analyzes faint radiation that a container might emit during a theft and quickly determines whether nuclear fuel is concealed inside. Once an alarm has been set off in a secure area, doors and gates slam closed and a variety of automatic defenses may begin to operate. In a few of the most sensitive installations where nuclear explosives are kept, special dispensers can quickly fill an entire room or corridor with dense foam that hampers movement and prevents an intruder from seeing anything.

New technology notwithstanding, human guards remain indispensable to security. Training at the Energy Department's academy now includes real weapons that fire laser simulators instead of bullets. A computer keeps track of the numbers and types of hits, letting referees gauge how effectively the students would react to a real threat. "It's a never-ending campaign," Mr. Wilson says, "but the stakes are too high to relax our guard even for a moment." [MWB]

Q&A

Q. *Can fiber-optic telephone lines be tapped?*

A. Fiber-optic telephone lines—microscopic filaments of glass fiber through which voice and data are sent on beams of laser light—can be tapped, but only with great difficulty. And the party who receives the call can detect the eavesdropping. According to Dr. Edmond Murphy, a researcher at the integrated optic components division of Bell Laboratories, the least complicated tapping mthod is to cut open the cable and find a single optical fiber. This fiber is wrapped tightly around a narrow rod. The light flowing through it migrates to the outer portion of the fiber where it can be picked up with a light detector and its information deciphered. Tapping a fiber-optic line reduces the strength of the signal and the receiver would know almost immediately that the line was being monitored.

GET SMART

A NEW generation of elevator technology is working its way into the world's office buildings, bringing with it a promise that many will find irresistible: the savings of dozens of wasted seconds each day. The result is, in effect, a thinking elevator—a hybrid of programmable computer chips and high-technology sensors. Advanced versions are appearing in new skyscrapers, and in many existing buildings the technologies are being grafted onto old cars and and motors. Elevator designers hope to save either passengers' waiting time or, by carrying more people with less equipment, building owners' money. Clever programming alone—apart from increases in elevator speed—can serve as many as fifteen hundred people an hour with elevators that formerly reached their limit at a thousand.

The smart elevator will sense not only how many people are riding but also how many are waiting at each floor. In buildings with just one or two elevators, the microprocessor has little to add, but where banks of multiple elevators handle heavy, complex traffic, the computer comes into its own. Using techniques borrowed from game theory, it makes rapid choices, sending cars to where they can do the most good, considering the system as a whole. The next leap forward will be learning ability, enabling elevators actually to anticipate riding patterns.

The idea is to stamp out the various bugbears of elevator traffic. There is the phantom pickup, for example—the door of car number two opens to greet a rider who is no longer there, having just boarded car number one. There is car-bunching—elevators all heading upward in packs around the twentieth floor while impatient crowds wait at the lobby. "In today's world an elevator system gets overloaded and overpopulated, and the elevators tend to bunch together" even when passengers are distributed more or less evenly, says William S. Lewis of Jaros, Baum & Bolles, a consulting engineering firm in New York. Any system with many imperfectly coordinated moving parts can fall into a undesirable pattern. To overcome that, the computer has to overrule the tendency of every car to try to answer the nearest call.

In the past, when elevators were controlled by relays, simple electromagnetic switches, complex logic was impossible, and individual cars were mostly on their own. Now the computer lets elevators look at one another and adjust to changing circumstances. "It can say, 'Skip these calls and run express to the lobby, even though someone is going to wait longer,' " Mr. Lewis explains. "That gets to be a very esoteric algorithm."

Over the past few years, some elevator riders have already had to adjust to the first signs of the new technologies, such as graphic color television displays of schematic elevators gliding up and down the screen. Some riders have had to suppress their annoyance at the sound of synthesized voices asking them to "Clear the door" or explaining dolefully, "There is no cause for alarm." Elevator systems have long been able to adjust somewhat to periods of peak up or down traffic, by automatically sending extra cars to the lobby in the morning, for example. With learning computers, a building's elevator system will act like a sort of anthropologist, studying rider habits until it can anticipate what people are going to do before they do it. This capability allows a system to adjust when it discovers that the law firm on the fourteenth floor begins to empty out early Friday afternoons during the summer, say, or to recognize cafeteria-bound traffic at lunchtime.

"React is the key word here," says Merton Meaker, director of technical marketing for United Technologies Otis Elevators. "The elevator control systems will recognize different traffic patterns in a building, day to day or week to week, and respond accordingly." Mathematical ingenuity is required for banks of cars to make the optimal decisions for moving people around large office buildings. Manufacturers have worked out programs that calculate different hypothetical futures and compare their desirability, just as a chess-playing computer does, using penalties and bonuses to help weigh the alternatives.

Bonuses can be awarded for assigning a call to a car that has a rider who wants to stop at that floor anyway—a seemingly obvious strategy that was beyond the capability of old-fashioned systems. Penalties are awarded for decisions that lead to too many stops. And, most important, penalties are awarded for keeping people waiting; long waits get extra heavy penalties. The appropriate threshold is a matter of debate. Officials of Fujitec America, for example, cite a psychological study that found "great irritation" after sixty seconds or more. That study was conducted in Osaka. "Thirty seconds is the rule of thumb in New York," says James J. Mancuso, area manager for Fujitec.

Fujitec plans called for installing the first infrared sensing devices to detect the body heat of waiting riders. Other companies, too, are preparing such devices—"people sensors"—to let their computers recognize a nascent mob before it is too late. In some ways, that information is even more important than it would have been in the era when every elevator was on its own. The new computer stategies can outsmart themselves, programmers have discovered.

For example, suppose twenty people are waiting on the eleventh floor. After twenty-nine seconds, one car arrives. Eight people board. The remaining people press the button again. Without people sensors, the computer will start its waiting-time count all over again at zero, giving the floor a low priority. Many systems already have the ability to estimate how many riders are in a car by weighing it, and that information, too, is being added to computer strategies. Cars that feel full can be instructed to stop trying to pick up new riders. And as an "antinuisance feature," when a dozen buttons are pushed in a car that is suspiciously light, the computer can automatically turn them all off.

The first generation of operatorless, automatic elevators is now more than thirty years old, so modernization of old systems forms the largest share of the market. In a new building, however, the potential for saving money is vast. The central elevator space of a building represents a tremendous cost, and allowing eight elevators to do the work of twelve can mean millions of dollars in rents.

Raw speed alone seems to have reached its useful limit. Said Mr. Meaker of Otis: "There's a physiological restriction that's going to limit us to about two thousand feet per minute: the comfort level on the human body when you change cabin pressure." In other words, ears start popping. "If you want to go faster your only alternative is to pressurize the cab," Mr. Meaker says.

Computer strategies must deal with tremendous variation and randomness in rider behavior. On average, some manufacturers estimate that an office worker will make six trips a day, but that figure can be much higher when a company extends over several floors. In a building with 10 floors and 1,000 people, elevators might need to handle 250 rides an hour. In the World Trade Center, more than 200 elevators carry 40,000 people an hour.

The millions of seconds freed by the technologies will not necessarily be distributed democratically. "We do have what is known as executive service," says Joel Tuman of Millar Elevator Industries in New York, which has updated elevator systems in many existing buildings. Certain New York City executives—they know who they are—can press a button at their desk, causing an elevator to speed toward their floor. Or an executive can dial a number from a car telephone to make sure an elevator will be standing by at garage level.

The microchip has made other special services available, as well. In the event of a terrorist attack, for example, an operator can instruct the computer to send all the elevators in a building to a particular floor and open the doors one by one, presumably delivering the occupants into the hands of the waiting security forces. For now, human operators do control even the computerized systems. Consoles with elaborate color-coded graphic displays have been set up at lobby security desks or in special elevator control centers.

In some buildings, the designers have begun to note an unexpected and possibly unsettling phenomenon. The operators develop a tendency to involve themselves in their work. "Particularly in New York, certain individuals like to play with the system," says Mr. Mancuso of Fujitec. "They make it like a video game."   [JG]

MICROMOTORS

RESEARCHERS at the University of California have fabricated an electric motor that is no larger than the width of a human hair. The motor, believed to be the smallest ever made, is one of a new class of microscopic machines that scientists envision will one day allow medical, manufacturing, and other operations to take place on scales heretofore thought impossible.

Scientists here and at other research centers have also created gears with teeth the size of blood cells, as well as springs, cranks, and tongs that are so small and light they are prone to being accidentally inhaled. Just what applications these micromachines might eventually have is still open to the imagination. But some engineers are convinced that mechanics is on the brink of the same kind of revolution that occurred in electronics with the development of tiny integrated circuits. Computers that took up entire rooms became small enough and cheap enough to put onto desktops and into wristwatches. "We'll have a whole new class of micromechanical systems," says Richard S. Muller, a professor of electrical engineering and director of the Berkeley Sensor and Actuator Center. "They will provoke a whole new line of products and a whole new category of capabilities."

A report, issued after a workshop sponsored by the National Science Foundation, lists many applications that seem feasible. Tiny scissors and even electric buzz saws could be used for delicate microsurgery, such as cutting scar tissue away from retinas. Tiny machines could also travel down arteries, scraping away fatty deposits that could lead to heart attacks. "Smart pills" could be developed that would be implanted or swallowed and would dispense precisely the right amount of medication through microscopic valves.

Micromachines could also be used for extremely precise manufacturing tasks, such as the exacting alignment of lasers, light detectors, and thin optical fibers needed in fiber-optic communications systems. Instruments used on spacecraft to take measurements could become extremely small and light, while consumer electronics devices like tape recorders could become even smaller and more versatile.

To make the tiny devices, engineers are utilizing the same techniques that have long been used to make integrated circuits. Precise structures are created on silicon chips by depositing ultrathin layers of materials in some areas and etching materials away in others. Engineers hope that by taking advantage of silicon chip technology, they will be able to manufacture micromachines as cheaply and as uniformly as computer chips are produced. "You're not just reducing things, you're making lots of them," says Kaigham J. Gabriel, a scientist at Bell Laboratories who has made several micromachines with his colleagues, William S. Trimmer and Mehran Mehregany.

George Hazelrigg, an official of the National Science Foundation who oversees research in micromechanics, says: "We would expect to be able to make motors for a tenth of a cent apiece, maybe less. You can talk about applications with a hundred thousand or a million motors. We haven't the foggiest notion of what we can do with that." It might be possible, for instance, to build a flat television in which the intensity of light at each point on the screen would be controlled by its own micromotor or microshutter, he said. At the Massachusetts Institute of Technology (MIT), engineers talk of developing armies of "gnat robots" that could perform some jobs, like cleaning a floor, more efficiently and cheaply than one human-sized robot.

While tiny gears, turbines, motors, and other moving parts are still experimental, nonmoving parts fashioned with the same "micromachining" techniques have already found commercial uses, mainly as sensors. A tiny pressure sensor, for example, can be made by etching a thin diaphragm in the middle of a silicon chip that bends in response to pressure. Circuits embedded in the silicon on the rim of the diaphragm measure the amount of deflection. Millions of such sensors are produced each year for measuring blood pressure and for gauging air intake pressure in electronic automobile engine control systems, says Kurt Petersen, executive vice president of technology for Novasensor, a Fremont, California, company that makes such sensors. Tiny accelerometers are also likely to find commercial use in the next few years, possibly in automobile airbag systems to detect the rapid deceleration caused by a collision, or in active suspension systems. An accelerometer can be made by building what is essentially a tiny diving board above the silicon chip. The greater the acceleration, the more the diving board bends.

Micromachining can also be used to make holes and grooves and other nonmoving parts for high-precision machinery. German scientists have made tiny nozzles that bend gas molecules through a sharp curvature. Since heavier gases will curve less easily, the nozzle is used to separate the lighter form of uranium, which is useful in nuclear reactors, from the heavier form, which is useless, Mr. Hazelrigg said. The development of moving parts has occurred only recently, and the devices built so far have only been for demonstration. The Berkeley motor, designed by Roger T. Howe, has turned when a voltage was applied, but in early tests it was not capable of sustained motion. The rotor in the device has a diameter of 60 microns, or 60 millionths of a meter. By contrast, a human hair is 70 to 100 microns thick.

To produce a moving part free from its silicon base, scientists have built some structures on "sacrificial" layers of a special kind of glass that is etched away at the end. An analogy would be a mechanic who builds a structure of alternating layers of metal and ice and then melts the ice, leaving only the metal. The ultimate objective in many cases would be to combine sensors, the electronic control circuits, and the actuator—the moving part—on the same chip. That would allow, for example, the development of a robot-on-a-chip that would sense its environment, determine how to react, and then do so. A tiny insulin reservoir implanted inside a person with diabetes could sense the glucose levels in the blood and continuously adjust the amount of insulin it dispensed.

Still, scientists say they are far from making such applications practical. One obstacle is that the structures built on silicon chips are essentially two-dimensional, while the real world works in three dimensions. And while engineers have long worked with silicon for electronics, they know very little about its mechanical properties, such as its strength, brittleness, and elasticity. Some think that silicon might prove unsuitable and that more durable materials, such as metal alloys, might have to be used. But in some initial tests, silicon has proven quite robust. "No one had realized the mechanical strength that this stuff has," says Dr. Muller of Berkeley. "It's comparable to some of the steels."

Working with micromachines will also introduce entirely new design challenges for mechanical engineers because physical forces that are insignificant at large size become much stronger at the microscopic level. "The prediction rules that we have don't work well at the micro level," Mr. Hazelrigg says. "We don't even know what the word 'friction' means when you get down to that scale." The resistance of air to movement, for instance, seems much greater to tiny spinning objects. "Air looks like molasses to these turbines," says Dr. Gabriel at Bell Labs. As another example, the motors being designed at Berkeley and MIT use the force responsible for static electricity, rather than the magnetic forces that usually drive motors. Static electricity can be used to pick up a tiny piece of paper with a comb rubbed in fur, but it is overcome by gravity for anything larger. But for objects as small and light as the micromotors, static electricity is the dominant force.

Roger Brockett, professor of applied mathematics at Harvard University, says that while micromachines are promising for some applications like microsurgery, they might be too small and light to accomplish some tasks for which they are envisioned, such as the assembly of electronic equipment, which involves moving things that are small, but not microscopic. Applications envisioned for micromachines are "not all equally believable," he adds.

Proponents point out, however, that many applications might arise that scientists cannot envision now, just as no one was able to predict all the uses of integrated circuits when they were first developed. "Who would have ever thought that a major use of computers would have been in wristwatches?" Mr. Hazelrigg says. Micromechanics "is going to be a multibillion-dollar industry in a few years." [AP]

Q&A

Q. *How do liquid crystal displays work?*

A. Liquid crystal displays, or LCDs, used in products like forehead thermometers and watches, rely on a kind of matter that has characteristics of both a solid and a liquid, explains Dr. John L. Lewis, a chemist at the Liquid Crystal Institute at Kent State University.

The substance can be sandwiched under a transparent plate, he said, and when an electrical field from a transparent electrode acts on the molecules of the substance, they change alignment. Like large crystals, the molecules have optical properties that change depending on which way they are pointing, so that a pattern of light and dark can be seen, he explained. To display numbers, for example, the electrodes can be arranged in the familiar seven-segment design (four vertical and three horizontal), and if the correct segments are turned on, a number appears under them.

A derivative of cholesterol was the first compound found to have liquid crystal properties, in the mid-nineteenth century, but many such compounds, found in soaps, tissue membranes, the oily-appearing substance in mood rings, and so forth, tended to be unstable. The chemicals most commonly used for displays today belong to a class called cyanobiphenyls, Dr. Lewis said. These molecules have two benzene rings in a plane with a cyano group attached and an alkyl chain at the end.

Dr. Lewis said that the molecules are roughly analogous to a box of pencils with strings on the end. The "pencil" is the lined-up benzene rings. The "string" is the alkyl chain. "The two rings want to remain linear relative to one another and more or less in the same plane," he said, but the string can twist.

Like a shaken box of pencils, the molecules tend to end up in the same alignment. When an electric field is applied, then removed, the molecules switch back and forth between untwisted and twisted states.

TRAVELING LIGHT

IT seems more like magic than a laboratory demonstration. Two transparent crystals are placed at random a few feet apart, and a separate laser beam is pointed at each one. At first the beams pass straight through the crystals as if they were glass, but after a few seconds the emerging beams fan out, and the fans from the two crystals begin to merge. A moment later the fans fade away and a new beam pops into existence between the two crystals. Incredibly, the crystals have found each other and created a two-way optical link between themselves in which each crystal not only propagates its own beam but acts as a mirror for the other's beam.

The crystals in this demonstration by Dr. Jack Feinberg, a professor of physics at the University of California at Los Angeles (UCLA), are made of the compound barium titanate. They are not endowed with intelligence or will, but merely exhibit the extraordinary properties of a new class of materials called photorefractive nonlinear optics. When light shines through these peculiar substances, they respond by rearranging the electric charges in their microscopic structures so as to increase their ability to bend the light, to change its color, or even to act as mirrors. Some photorefractive crystals can "see" weak light coming from any source in their vicinity and modify themselves to reflect the light directly back. This odd behavior is more than a laboratory curiosity. It is, in the view of many physicists, the dawn of a vitally important new family of technologies.

These materials are expected to make telephone systems more efficient and cheaper, computers more flexible and intelligent, and lasers far more powerful. Photorefractive crystals are already under development as devices for storing holograms. The many organizations backing research in the field include large communication companies and the Defense Department, and discoveries and inventions are coming thick and fast. Although it has been known for decades that the refractive indexes (light-bending abilities) of some materials can be slightly changed by the passage of light, only in recent years have substances been invented or discovered whose light-bending ability changes markedly under the influence of light or electric fields. Communication engineers say that when these substances are eventually incorporated in new telephone systems based on optical fibers, the transmission of voices and information will become vastly more efficient.

Because light travels faster than electrons (and has some other advantages), optical fibers have already replaced electric wires in many telephone systems. Optical connections have begun to replace electricity in computer components as well, reducing the time it takes information to travel from one device to another. Scientists believe that optical on-and-off switches and even optical switchboards will one day replace their electronic counterparts in computers, though there is general agreement that practical optical computers are still many years away. One problem that remains to be solved is size; the crude optical switches devised so far are vastly larger than their electronic counterparts, which can be packed by the million on a thumbnail-size semiconducting chip.

Scientists at AT&T Bell Laboratories have unveiled a prototype, a hybrid computer based on chips containing labyrinths of microscopic lasers and tiny optical on-off switches ("gates") that respond to electric fields by becoming either transparent or opaque. The comparatively large devices in the Bell prototype serve the same functions that some electronic transistor switches do in conventional computers. Dr. Alan Huang, who designed the machine, says the chips containing electro-optical switches have as many as fifteen hundred layers only a few atoms thick, deposited by beams of molecules hurled at a wafer of gallium arsenide.

Many experts believe that optical materials will eventually take over many tasks now performed by semiconducting electronic devices. Dr. Peter W. E. Smith, a science director of Bellcore, a research institution in Red Bank, New Jersey, financed by telephone companies, says that if computers can ever function in ways approximating those of the human brain, photorefractive optics will be important components: "When we see a familiar face we are able to recognize it instantly, not at the end of a sequence of digital computations. The image need not even be very clear for the brain to recognize it." If a computer could store entire images as holographic patterns that could be retrieved whole rather than piece by digital piece, it might approach some of the brain's remarkable ability to recognize patterns almost instantly, he said.

Bellcore recently devised a holographic system, based partly on photorefractive substances, that can not only retain a memory of images but also recognize them and learn from its own mistakes. A principal object of the research is to develop "neural networks" of computer interconnections with some of the same associative abilities that a brain has. Photorefractive materials differ from ordinary glass or plastic, says Dr. Feinberg of UCLA, in that their microscopic structures are changed by the light passing through them, and their altered structures, in turn, change the light. All transparent materials bend beams of light by varying amounts; the amounts are expressed in terms of "refractive index." Diamond has the highest refractive index of all known substances, which accounts for the sparkle of a faceted stone. In most substances the refractive index remains fairly constant, but in photorefractive materials, light-bending ability changes radically when either an electric field or a strong beam of light is applied.

Electric fields and light beams can rearrange the positive and negative charges in photorefractive materials, and this changes their refractive indexes. In some cases two or more beams of light entering a photorefractive material interfere with each other to create microscopic bands of positive and negative charges, which form "gratings." Optical gratings, rows of microscopic lines separated by spaces the widths of the wavelengths of light, can be etched, printed, or electronically implanted on transparent material. They bend and break up light in much the way prisms diffract white light into a rainbow of colors. When such patterns are created by the interference of light beams, they can be used to store complex images encoded as holograms.

Dr. Smith and other leaders in the field say the rush is on to discover photorefractive substances with new and useful properties. Dr. Samson A. Jenekhe of the University of Rochester announced the invention of photorefractive polymer that may be the fastest optical switch yet, one whose refractive index changes radically in less than one trillionth of a second after exposure to strong laser light. The material, Dr. Jenekhe says, is a derivative of polythiophene, a chain of carbon-atom rings linked to sulfur atoms. Dr. Jenekhe, whose research is supported by the Naval Air Development Center and the Honeywell Corporation, said he believes the substance may eventually be used in optical computers, but a more immediate application will be in powerful lasers.

A problem with lasers, including those under investigation by the Strategic Defense Initiative program as antimissile weapons, is that at very high power they generate intense heat that distorts their internal light paths, dissipating and wasting energy. While a beam is building up inside a laser device, the light is reflected back and forth between two mirrors many times before it gains sufficient energy to penetrate the thin mirror at one end and emerge from the lasing device. As the beam gains energy, heat and other factors tend to distort the lasing medium inside the cavity and rob the beam of energy.

But by replacing one or both of the mirrors in the laser device with photorefractive material, the distortions can be corrected. Instead of acting as simple mirrors that reflect the beam back and forth to build its energy, the photorefractive materials change their structures and create a distorted reflection that exactly cancels the distortion within the beam itself. Photorefractive mirrors thereby "heal" the laser beams they contain, permitting the beams to reach maximum intensity.

An application of vital importance to communication companies is in optical fiber telephone lines. At present, the laser light signals carried by these fibers can travel only relatively short distances before being absorbed and weakened by their glass carriers. The signals must therefore be strengthened at frequent intervals by converting pulses of light into electric pulses, then amplifying the electric pulses electronically and reconverting them into light pulses to be sent on their way to the next amplifying station. But if signal strengthening could be achieved optically without electronic amplification, telephone systems would become vastly simpler, more efficient, and cheaper to operate. In such a system, Dr. Feinberg explains, the signal beam would draw fresh energy from a "pumping beam" directed at the medium through which the signal was passing. Thus fortified, the signal would continue on at the speed of light without having to pause for conversion and electronic amplification.

An astonishing new blend of photorefractive plastic and glass has been patented by Dr. Paras N. Prasad of the State University of New York at Buffalo, working under contract to the Air Force Office of Scientific Research. Such materials are likely to be critical elements in future communication systems and many other applications, according to Dr. Donald R. Ulrich, program manager at the Air Force Office of Scientific Research. Glass is ordinarily made by melting silica or similar compounds at high temperatures. But Dr. Prasad explains that in the "sol-gel" method, glass can be cast at room temperature from a water solution of tetramethoxysilane containing a catalyst. In his research Dr. Prasad discovered a photorefractive polymer, poly-p-phenylenevinylene, that can dissolve in the same water-based mixture from which glass is cast by the sol-gel method. When the precursors of the polymer and glass are mixed in the same solution, he found, they solidify into a glasslike substance that combines the photorefractive qualities of the polymer with the optical clarity of glass.

"We've cast it into thin films that demonstrate its potential as an optical-fiber wave guide, and it has a very fast switching time if used as an optical switch," he says. "It seems to have great potential for communications." [MWB]

OPTICAL TRICKERY

An optical system that concentrates sunlight at least 56,000-fold has been developed by scientists at the University of Chicago. A wide range of applications is envisioned, including production of electrical energy, disposal of hazardous wastes, creation of high-strength fibers, and generation of intense laser beams for industrial applications and space communications.

The system arises from the new field of "nonimaging optics." While a telescope produces an image at its focus, nonimaging optics avoids the limitations inherent in image production. It uses optical tricks, involving both reflection and refraction, to achieve a high concentration of light.

Refraction is the bending of light when it enters a material that alters its velocity. For instance, refraction makes a straight pole seem bent when partially immersed in water. In the device tested so far, a parabolic mirror sixteen inches wide focuses sunlight on a "secondary concentrator" that concentrates the energy close to the theoretical limit. The concentrator is walled with silver, an efficient reflector, and filled with a highly refractive oil. Even if the sunlight, after focus by the mirror, is not perfectly aligned with the concentrator, it is refracted by the oil and then reflected by the silver walls, which are shaped to produce extreme concentration at the bottom. Because the surface of this concentrator is convex, its edges divert light that might otherwise be lost toward the point of extreme concentration at the bottom.

Tests have shown the resulting light to be 56,000 times as intense as that of sunlight hitting the earth on a clear day. The resulting temperatures are far beyond the tolerance of an electricity-generating cell. Nevertheless, Richard Winston, a professor of physics at the University of Chicago and an author of the report, said that Stanford University and the Electric Power Research Institute are developing a similar but less intense concentrator for electricity production.

Furthermore, the Solar Energy Research Institute in Golden, Colorado, is investigating ways such beams might convert toxic industrial products, like polychlorinated biphenyls (PCBs), to harmless substances. Professor Winston also said that a group in Israel is exploring a variety of industrial applications of lasers powered by solar energy from such a concentrator. They could also be used to relay communications between satellites in orbit, where sunlight is undiminished by the atmosphere.

HOT RODS

A NEW type of small nuclear reactor is being seen by experts as a safer alternative to conventional reactors and coal- and gas-fired plants whose emissions have been linked to environmental ills like acid rain and global warming. Despite setbacks, American and foreign companies have been pushing to bring the innovative design to market. The new machine is called a modular, high-temperature, gas-cooled reactor.

Designers of the new reactor say the machine is intrinsically safe because its physical characteristics make it immune to meltdown, the most feared reactor accident. It relies on laws of nature rather than complicated machinery and error-prone caretakers to prevent major accidents.

All reactors rely on a chain reaction in which heavy atoms are split to release bursts of energy and heat. The strength of the reaction depends in part on the concentration of the fuel. The challenge is to keep the reaction under control. The key safety feature of the new design is that its fuel is more diluted than usual. It is made by forming uranium into billions of tiny grains and covering each of these with a tough ceramic shell that can withstand unusually high temperatures. This arrangement means that the fuel and its radioactive by-products are tightly sealed off from the environment. Most important, it also means that the fuel is less dense and reactive than the concentrated fuel rods in conventional reactors.

Indeed, the high-temperature, gas-cooled reactor is considered so safe that it can withstand the simultaneous failure of the rods that control the nuclear reaction, the pumps, and all cooling systems. "It can survive that accident easily," says Lawrence M. Lidsky, a professor of nuclear engineering at the Massachusetts Institute of Technology (MIT) who is investigating the new design. Robert D. Pollard, a nuclear safety engineer with the Union of Concerned Scientists, a private group critical of the nuclear power industry, responds that Mr. Lidsky's statement is too sweeping, but he calls the reactor "clearly safer" than the existing generation of nuclear power plants. "We've got to change the industry so safety is dependent on physical laws rather than complex machinery," he says, adding that the new design is a step in the right direction.

While many skeptics remain cool to nuclear power, calling it inherently dangerous, growing ranks of scientists, federal officials, and even environmental groups are reexamining nuclear power to see whether it might be more environmentally benign than fossil-fuel power plants. The burning of coal and oil produces gases that collect in the atmosphere, and most experts believe the gases trap heat as in a greenhouse, warming the earth. The allure of nuclear reactors is that they produce none of these gases. "The greenhouse effect has changed a major parameter in the antinuclear equation," says John F. Ahearne, a former chairman of the Nuclear Regulatory Commission who became vice president of Resources for the Future, a Washington-based environmental group.

So too, Senator Timothy E. Wirth, a Colorado Democrat who has introduced legislation intended to deal with the new environmental ills, called for "a fresh look at nuclear power," citing the threat of climatic destruction. The new interest in nuclear power comes a decade after the reactor business all but died in the United States. Since 1978, no American utility has ordered a new reactor.

The nuclear power plants now operating in the United States are generally water-cooled reactors. Like high-powered racing cars, they are temperamental and require constant attention. Most important, their very hot nuclear fuel has to be carefully and continuously cooled to avoid accidents.

In such reactors, half-inch-thick rods of uranium-oxide fuel get yellow-hot at their core, reaching temperatures of about 4,100 degrees Fahrenheit, according to *A Guidebook to Nuclear Reactors*, published by University of California Press. The metal casing around the fuel is cooled to the much lower temperature of about 650 degrees by circulating water. This water, which is under immense pressure, goes to a steam generator that drives turbines to generate electricity. If the water disappears for just seconds, hot fuel can rapidly destroy the metal casing. At 1,800 degrees the casing starts to break down and at 3,370 degrees it melts. The danger comes when fuel starts to lump together in a molten mass that can damage or destroy the reactor. To avoid meltdowns, engineers have devised a series of safety systems. One major line of defense is emergency cooling loops of water to keep the fuel core from melting. Another is a series of concrete containment vessels that surround the fuel core so that, even during a serious loss-of-coolant accident, no radioactivity will escape.

This "defense in depth" philosophy has proven to have serious drawbacks. "Complexity is the reason light-water reactors are so hideously expensive and hard to run," says Dr. Lidsky of MIT. The complexity and expense of safety systems have forced utilities to build very large plants that can produce more electricity and thus more revenue. Yet all the precautions are sometimes insufficient, as shown by the 1979 accident at the Three Mile Island reactor in Pennsylvania. In that accident a partial meltdown occurred and some radioactive gas was released into the environment.

Critics have long called on industry to renounce the current generation of reactors, saying that no matter how extensive the safety measures are, the machines are disasters waiting to happen. "It's bad energy policy to rely on reactors that have the potential for accidents with enormous off-site consequences on the grounds that the probability is low," says Mr. Pollard of the Union of Concerned Scientists.

The main attraction of the modular, high-temperature, gas-cooled reactor is that it is far less sensitive to changes in fuel temperatures than the conventional reactors, a safety feature that industry experts say will aid its marketplace success. "We see a very bright future," says Richard A. Dean, senior vice president of General Atomics, a San Diego–based company that helped pioneer the technology.

The design's secret lies in its tiny fuel particles, each the size of a grain of sand. These individual grains are encapsulated in multiple spheres of glassy carbon-based materials, which trap radioactive fission products and can transmit heat but remain intact up to temperatures of 3,300 degrees. The capsules each have a diameter of only one millimeter, dozens of them fitting atop a dime. For ease of handling, they are bound together into small rods or containers the size of billiard balls. Since the top temperature that fuel grains can achieve in such a matrix is about 3,000 degrees, there is no way for them to melt through the protective capsules, no matter what the accident, nuclear engineers say. In effect, the billions of tiny capsules act like the far more massive containment systems of conventional reactors.

Instead of water, the coolant in the new reactor is helium, an inert gas that transfers heat from the reactor core to electricity-making turbines. The helium can be heated to much higher temperatures than water, the "high temperature" in the reactor's name referring to coolant temperature rather than fuel temperature. This relatively high coolant temperature allows the new reactor to operate with greater efficiency than water-cooled reactors. Typically, water-cooled reactors convert nuclear heat into electricity with an efficiency of about 33 percent, whereas gas-cooled reactors in theory can do the same at efficiencies of 40 to 50 percent.

But there is a tradeoff in all this. Since the new fuel is less concentrated and safer, the reactor's core would have to be quite large to achieve the same power output as conventional reactors, which typically produce about 1,100 megawatts of electricity. However, increasing the core size would also increase its top temperatures. So for safety's sake, designers limit the core size so that reactor output is about 140 megawatts. Designs typically call for three or four of these small reactors to be strung together to make up one power plant, thus the term "modular."

"We were concerned it wouldn't be economic," says Dr. Dean, whose company has studied larger high-temperature, gas-cooled reactors for decades. "But when we saw that a lot of the engineered safety systems could be eliminated, and saw the cost savings, we realized it was very competitive."

For two decades, a very small, high-temperature, gas-cooled reactor has operated successfully in West Germany. Although its fifteen-megawatt output is too small to be of commercial interest, the reactor is ideal for research.

In a test in the fall of 1988, it was subjected to the worst possible accident: the withdrawal of all the rods that control the rate of the nuclear reaction as well as the simultaneous total loss of all coolant, actions that would destroy a conventional reactor in seconds. Instead, over the course of days, the natural processes of heat conduction and radiation kept the hottest part of the reactor core below the temperature at which the fuel would fail. "We'd play tennis, and then go back and see how the temperatures were doing," recalls John C. Cleveland, a nuclear engineer at the Oak Ridge National Laboratory in Tennessee who monitored the experiment in Germany. "The test didn't damage a thing."

In the United States, a gas-cooled (but not modular) reactor operating in Fort St. Vrain, Colorado, ran into problems unrelated to the safety aspects of its design and was forced to close, and another of the reactors, proposed for Idaho, was supposed to be used in weapons production but the plan was cancelled in the face of resistance from Idahoans who no longer saw weapons production in their backyard as desirable.   [WJB]

GOLDEN ARMOR

Gold, one of the heaviest chemical elements, is the basis of a new lightweight plastic foam to be used as a radiation shield. Scientists at Texas A&M University have found a way to intersperse gold atoms with other atoms in the long molecular chains that make up polymers. By bubbling gas through the gold polymer, it can be expanded into a light foam that reportedly shows great promise as a shielding agent against neutrons and other types of radiation.

Dr. John Fackler, director of the program, says the new polymer combines gold with triphenylphosphine, a compound of carbon, hydrogen, and phosphorus, in a form that may be suitable for making antiradiation garments. The polymer is 11 percent gold by weight, and the gold atoms in the substance efficiently scatter or absorb most forms of radiation, including X rays. Chemically incorporated into a polymer, gold is less poisonous than other heavy metals that also block radiation. (Metallic gold is not poisonous, but when incorporated into compounds it may be.) Dr. Fackler says that because gold is chemically very stable, it tends to revert to its native state from some of the compounds it forms. "We often have trouble with shiny, yellow metallic gold precipitating out of liquid compounds—just what refiners want, but the opposite of what we want."

BRILLIANT PEBBLES

RAPID advances in the miniaturization of parts for high-technology weapons have allowed the Pentagon to give its "Star Wars" anti-missile program a major new emphasis. Instead of focusing exclusively on big, bulky, costly weapons, it would now create swarms of small, cheap, brainy rockets to hurl at enemy missiles with deadly accuracy. The proposal at the center of growing excitement and debate is "Brilliant Pebbles." The idea is to sow space with ten thousand to a hundred thousand small weapons that would home in on enemy missiles and destroy them by force of impact.

Compact and smart, each weapon would be three feet long and weigh about a hundred pounds, with its "brains" in a silicon chip said to be as powerful as a supercomputer and its "eyes" in an innovative wide-angle optical sensor. These devices would track the fiery exhaust of missiles and pick targets, eliminating much of the need for outside guidance from sensor satellites and ground stations. Not coincidentally, the system would also be far cheaper than those usually envisioned for space deployment by the Strategic Defense Initiative (SDI) program. In spirit, if not form, the weapons of earlier systems resembled battle stations from the movie *Star Wars*.

"SDI is alive and well, but like everything else, it has to fit into a reduced budget," Defense Secretary Dick Cheney said in announcing the Star Wars reorientation. The emergence of miniature space weapons is seen as creating new technical and diplomatic options for the Bush administration as it ponders what stance to take on the disputed anti-missile program—and new targets for Star Wars critics as well. Vice President Dan Quayle, who has long had an interest in Star Wars issues, hailed Brilliant Pebbles as having the potential to make anti-missile defenses more affordable, more deployable in the near future, and more easily lofted into space than older technologies being investigated by the anti-missile program. "It could revolutionize much of our thinking about strategic defense," Mr. Quayle said in a speech to the Navy League of the United States, noting that more research and testing were needed to prove the feasibility of the idea.

Brilliant Pebbles is the brainchild of Lowell L. Wood, a physicist at the Lawrence Livermore National Laboratory in California who is a protégé of Edward Teller, the elder statesman of the laboratory whose early support was crucial for the birth of the anti-missile program. The two scientists have extensively lobbied both the Reagan and Bush administrations to promote the idea, on which a few million dollars have so far been spent. But some experts criticize such high-level attention for an idea that has undergone relatively little study and evaluation. "It's premature for this program to receive the kind of advertising and prominence it's gotten," says Ashton B. Carter, a physicist and professor of public policy at the Kennedy School of Government at Harvard.

Other experts say the technology is doomed to fail in ways similar to older Star Wars concepts. "It is wrong to imagine that this is a novel idea, with its flaws, if any, remaining to be discovered," says Richard L. Garwin, a prominent critic of the anti-missile program.

Whatever its faults or merits, or obscure history, the small size of Brilliant Pebbles makes it a major departure from orthodox Star Wars plans. For instance, the Pentagon's $1.5 billion Zenith Star laser is so big that in the mid-1990s it is scheduled to be launched into space in pieces and assembled there for testing. So too, the Pentagon's official plan for rudimentary deployment of space weapons calls for large homing rockets to be clustered in orbiting "garages" from which they would be fired in time of war. Over the years, critics have repeatedly attacked proposals for large space-based weapons, saying they would be extremely vulnerable to a preemptive enemy strike and dangerously dependent on complex systems of sensor and communication satellites.

In contrast, advocates say Brilliant Pebbles would be much harder to attack effectively because of the sheer number of pebbles (especially if decoys were added) and less vulnerable to disruption because it has fewer supporting links. They say its technical feasibility is intuitively understandable to anyone who has witnessed the startling size reduction of computers, radios, televisions, and tape players in recent years. "Today you get more power in a laptop computer than you used to get in a whole building's mainframe," says Robert C. Richardson 3d, a retired brigadier general in the Air Force who is a proponent of Brilliant Pebbles and an adviser to High Frontier, a group in Washington that lobbies for the anti-missile system. Such miniaturization is also happening in weaponry, advocates say. Navigation systems, lenses, computers, engines, and laser communication systems are all shrinking fast in size and expense. The upshot is dramatic projected cuts in the expected cost of a Star Wars system.

Lieutenant General James A. Abrahamson of the Air Force, former director of the Strategic Defense Initiative, said in his final Pentagon report that Brilliant Pebbles could cut the cost to $25 billion. To put that in perspective, the Star Wars program to date has consumed about $16 billion for research. Brilliant Pebbles could be perfected in two years and deployed in five, General Abrahamson said, calling it "the most compelling and immediate" way to achieve a space-based defense. Other experts call such schedules naively optimistic.

The most novel feature of Brilliant Pebbles is its optical sensor, which was recently explained in a Livermore publication, *Energy and Technology Review.* It has a spherical focal plane, fiber optics, and multiple charge-coupled devices, giving the compact, wide-angle camera extremely high resolution. "Elimination of the flat focal plane enables us to design optical systems capable of generating between one billion and 10 billion resolvable spots, hundreds of times better than the current state-of-the-art photographic lenses," the Livermore report said. For example, it noted that from an altitude of about six hundred miles "such a wide-field-of-view camera could image a land area the size of Virginia and resolve individual buildings."

The sensor's data would be digested by a computer about the size of a cigarette package that would, according to Dr. Wood of Livermore, have the processing power of a Cray-1 supercomputer. "Each pebble carries so much prior knowledge and detailed battle strategy and tactics, computes so swiftly, and sees so well that it can perform its purely defensive mission adequately with no external supervision or coaching," Dr. Wood said in a speech. The swarm might theoretically be so smart, designers say, that no two pebbles would be likely to attack the same target.

Speeding through space a few hundred miles above earth, a pebble would need only a nudge from its engines to send it hurtling toward a rising missile. During peacetime, a pebble would be housed in a protective "life jacket" equipped with solar-powered panels to keep it ready for war. Designers say pebbles would be put on war alert when ground controllers sent them a special encrypted signal.

While conceding that the idea has some intriguing features, critics say Brilliant Pebbles is nonetheless fatally flawed. The system's low cost is pure fantasy, skeptics say, as is the prospect of achieving such high levels of processing power with an on-board computer so compact and inexpensive. Years earlier, they note, Dr. Wood and Dr. Teller promised President Reagan spectacular results from the X-ray laser, an innovative Star Wars weapon that has so far failed to materialize. "Lowell Wood has a terrible track record as a technological prognosticator," says John E. Pike, a Star Wars skeptic who is head of space policy for the Federation of American Scientists, a private group based in Washington.

Jeff Garberson, a Livermore spokesman, responds that Brilliant Pebbles was not a "personal project" of Dr. Wood that relied solely on his analyses for support, but that it had undergone careful scrutiny by other Livermore scientists and the nation's defense experts. Mr. Pike disagrees, saying the proposal is riddled with obvious flaws. For instance, the sheer numbers of Brilliant Pebbles were likely to produce "traffic jams" and collision "disasters" in space. "The total number of satellites since Sputnik is about five thousand," he says. "And they're talking about putting up, at the high end, many times that."

Indeed, experts have repeatedly warned that orbiting debris already poses a hazard for astronauts. It is becoming so abundant that there could be a chain reaction in space where one collision and fragmentation would start a rising tide of destruction among billions of dollars' worth of spacecraft. Perhaps the ultimate challenge to Brilliant Pebbles, experts say, is what the Soviet Union might do. If the Soviets simply made their rockets burn faster and finish powered flight more quickly, that would dramatically cut the time a defensive weapon had to lock onto bright rocket flames, in most instances outwitting the defense.

Ray E. Kidder, a Livermore physicist who is critical of Star Wars, says Brilliant Pebbles might be effective against the current generation of slow-burning SS-18 missiles, which fire their engines for six minutes. "But," he says, "it couldn't possibly work against a fast-burn booster" that fired its engines for sixty seconds or less. Major Bill O'Connell of the Army, a spokesman for the Strategic Defense Initiative in Washington, says SDI officials felt Brilliant Pebbles would not be "seriously degraded" by fast-burn boosters, but the details of how they might outwit fast Soviet missiles are secret and cannot be publicly released. [WJB]

METEORS IN WARTIME

Experiments have demonstrated that meteors flashing through the upper atmosphere could be used for emergency communication if nuclear explosions should disrupt critical military radio links in wartime. In tests conducted by GTE Government Systems Corporation, voice messages were transmitted up to eight hundred miles with the help of meteors. Meteor-assisted communication is possible up to a range of 1,240 miles, the company said, and might be useful even in peacetime. The technique could be used to talk to aircraft crews during radio blackouts caused by such atmospheric disturbances as solar flares.

Meteor communication is based on the fact that a meteor blazing through the upper atmosphere leaves a trail of electrically charged atoms or ions from which radio waves broadcast at certain frequencies from an earth station can be reflected back to a distant earth station. An ionized meteor trail lasts only for about a third of a second, but during this brief interval GTE's radio equipment can send up and bounce back to earth a digitized and highly compressed spoken message about seventy words long.

Meteors hit the atmosphere sporadically and unpredictably at intervals ranging from a few seconds to about twelve minutes. The broadcasting system stores each message it is to transmit until the instant a radio probe detects the arrival of a suitable meteor. Then the transmitter fires its message, and a decoder at the receiving station restores the digitized information to spoken words. The Defense Department has sponsored occasional experiments with meteor communications ever since the advent of nuclear weapons, when it was discovered that nuclear blasts can disrupt normal radio communications for twenty-four hours or more. Military planners lost interest in meteors when communications satellites began operating in the 1960s, but recent developments in antisatellite weaponry have made communications satellites vulnerable.

THE BLIMP WILL RISE AGAIN

A QUARTER century after the United States Navy consigned its last dirigible airship to the scrap heap of antiquated weaponry, a gigantic new blimp is taking shape to defend America against cruise missiles, drug smugglers, and a host of other modern threats. The airship, designated by the Defense Department as the YEZ-2A, will be 425 feet long and filled with 2.5 million cubic feet of helium, and will thus be the largest blimp ever built. Equipped with powerful radar and capable of patrolling for up to five days without refueling, the YEZ-2A is intended to become a formidable sentry for the Navy, Air Force, Coast Guard, and civilian police agencies.

Supporters of the ship say that it will combine some of the most useful surveillance features of satellites, airplanes, and surface ships, and that because it is made mostly of nonmetal substances, it will be nearly invisible to enemy radar. Its own big radar antenna, mounted inside the gas bag, will be capable of giving timely warning of the approach not only of smugglers' low-flying aircraft, but also of bombers, ships, and even supersonic cruise missiles hugging the waves.

While flying one hundred feet or so above the ocean, the ship will be able to tow mine-sweeping devices that could detonate mines without endangering the blimp. Unlike helicopters, which use large amounts of fuel and power to remain stationary, the blimp could hover indefinitely over a spot where its towed listening devices detected a submarine. Military planners hope the new blimp will begin flying in 1992.

The great dirigibles of the 1930s, which had rigid aluminum skeletons to maintain their shape and to support an internal string of cylindrical gas bags, were much larger than blimps. The German rigid airship *Hindenburg,* one of the two largest airships ever built, was 804 feet long and contained more than 7 million cubic feet of hydrogen, an explosive gas. Most of the big rigid airships of the past, including the *Hindenburg,* came to grief in explosions or crashes. By contrast, helium blimps like the familiar Goodyear airships have no internal skeletons and their impermeable skins are supported entirely by the pressure of the nonflammable gas inside them. This pressure is maintained by pumping air into two or more "ballonets," large air sacks inside the blimp that can expand to fill the space left by contracting helium. (The gas that supports an airship, helium or hydrogen, contracts when the craft descends to a lower altitude or when it is cooled by passing under a cloud that blocks warming sunlight. This reduces the lift of the gas.) The four largest blimps ever built were the Navy's ZPG-3Ws, built in the late 1950s, which were 403 feet long and contained 1.5 million cubic feet of helium.

Most blimps filled with helium have proved to be safe and reliable. During World War II, only one of the Navy's 168 oceangoing blimps was destroyed by fire from an enemy submarine, and of the 89,000 ships escorted by anti-submarine blimps, none was sunk. The Navy, nonetheless, ended its airship program in 1961.

Although the design of the new YEZ-2A descends directly from that of the French airship *La France,* which first flew in 1884, it will embody technology undreamed of until recent years. Its three-deck gondola, with accommodations for a fifteen-member crew, minesweeping and submarine-detection gear, refueling hose, and other equipment, will be the first pressurized airship cabin ever built. Even at the blimp's operating ceiling near ten thousand feet, crew members will not need oxygen masks.

The British-American consortium developing the new ship, Westinghouse-Airship Industries, Inc., believes the YEZ-2A could herald a renaissance for commercial and military airships. The main obstacle, according to the company's president, Edward J. Hogan, Jr., will be "getting entrepreneurs to overcome the giggle factor." Mr. Hogan, a retired United States Navy rear admiral and former fighter pilot, says that many business executives fail to take blimps seriously as efficient cargo or passenger carriers until they have witnessed demonstrations.

Meanwhile, however, the company is backed by a $170 million Defense Department contract signed in 1987, which provides for the development and construction of the YEZ-2A, a prototype that could be the first of a series. Mr. Hogan says some aspects of the program have subsequently bogged down because of a lack of agreement between the Navy, the Air Force, and the Defense Advanced Research Projects Agency regarding their respective management roles in the blimp project.

A ground-test replica of the gondola was being built at the company's hangar in North Carolina, a former Navy blimp hangar built during World War II. The cavernous building, whose height of 220 feet will barely accommodate the YEZ-2A, is nearly the length of three football fields placed end to end. The four normal-sized Airship Industries blimps usually housed in it seem diminutive in the vast space. But these blimps, designed in England and assembled in the United States, embody features essential to future high-technology airships, including the YEZ-2A. Flight operations for Airship Industries' advanced projects office are managed by Peter A. Buckley, a former Royal Air Force pilot who learned to fly blimps while working for the Goodyear Company in Europe. He believes his company's Skyship 600 blimps, which carry sightseeing passengers, display advertising, and search for ocean pollution, are also test beds for revolutionary improvements in lighter-than-air technology.

In October of 1989, one of these blimps became the first aircraft to fly using a "fly-by-light" control system. Older blimps are controlled manually with a large wheel at the pilot's side to move the elevators and two rudder pedals, all linked by long cables to the control fins at the tail of the airship. But the latest Skyship 600 has a side-stick controller like the flight control used in the newest fighter planes and airliners. By manipulating a small joystick resembling those used in computer games, the pilot sends instructions to a computer that translates them into coded light beams sent along optical fibers. These fibers are connected to receivers and control motors at the tail of the airship that move the rudders and elevators. The system, which is much less fatiguing to a pilot than traditional controls and which can be mated to an automatic pilot, will be incorporated in the YEZ-2A. But the most important feature of Airship Industries' blimps is their ability to move straight up or down, backwards or forwards, or even to turn around while hovering. This is made possible by "vectored-thrust" engines, whose propellers can be swiveled to any up or down angle.

"Experts are always amazed at the maneuverability of these new airships," Mr. Buckley says. "Recently I was demonstrating one of them at Charles de Gaulle Airport in Paris, and I was maneuvering at low altitude between two busy runways. The traffic controller suddenly told me to stop, probably to see how rapidly I could comply in an emergency. I stopped instantly by reversing thrust, and he was duly impressed." The French-made, plastic-lined fabric from which the blimp hulls, or "envelopes," are made is nearly impermeable, and the helium in an Airship Industries blimp has to be purified or partially replenished only about twice a year, pilots said. The reinforced noses of the airships are strong enough that the blimps can fly up to ninety miles an hour without risk of damage.

Since 1980 Airship Industries has built sixteen blimps, fourteen of which are now flying: five in the United States, four in Europe, one in Korea, two in Australia, and two in Japan. The company is thus the largest airship manufacturer in the world. Goodyear continues to fly its four remaining blimps, the only survivors of more than three hundred airships Goodyear built since 1911. (In 1987, Goodyear sold its aerospace division to the Loral Corporation and thus ended its lighter-than-air tradition.)

Mr. Buckley and his associates have been testing a section of the YEZ-2A gondola that will be used for refueling. A motor-driven spool in the compartment contains a collapsible hose that can be lowered more than six hundred feet to a surface ship at sea, which would pump fuel up to the blimp. The compartment is also fitted with doors and winches. "If we were out of range of a surface tanker," Mr. Buckley says, "we could have a C-130 cargo plane drop collapsible fuel cells into the ocean. Using our winches, we could retrieve the cells and haul them into our fuel bay. With refueling and resupply, the YEZ-2A could stay aloft almost indefinitely."

The airships of the 1930s were built to accommodate passengers in style. The two-deck dining room of Britain's R-100 featured Oriental carpeting, a colonnade, and a handsome double staircase. The *Hindenburg*'s salon was furnished with an aluminum grand piano covered with yellow pigskin. Does Airship Industries hope to attain such elegance? "That kind of thing is probably a long way off," Mr. Buckley says, "but meanwhile, airships on the scale of the YEZ-2A will be able to do things that nothing else can do. We should build and use them, develop the market, and then move on to better ones." [MWB]

Curiouser
and
Curiouser

HOW REMARKABLE IS A
COINCIDENCE?

COINCIDENCES, those surprising and often eerie events that add spice to everyday life, may not be so unusual after all. After spending ten years collecting thousands of stories of coincidences and analyzing them, two Harvard statisticians report that virtually all co-incidences can be explained by some simple rules. Some of the analyses performed by them or other statisticians showed that events that looked extremely unlikely were almost to be expected. When a woman won the New Jersey Lottery twice in four months, the event was widely reported as an amazing coincidence that beat odds of 1 in 17 trillion. But when carefully analyzed, it turned out that the chance that such an event could happen to someone somewhere in the United States was more like 1 in 30.

It was an example of what the authors, Dr. Persi Diaconis, a professor of mathematics at Harvard University, and Dr. Frederick Mosteller, an emeritus mathematics professor at Harvard, call "the law of very large numbers." That long-understood law of statistics states, in their formulation: "With a large enough sample, any outrageous thing is apt to happen."

Dr. Diaconis, whose work also led to the recent discovery that seven shuffles are needed to mix a fifty-two-card deck randomly, says the findings on coincidence were meeting mixed reactions. "Some people are enormously relieved, but others are furious." Not every-one who supplied an amazing coincidence to the researchers wanted to hear that a cherished dramatic story was really nothing special. "I think the whole subject is fascinating," says Dr. Erich Lehmann, a statistician at the University of California at Berkeley. Although it can be a difficult statistical problem to decide just how unlikely an event is, Dr. Lehmann says, there is no dispute about the validity of the findings of Dr. Diaconis and Dr. Mosteller, some of which have been discussed at statistical meetings in recent years. The two Harvard statisticians reviewed a large body of calculations and analyses of coincidences performed by other researchers, and they devised new techniques and approaches for studying the phenomenon in a wide range of circumstances. The research results, Dr. Diaconis says, "are aimed at very basic problems of inference that arise in messy, real statistical problems."

Dr. Bradley Efron, a statistician at Stanford University, says coincidences arise "all the time" in statistical work. When researchers find clusters of odd cancers or birth defects or other diseases, statisticians are asked "to decide which events are the luck of the draw" and which may reflect some underlying cause, Dr. Efron says. "That's what Persi and Fred are trying to unravel. I think it's a very interesting enterprise."

Dr. Diaconis says one application of the new analyses is in scrutinizing data from clinical trials of new drugs. "As you are looking around in that mass of data, you find that in a certain subgroup, there are twice as many deaths in people taking drug A as there are in people taking drug B." Is it a coincidence or an indication that drug A is so dangerous to some patients that the trial must be stopped?

Dr. Diaconis and Dr. Mosteller say they decided to study coincidences because they were fascinated by the role these odd events play in everyone's lives. "All of us feel that our lives are driven by coincidences," Dr. Diaconis says. "Who we live with and where we work, why we do the things we do often rest on slim coincidences." These chance events "touch us very deeply." The two statisticians defined a coincidence as "a surprising concurrence of events, perceived as meaningfully related, with no apparent causal connection." Dr. Diaconis and Dr. Mosteller began with the presumption that there are no extraordinary forces outside the realm of science that are acting to produce coincidences. But they also recognized that seeming coincidences are an important source of insight in science and so should not be dismissed out of hand. What looks like a coincidence may in fact have a hidden cause that can lead to a new understanding of a phenomenon. A sequence of odd blips on a chart or a clustering of cases of a rare disease can tell researchers that a new event is occurring.

A decade ago, Dr. Diaconis and Dr. Mosteller started asking their colleagues, friends, and friends of friends to send them examples of surprising coincidences. The collection quickly mushroomed. Dr. Mosteller said he had thirteen notebooks, each three and a half inches thick, full of coincidences. "These notebooks are eating up the shelves in my den." Dr. Diaconis said he had two hundred file folders full of coincidences. When they began to study these coincidences, they learned that they fell into several distinct groups. Some coincidences have hidden causes and are thus not really coincidences at all. Others arise from psychological factors, like selective memory or sensitivities, that make people think particular events are unusual whether they are or not. But many coincidences are simply chance events that turn out to be far more likely statistically than most people imagine. The analyses often required the researchers to develop new statistical methods, but in the end almost all coincidences could be analyzed.

The law of truly large numbers, which explains the double winner of the New Jersey Lottery, says that even if there is only a one-in-a-million chance that something will happen, it will happen eventually given enough time or enough people. "It's the blade-of-grass paradox," Dr. Diaconis says. "Suppose I'm standing in a large field and I put my finger on a blade of grass. The chance that I would choose that particular blade may be one in a million. But it is certain that I will choose a blade." So if something happens to only one in a million people per day and the population of the United States is 250 million, "you expect two hundred and fifty amazing coincidences every day. If a one-in-a-million thing happens to you, you start telling people about it. You might say to me, 'So what do you think of that, wise guy?' And I say, 'It's an example of the law of truly large numbers.' "

When a New Jersey woman won the lottery twice in a four-month period, it was reported as a 1-in-17-trillion long shot. Narrowly speaking, that is correct. But as Dr. Diaconis and Dr. Mosteller reported, 1 in 17 trillion is the odds that a given person who buys a single ticket for exactly two New Jersey lotteries will win both times. The true question, they say, is, "What is the chance that some person, out of all the millions and millions of people who buy lottery tickets in the United States, hits a lottery twice in a lifetime?" That event was called "practically a sure thing" by Dr. Stephen Samuels and Dr. George McCabe, two statisticians at Purdue University. Over a seven-year period, they concluded, the odds are better than even that there will be a double lottery winner somewhere in the United States. Even over a four-month period, the odds of a double winner somewhere in the country are better than 1 in 30.

Another principle that demystifies many coincidences is what the researchers call "multiple end points"—occasions when what might qualify as a coincidence is not spelled out ahead of time and when many chance events would qualify. This could apply, for example, at a party where two people might discover that they come from the same town. This may seem a surprising coincidence. But the truth is that almost anything two strangers have in common would count as a coincidence—the same first or last names, the same birthday, the same item of clothing. "Clearly, the chances of getting a match in any of several things is bigger than if you look at just one thing," Dr. Diaconis says. He and Dr. Mosteller have developed a formula that can evaluate such problems.

Multiple-end-point coincidences often sound amazing on the surface. For example, Dr. Diaconis says, "I had a friend who said, 'Gee, I was watching a James Bond movie and there was a four-digit code on a bomb that was exactly the same as the code on my Israeli bank account.' " It sounded extraordinary, since there are 10,000 possible four-digit numbers. But Dr. Diaconis went on: "If you know a hundred and twenty numbers—Social Security numbers, bank codes, telephone numbers of friends—there are even odds that four digits of two of them will match," he says.

A third category of coincidences is those that are close but not exact. The odds of a coincidence then go up enormously. Once again, the two Harvard statisticians have developed a formula to analyze these problems. For example, with twenty-three people in a room, the chances are even that two of them will have the same birthday. But with only fourteen, there is an even chance of finding two people born within a day of each other, and if you ask instead that the birthdays match within a week, you need only seven people. "I have a friend who said, 'My daughter, myself, and my husband all have a birthday on the eleventh of a month,' " Dr. Diaconis recalls. "O.K., there are thirty categories, thirty days of the month. How many birthdays do we have to know so that three are on the same day of the month?" The answer, he calculated, is eighteen. "So if you know eighteen people, it's even odds that three will be born on the same day of the month." So the friend's birthday coincidence, he said, "is not so unusual."

By analyzing coincidences with these three principles in mind, Dr. Diaconis and Dr. Mosteller find that what looks unexpected usually turns out to be expected. No strange forces outside the realm of science are needed to explain coincidences. "Why does an educated person think there might really be something in coincidences?" Dr. Diaconis asks, and answers his own question: "No one story holds up on its own. But taken together, they mean something." If you put all these near coincidences together, doesn't it indicate that something strange is going on?

It does not, Dr. Diaconis replies, and quotes another fundamental law of logic: "A lot of flawed arguments don't produce a sound conclusion." [GK]

Q&A

Q. *Why do people often sweat after eating spicy foods such as chili peppers?*

A. Spicy foods such as chili peppers contain capsaicin, a chemical that activates nerve endings in the mouth and tongue that are normally stimulated when body temperature rises, according to Dr. Barry Green, an associate member of the Monnel Chemical Senses Center at the University of Pennsylvania. The impulses from these nerves signal the brain that the temperature has risen, which sets off a chain of physiological events that induce facial sweating. Sweating usually occurs near the stimulated area, such as the facial region, Dr. Green said.

BIRDS OF A FEATHER

NOTHING in the motion of a single bird or a single fish, no matter how graceful, can prepare a scientist for the sight of ten thousand starlings wheeling in formation over a cornfield, or a million minnows, threatened by a predator, snapping into a tight, polarized array. Yet somehow the actions of individual animals sum together in ways that researchers are only beginning to understand, creating patterns of motion so complex that they seem to have been choreographed from above. Flocks and schools have a distinctive style of behavior, acting with a fluidity and a seeming intelligence that far transcend the abilities of their members.

Vast congregations of birds, for example, are capable of turning sharply and suddenly en masse, always avoiding collisions within the flock, and zoologists now believe that such movements take place without guidance from a leader. Fish, too—their vision limited in murky seas—manage complex, seemingly instantaneous maneuvers when alarmed by an intruder. Flocking and schooling create some of life's most breathtaking spectacles, and they have been among the most difficult to explain. Zoologists, studying bird and fish behavior with the help of miles of high-speed film, have often assumed some high level of coordination. But now, gaining insight from new computer models, they see synchronized maneuvers as a surprising product of the actions of individual fish following individual rules for fleeing predators and staying clear of their neighbors. Thousands of simulated animals are programmed to fly or swim independently, and flock- and school-like behavior emerges on its own.

Despite recent progress, the science of analyzing the movements of flocks and schools is at an early stage, in which scientists have identified many of these individual patterns but have yet to understand precisely how they emerge as movements of the group. Perhaps no conventional answer will ever emerge. "The complexity is fascinating because there is so much going on, but it isn't unstructured—it isn't like dropping a thousand Superballs into a tank," says Craig W. Reynolds of Symbolics Corporation, who has modeled flocking and schooling. "The synchronization speed is pretty astounding. And since birds aren't mental giants, they can't be doing deep thinking as they fly along. They must use fairly simple rules."

Flocks and schools are more than just groups of animals clumped together, and not all species display flocking and schooling behavior. Those that do, often in response to hungry predators, are capable of high-speed motion, flying or swimming around obstacles, and abrupt course changes. Herding can be a similar phenomenon, but land-bound animals cannot match the flexibility of birds and fish liberated in the three-dimensional space of air and water. The seemingly effortless rapport of thousands of animals has driven otherwise sober scientists to talk of "thought transference" and "magnetic field perturbation."

"All kinds of crazy things have been proposed," says Frank Heppner, a University of Rhode Island zoologist. "I once put forward my own bit of insanity by proposing that there had to be some biological radio." Ornithologists have also traditionally believed that birds must be responding to cues from a leader—"because how else could you account for the near-simultaneous movement," Dr. Heppner says. Now, however, he and other zoologists believe that the motion can be explained along the lines of computer models in which no individual is a leader, or, in another sense, every individual is. "Until very, very recently, there's been no conceptual model that permitted this. You have a situation where you have some simple rules, but growing out of those simple rules, you have emergent properties that have nothing to do with the original rules."

Wingless and finless creatures are also capable of coordinated motion with no leaders, as Dr. Heppner demonstrates with groups of hundreds of students. Asked to begin applauding en masse but in synchronization, they start with a random scattering of claps, yet manage to find a coordinated tempo almost instantly, usually within two beats. Another kind of human behavior, the "wave" that rolls through masses of fans rising and sitting at a baseball stadium, may resemble the dynamics of a turning flock of birds. One zoologist, Wayne K. Potts, filmed the maneuvers of thousands of dunlin in Puget Sound, Washington, in 1983 and found that a turn propagates from one side of a flock to the other like a wave through a fluid.

The wave is fast. In the dunlin flocks, the turn spreads from bird to bird in about a seventieth of a second—three times faster than a bird's reaction time. If a bird waited until its neighbors turned, it would be much too late. Dr. Potts saw this as a perplexing challenge, in part because he had experience flying military aircraft in formation. Even simple formations, he felt, had a strong tendency to break up because of a time lag that tends to be magnified from pilot to pilot. "If the first does something abrupt, the second is already behind. If anything like that was going on in a flock of a thousand birds, it would just be total chaos immediately. It's still underappreciated how difficult a problem it is." Dr. Potts decided that birds must sense the approach of the wave from a distance and time their reactions accordingly. To test this idea—the "chorus-line hypothesis"—he provoked sudden maneuvers by shooting arrows near the edge of a flock. As expected, a turn began slowly with one or two birds and then accelerated rapidly through the rest.

Flocks and schools both execute far more complex maneuvers in the presence of a predator, and that is thought to be their evolutionary reason for being. Fish swimming in a more or less random way can instantly polarize—forming a dense school, aligned in parallel formation. And when a predator attacks, they can perform a fast expansion called the "fountain," quickly splaying outward from the intruder.

Traditionally, though their problems seem increasingly similar, flock experts and school experts have not worked together. Predators act differently in water and air, and the problems of how individual creatures perceive one another tend to be quite different. While birds rely on vision, for example, fish have another sense organ, known as the lateral line—"a sense of distant touch," as Julia Parrish of the University of California at Los Angeles (UCLA) puts it. The lateral line allows them to respond sensitively to changes in water pressure caused by nearby motion.

Experimentally, those studying schools have had an advantage. Fish can be observed in laboratory tanks. "Whereas we're kind of stuck with a field situation," says Dr. Heppner, "because nobody has figured out how to get ten thousand starlings in a big cage and get them to do anything." Even school experts, though, face daunting levels of ignorance. Researchers need to find out how much switching and shifting places of individuals takes place in a smoothly swimming school, for example. They wonder how different types and sizes of fish sort themselves out within a school. Technologically, such questions pose a challenge. Watching the overall shape of a school or flock is one thing; picking out the paths of individual members is quite another, only now coming within reach of computer tracking systems. "I think there's going to be a renaissance in studies of these groups," says William M. Hamner, a zoologist at UCLA. "One needs to model them from a series of individual behaviors. One wants to take those individual behaviors and construct a whole from them."

Neither zoologists nor computer modelers know exactly what rules guide the motions of individual animals. In their models, they experiment with different possibilities. They do know that birds and fish must try to keep some minimal distance away from their neighbors, and they know that stragglers on the outskirts of a flock put themselves at risk if they stray. Dr. Potts found from analyzing his films of dunlin that a left turn, say, starts not with the birds on the left side leading away, but rather with one or more birds turning inward from the right flank. That makes sense, he said, for individual birds fearing falcons and other raptors. "The time when a bird is most vulnerable is when he's away," Dr. Potts says, "so the last thing he wants to do is turn outward."

Dr. Reynolds used a computer model to mimic such maneuvers. Instead of directing the motion of the whole flock, he allowed each bird to fly at certain speeds, moving on the basis of information about a limited number of neighbors. "You get this fluid-seeming motion," he says. "The flock becomes a big amorphous thing that changes shape like a jellyfish, but there isn't a central mind, there isn't a flock mind." The human tendency to see the whole as a coherent, willful entity has been misleading, Dr. Hamner believes. The coordinated motion of a school or flock does not imply purposeful coordination on the part of individuals, and the feeling of purpose may be deceptive.

Looking down at freeway traffic from atop a skyscraper with cars smoothly weaving in and out, one has to fight the illusion that all the cars are cooperating. On a more intimate scale, traffic does not seem quite so well planned. "Flocks form patterns and the patterns entrain our brain," Dr. Hamner says. "We like patterns—we like patterns in waves, and we like patterns in a fire, and we see a flock of birds in the sky and we see a pattern in the overall movement. That's the beauty of the whole system, but it's also the thing that screws up human investigators." [JG]

Q&A

Q. *How do people who have been deaf from birth articulate their thoughts internally?*

A. "Since Aristotle, there have been arguments on all kinds of fronts related to the need for language in order to think," says Dr. Howard Busby, an administrator at Gallaudet College in Washington, D.C., which serves deaf students.

"Aristotle himself said deaf people would never learn language or to think," says Dr. Busby, who has been deaf since birth. "Some people still think that deaf people don't think. But deaf people do think and articulate their thoughts with a variety of languages ranging from English to American Sign Language. Most people visualize, in their mind's eye, so to speak. What actually makes them think that they're thinking in words is when they try to articulate it. The description of those thoughts gives language to them. The language you choose to describe them forms the shapes of those thoughts. If you articulate them in English, you assume that you thought them in English.

"Think of a high jumper, for example. He's standing back, he's looking at where he's going to jump; he doesn't tell himself, 'I am going to run this way, I am going to stop right here, I am going to spring on my right foot.' He doesn't really say it. He pictures it instead.

"I am signing to you in English," says Dr. Busby, whose remarks in a telephone interview were translated by Susan Newburger, a member of the college staff. "I can probably convey to you in English what I am thinking. If I were using Ameslan [American Sign Language, or ASL], maybe it would be a little different. It doesn't mean I was thinking in English or thinking in ASL. It means I had categories of concepts that I drew on. I put it into appropriate language."

SECRETS OF THE SUCCESSFUL CAT

THE inscrutability of the cat, long a creature of mystery to humans, is slowly yielding to a vigorous new effort among scientists to fathom this wildest and most independent of domestic animals. The scientists are finding, among other things, that household cats display an extraordinary flexibility that enables them to deal with the potentially schizoid state in which they find themselves: domesticated on the outside, but wild at heart. On one side of that duality, the experts are finding that cats are more sociable than popular myth would have it. Kittens can easily be conditioned to become friendly and affectionate rather than aloof. But on the other side, housecats are being revealed as such efficient killers, even when well-fed, that they often have a major effect on their surrounding ecosystems.

A cat can instantly change from a purring ball of fur to a merciless stalker of prey, animal behaviorists say, because its nervous system is equipped for a range of divergent responses to the world around it, any of which can be tripped at any moment by the right stimulus. This characteristic has made cats unusually adaptable, able to live successfully in a one-room apartment and on a 150-acre farm. Some animal behaviorists have also concluded that cats display as much range and variety in personality and temperament for their species as humans do for theirs, and as much quirkiness, both maddening and delightful. Cats, for instance, are highly sensitive to the actions and moods of people and are capable of sulks and snits when they feel slighted or deserted.

The experts have also developed startling evidence of the cat's renowned ability to survive, this time in the particular setting of New York City, where cats, particularly in the summer, are prone to fall from open windows in tall buildings. Researchers call this phenomenon feline high-rise syndrome. One study involved 132 cats that accidentally fell distances from two to thirty-two stories; it found that 107 of the cats lived to purr and pounce another day. The cat that fell thirty-two stories was one of the survivors.

Cats are now believed to outnumber dogs as pets in the United States and some other Western countries, reversing a historical trend. With the rise of the two-income family, cats have surpassed dogs as the pet of choice for many city dwellers; they do not have to be walked and can be left alone indoors all day. "This is the genetically engineered pet for working people," says Phyllis Wright, the vice president for companion animals of the Humane Society of the United States. The trend, in turn, has stimulated the new interest in learning more about the cat's nature, a subject long treated in literature and song but less often a subject of science. "For a long time the focus was on dogs; now it's turned around," says Dr. Benjamin L. Hart, a professor of physiology and behavior at the University of California at Davis who is an authority on both cats and dogs. He and his wife, Lynette A. Hart, the director of the university's human-animal program, have written a textbook for feline and canine behavioral therapists.

The number of therapists who deal with cats is growing, and they are finding it a lucrative field. Because of the cat's inscrutability, many first-time cat owners (and even veteran owners) misunderstand their pets, the experts say. Often, conflict arises because the cat is simply being a cat. "A cat playing is normal behavior; a cat playing on your head at four o'clock in the morning is not appropriate behavior" from a human point of view, says Dr. Dale Olm, a veterinarian and animal behavioral consultant at the New York hospital of the American Society for the Prevention of Cruelty to Animals.

Misunderstandings about the cat's nature can create false expectations that lead the owner to abandon the animal or take it to a shelter. Half the animals in many shelters today are cats, about double the proportion in the early 1970s. Of all the domesticated species, only the cat lived a solitary life in the wild. And it remains fundamentally an asocial animal incapable of being dominated by humans. This characteristic is widely seen as a major reason why so many people dislike cats—and also why others value them. "A dog or a horse can be admonished or struck and it will become submissive," Dr. Hart says. "You can't do that with a cat. They do not go into a subordinate role as a dog or a horse does. They just fight back."

Cats also differ in another major aspect of their relationship with people, and it is here that scientists have discovered a way to break down the wall of aloofness. Dogs see humans as members of the pack, as companions, scientists say, whereas cats, with their solitary heritage, do not. A common view among scientists is that what cats see in humans, if they are going to see anything, is a stand-in for the mother. The humans are perceived as fulfilling the mother's role as provider and care-giver, especially when a cat comes to live in a house when it is young. One small sign of this is the characteristic bit of behavior in which a cat "kneads" its owner's lap with its paws. It is the same behavior that accompanies nursing. Another is the fact that cats purr when humans pet them, "which is what a kitten does when it's around its mother," Dr. Hart says. "But two adult cats don't normally get together and purr with one another," he said. Adult dogs, on the other hand, use the same signals on humans that they use on each other, like wagging the tail.

Animal behaviorists have found that most cats can be shaped into affectionate, friendly creatures by simply petting and handling them a lot from the age of two to seven weeks, a sensitive time in the development of social relations. This should occur before a kitten is separated from its mother, and the handling usually has to be done by the mother's owner. "Once a cat has had this experience, it will remain a cat that is friendly to humans for its entire life," says Dennis C. Turner, a lecturer in ethology and animal research at the University of Zurich-Irchel, Switzerland. "And if the kitten has missed this experience, it will remain shy of people the rest of its life."

Dr. Turner is the co-editor with Patrick Bateson, a professor of ethology at Cambridge University, of *The Domestic Cat: The Biology of Its Behavior*. The cat "is exquisitely sensitive to the behavior of other individuals and, when it is kept as a pet, to the actions and moods of its human owner," Dr. Turner and Dr. Bateson wrote. Because of this, problems can result when a major change in the relationship takes place. "When our behavior does change from what they expect, it can affect the cat," Dr. Hart says. Leaving for a trip or changing working hours, "whatever it is, the cat is going to see it as a major change in its environment. It becomes upset." The resulting problems can include urinating outside the litter box—on the owner's shoes, for instance. Some cats mope and later punish absent owners by ignoring them when they return. Not all cats would react this way. "Some cats adapt to people coming and going and pretty soon it doesn't bother them," Dr. Hart says. This is another way of saying that where particulars are concerned, it is as difficult to generalize about cats as it is about people.

Some differences among cats are genetic. Siamese, for instance, are friendlier than the Abyssinian, which is more retiring, fearful, and nervous. Researchers studying cats living in the wild have found that the characteristics of a cat's father partly determine those of its offspring. Researchers have also identified at least three broad personality types among cats. One is described as sociable, confident, easygoing, and trusting. The second is characterized as timid, nervous, shy, and unfriendly, while the third is active and aggressive. There are countless varieties of feline temperament and style in these categories, all reflecting the cat's flexibility. Most people, experts suspect, select a cat based on coloring and general appearance. Considerations of personality are seldom involved. Careful investigation, including the circumstances of the cat's early rearing, is recommended before making a choice.

One of the hottest areas of behavioral research involves the interaction between cats and humans, and some of the findings are being converted to practical use by counselors. Researchers have found, for instance, that problems can develop between cat and human even in the best of circumstances. Often the problems are soluble. The cat that jumps on its owner's face at four A.M., for instance, might be used to being fed in the morning and is trying to rush the owner into getting up and putting out the food. The solution: change the cat's feeding time. "The cat goes crazy for about a week," Dr. Olm says, but it will usually adjust.

The most frequent behavioral problem involves the litter box. The experts have learned that a cat simply will not use a litter box that it thinks is dirty. If the litter box is not changed often enough to suit the cat, it will do its business elsewhere. Cats at times balk at the kind of litter used. Some litter is meant to be aromatic, but cats can reject it. Sometimes there are too many cats using one litter box. In one case, a cat would not use the box because the family dog sat by the box and kept staring. Two other common problems are the spraying of urine and clawing of furniture. Therapists have devised ways to keep cats from clawing furniture, usually by manipulating the cat into transferring its activities to a scratching post. Scratching is an innate behavior intended not so much to sharpen claws, some scientists say, as to leave marks and odors that advertise a cat's presence for the benefit of other cats. Experts are divided on declawing. Some say that it amounts to abuse and should be outlawed, while others say that it is an acceptable last resort that does not, as is sometimes believed, seriously impair a cat's ability to defend itself.

Urine spraying, likewise an innate behavioral pattern of males and females, is also believed to be a marking maneuver. But contrary to conventional thought, some experts believe the cat is not staking out territory to defend, but is merely letting other cats know of its presence. Spraying is not easy to control by punishment or management. As one measure, the Harts recommend booby-trapping the spray area with upside-down mousetraps. Injections or oral doses of the hormone progestin are sometimes used to eliminate the behavior, although the effectiveness is limited.

Problems often arise because owners treat cats as if they had human personalities. "An anthropomorphic view of the cat is very common among owners," two Swiss ethologists, Claudia Mertens and Rosemarie Schar, wrote in the Turner-Bateson book, "but this fosters misunderstanding. A cat should always be seen as a cat, not a human being." [WKS]

Q&A

Q. *Why are days in January colder than those in December, though the amount of sunlight (and hence heat) is steadily increasing as the new year progresses?*

A. "Most of the heat in the atmosphere is dependent on the oceans," says Gary Petti, a meteorologist at the National Weather Service. "The oceans store most of the heat that comes from the sun. But water changes temperature much more slowly than land does. As the sun moves north, the water heats up, but the atmosphere lags behind the sun." Moreover, by midwinter there has been very little sunlight in the Arctic areas for several months. "Cold air that just sits over that area builds and builds and builds," he says. "If the jet stream takes a dip, that allows the cold air to come down." Until the oceans warm up and lengthening daily sunlight begins to dissipate the Arctic cold, average temperatures in North America continue to drop. The influence of the oceans on temperatures is also seen in other seasons, according to Mr. Petti. "The same thing happens in the fall. It's not unusual to have warm weather in September, which is an equinox, just as March is. But we don't have October weather in February."

THE ENIGMA OF GENIUS

IN some ways, mathematicians are finally beginning to penetrate the mind of Srinivasa Ramanujan. A little more than a century has passed since Ramanujan was born in Erode, near the small city of Kumbakonam in southern India. When he died thirty-two years later, he left a strange, raw legacy, about four thousand formulas written on the pages of three notebooks and some scrap paper. Some of the power and originality of Ramanujan's mathematics was understood a few years before his death. His contemporaries saw from the theorems scrawled across his pages that he possessed a genius for calculating the hidden laws and relationships that govern the wilderness of numbers. But Ramanujan was uneducated in standard mathematics and isolated by geography for most of his productive life. Often his formulas seemed as obscure as they were elegant. He worked in a place of his own and a way of his own, drawing his formulas and theorems from a mental landscape that remained far from the frontier of mathematics as it was seen in his day.

Now his work is flowing into mathematics and science more deeply than could have been imagined a generation ago. Computers, with special programs to manipulate algebraic quantities, have made it possible for more ordinary mathematicians to pick up the trail of his thought. And modern physics, from the superstring theory of cosmology to the statistical mechanics of complicated molecular systems, finds itself turning more and more often to the pure findings of number theory and complex analysis—the worlds of Ramanujan. So researchers are intensifying a process of forensic mathematics, or mathematical archaeology—poring over the rough pages, trying to understand the formulas and prove them. As they learn more of why Ramanujan chose particular paths, they sense a foundation that has not yet been revealed. "When he pulled extraordinary objects out of the air, they weren't just curiosities but they were the right things," says Jonathan M. Borwein of Dalhousie University in Halifax, Nova Scotia, one of many mathematicians who have lately found themselves turning to Ramanujan's formulas. "They are elusive evidence of a theory that's lurking around somewhere that he never made explicit."

The trail is hard to follow. Out of necessity and then perhaps out of habit, Ramanujan worked in a style that awes and frustrates modern mathematicians. He used a slate, jotting down formulas, erasing them with his elbow, jotting down more, and then recording a result in his precious notebook only when it had reached final form. The intermediate results, the links of the chain, are lost. Unlike mainstream mathematicians, he felt no need to prove that a result was true. His legacy is simply a set of discoveries. "He seems to have functioned in a way unlike anybody else we know of," Dr. Borwein says. "He had such a feel for things that they just flowed out of his brain. Perhaps he didn't see them in any way that's translatable. It's like watching somebody at a feast you haven't been invited to."

So mathematicians have spent years—often valuable and productive years—proving theorems that Ramanujan knew to be true. Deriving the formulas has often been more illuminating than the formulas themselves. Whole new subdisciplines within mathematics have blossomed around ideas that Ramanujan put forward in a peculiar, stark isolation. With the special excuse of his centennial year, mathematicians gathered in 1987 to discuss the implications of Ramanujan's work at meetings in the United States and India. They have far more raw material to work with than ever before, because recent years have brought a new effort to find and organize the pages that make up his legacy.

A University of Illinois mathematician, Bruce Berndt, has spent years editing the notebooks, tracking down sources and relationships, and, above all, proving as many of the unproved theorems as possible. A mathematician at Pennsylvania State University, George Andrews, has been performing the same task with the so-called Lost Notebook, 130 pages of scrap paper from the last year of Ramanujan's life. "The work of that one year, while he was dying, was the equivalent of a lifetime of work for a very great mathematician," says Richard Askey of the University of Wisconsin, who has collaborated with Dr. Andrews in trying to understand some of Ramanujan's work. "What he accomplished was unbelievable," Dr. Askey says. "If it were in a novel, nobody would believe it."

Ramanujan might have died in complete obscurity if he had not written a series of desperate, bold letters to English mathematicians in 1912 and 1913. By then he was twenty-five years old, working as a clerk after several years of unemployment, unwilling to put aside his slate and formulas. His family was Hindu, high-caste but poor. His father and grandfather before him worked as clerks for cloth merchants. Ramanujan was lucky enough to have a fairly good high school education in Kumbakonam, and he began his creative exploration of mathematics after discovering the few outdated and second-rank textbooks in the library there.

His intellect stood out clearly, but in college at Madras, about 150 miles north of his birthplace, he failed again and again to pass examinations in other subjects. In mathematics itself, he had no teacher. He worked, as the English mathematician Godfrey H. Hardy later said, "in practically complete ignorance of modern European mathematics." Hardy was not the first mathematician to receive a letter from this "unknown Hindu clerk," as he recalled—"at the best, a half-educated Indian." But he was the first to understand what the letter contained. Ramanujan's letters said, in effect, "I know the following . . . and I also know this . . . and, by the way, I have discovered this." He offered a carefully chosen selection of his theorems. Most were in the form of identities—statements that some familiar quantity, like pi, was equal to some unfamiliar quantity, or that two unfamiliar quantities were equal. Hardy examined them with bewilder- ment. A few struck chords of recognition, he said later; he thought he had proved similar statements himself. Some he thought he could prove if he tried—and he succeeded, although with surprising difficulty. Other theorems were already known. Still others, however, "defeated me completely," Hardy said in an essay years later. "I had never seen anything in the least like them before," he said. "A single look at them is enough to show that they could only be written down by a mathematician of the highest class. They must be true because, if they were not true, no one would have had the imagination to invent them."

Furthermore, Hardy could tell that Ramanujan was holding some things back, offering specific examples of theorems for which he surely must have discovered more general versions. He arranged an invitation to Cambridge University, and in 1913 Ramanujan arrived, leaving his wife behind. He stayed for just under five years. The two men collaborated often. Hardy remembered a slight man, of medium height, with eyes through which some light seemed to shine. Ramanujan remained a strict vegetarian, cooking all his own food in his rooms, and when he fell mysteriously ill in 1917, Hardy thought his vegetarianism contributed to his failing health.

Years later, Hardy took some pains to dispel the idea, perhaps a by-product of subtle English racism, that Ramanujan was some sort of Asian curiosity—either an "inspired idiot" or "some mysterious manifestation of the immemorial wisdom of the East." On the contrary, in Hardy's eyes he was a deliberate rationalist, often shrewd, and not nearly so religious as his dietary habits made him appear. They shared a fascination with numbers as almost living things or characters in a story.

One day after Ramanujan fell ill, Hardy visited him by taxicab and remarked that the cab's number had been rather uninteresting— 1729, or 7 × 13 × 19. "No, it is a very interesting number," Ramanujan responded, as Hardy later told the story. "It is the smallest number expressible as a sum of two cubes in two different ways." (It is the sum of 1 × 1 × 1 and 12 × 12 × 12, and it is also the sum of 9 × 9 × 9 and 10 × 10 × 10.) Hardy understood and appreciated Ramanujan more than any of his contemporaries. But even he could not see beyond the blinkers of his time and place. To him, Ramanujan's story was ultimately a tragedy—of inadequate education and of genius unguided. When he finally came to assess the younger mathematician's work and its likely influence on the future of his subject, he expressed disappointment. "It has not the simplicity and the inevitableness of the very greatest work," Hardy wrote in 1927. "It would be greater if it were less strange."

Few mathematicians accept that assessment today, as strangeness comes into the light and Hardy recedes into Ramanujan's greater shadow. "Hardy thought it was a shame that Ramanujan wasn't born a hundred years earlier," Dr. Askey says. That was the great age of formulas, the era of ground-laying work by such mathematicians as Leonhard Euler and Carl Gauss. "My comment is that it's a shame Ramanujan wasn't born a hundred years later. We're trying to do problems in several variables now—the problems are harder, and it would be marvelous to have somebody with his intuition to help us get started."

Not that his intuition was infallible. Ramanujan made some errors, once claiming to have found a formula for the approximate number of primes less than any given number. No such formula exists. He was too optimistic, and it was the optimism of an earlier time; by the nineteenth century, mathematicians had learned that some problems could never be solved, but Ramanujan's isolation shielded him from their doubts as much as from their knowledge.

In 1919, increasingly ill, having entered and left a nursing home and several sanitariums, Ramanujan returned to India. He continued to work feverishly, fighting the pain of his mysterious ailment, writing on whatever paper he could find. The next April, at the age of thirty-two, he died. The work of his last year, 130 unlabeled pages, came to rest at the library of Trinity College, Cambridge, where it lay in a box, along with assorted bills and letters, until Dr. Andrews of Pennsylvania State University found it in 1976. This was the Lost Notebook. "It's a bizarre term to use for something that was in the major library of the major college in England," Dr. Borwein says, "but in terms of people appreciating its contents, it was certainly true."

Dr. Andrews found that Ramanujan had cleared a path that mathematicians had not succeeded in matching in the intervening half century. Many discoveries concerned a family of identities he called mock theta functions—"simple assertions in arithmetic," as Dr. Andrews puts it, although "their implications are quite profound."

Such mathematics has helped drive one of the major new conceptions of theoretical physics, superstring theory, as the physicist Freeman Dyson told a Ramanujan conference. "As pure mathematics, it is as beautiful as any of the other flowers that grew from seeds that ripened in Ramanujan's garden," he said. Another identity was used to enable a computer to calculate millions of digits of pi. It converges on the exact value with far greater efficiency than any previous method. Yet, as always, Ramanujan had merely asserted his discovery; only later did Dr. Borwein and his brother, Peter B. Borwein, prove rigorously that those millions of digits really were pi. The applications of Ramanujan's magical-seeming formulas make mathematicians think that he was mining a deep vein of theory, the full outlines of which are not yet known. But many prefer not to dwell on just how Ramanujan was able to think as he did.

Hardy looked at Ramanujan's origins and saw a crippling neglect by an inadequate educational system cut off from European society. Still, as mathematicians realize now, Ramanujan had a decent high school, a handful of books, and the traditions of a culture that allowed him to aspire to a life as a scholar. Those looking for lessons in his brief, rich life sometimes note that now, one century later, much of the planet lacks that much. "Ramanujan is important not just as a mathematician but because of what he tells us that the human mind can do," Dr. Askey says. "Someone with his ability is so rare and so precious that we can't afford to lose them. A genius can arise anywhere in the world." [JG]

HOW TO REMEMBER PI

"How I want a drink, alcoholic of course, after the heavy lectures involving quantum mechanics!"

No, the foregoing plaint is not the latest evidence that faltering American physics students are succumbing to the lure of demon rum. It merely exemplifies a growing body of mnemonic phrases that are supposed to help people remember the value of pi—the ratio between the circumference and diameter of a circle. If the number of letters in each word is counted as a single digit, then the sentence reads: "3.14159265358979," the approximate value of pi.

Mathematicians and scientists readily acknowledge that such putative memory joggers are generally harder to remember than the numbers for which they stand. Moreover, in the case of pi, there is scarcely ever any need to know more than the first half dozen digits after the decimal point. Nevertheless, the writing of phrases, poems, and even songs embodying the numerical value of pi has become a kind of sport for pi enthusiasts. In an issue of *Mathematics Magazine* a Venezuelan mathematician issued an appeal for new pi mnemonics in languages other than English, French, German, Spanish, and Greek. He has already collected examples in those tongues.

In his article, Dr. Dario Castellanos of the University of Carabobo in Valencia, Venezuela, discusses the peculiar fascination pi has exerted on professional and amateur mathematicians over the centuries. Part of the number's appeal is that it is transcendental: the virtually random sequence of numbers following the decimal point is believed to be infinite.

Nevertheless, improved mathematical techniques have enabled researchers to calculate the value of pi to an immense degree of accuracy. With the help of a supercomputer and a mathematical tool called a quadratically converging algorithm, Dr. Yasumasa Kanada of the University of Tokyo established a record for pi. He calculated its value to 134,217,728 decimal places.

Mnemonics have yet to be devised that could stand for really long approximations of pi, but for reasons best known to themselves, people continue to compose pi prose. Dr. Castellanos noted that in 1985, A. K. Dewdney, author of the computer column in *Scientific American*, invited pi mnemonics from readers and received the following twenty-decimal-place example from Peter M. Brigham of Brighton, Massachusetts: "How I wish I could enumerate pi easily, since all these (censored) mnemonics prevent recalling any of pi's sequence more simply."

Pi mnemonics from Europe often honor the ancient Greek scientist Archimedes, who calculated pi to four decimal places before deciding that he had had enough. For example, Dr. Castellanos quotes a mnemonic poem in French encoding thirty decimal places of pi, which appeared in 1879 in a Belgian mathematical journal: *Que j'aime à faire apprendre un nombre utile aux sages! / Immortel Archimède, artiste ingenieur / Qui de ton jugement peut priser la valeur? / Pour moi ton problème eut de pareils avantages.* (How I love to learn a number useful to the sages! / Immortal Archimedes, artist engineer, / Who can put a value on thy judgment? / For me thy problem would have such advantages.)

WHY DOES THE KNUCKLEBALL BEHAVE THAT WAY?

RESEARCHERS armed with computer and wind tunnel are finally taking the measure of baseball's most demonic challenge to science, the knuckleball. Amid a crime wave of scuffed, jellied, and sandpapered baseballs, scientists find that the unassisted knuckleball—a perfectly legal, disarmingly slow sort of pitch—can fly in trajectories as bizarre as any of its illegal cousins. Recent simulations show, for example, that a knuckleball can veer more than once in flight, a piece of baseball lore that scientific tradition had long doubted. The simulations also appear to show why. Scientists are focusing on the role of the baseball's stitches in creating turbulence in the airflow, and they calculate that, under certain conditions, the knuckleball undergoes an aerodynamic "crisis" that warps its trajectory in midflight.

Enemy of athletic grace, saboteur of scientific predictability, fly in the aerodynamic ointment, this obstreperous pitch has long confounded the intuitions of those who study it, physicists as well as batters. The pitch is hard to throw: a pitcher must dig his first two fingers into the ball, his knuckles protruding, and launch it forward with as little spin as possible. Those who master it, though, have managed to prolong careers well into middle age. A properly hurled knuckleball seems to make up its own rules on its way to the plate. It floats, flutters, twitches, lurches, and dives. "It actually giggles at you as it goes by," a batter once said of Phil Niekro's knuckleball.

Scientists know that such phenomena must obey the laws of nature, but some physicists, unable to account for complex trajectories, have contended that the relevant laws are those of psychology. They suggested that frustrated batters and confused catchers—who lunge after knuckleballs with their special, comically oversized mitts—were victims of optical illusion. "Of course, anybody who's ever tried to hit one doesn't believe that," says Cliff Frohlich, a physicist at the University of Texas who has studied a variety of sports problems. "The reason you come up with such an idea is that you don't have a good physical explanation for it." Now science is catching up to reality. Improved computer simulations, combined with some clever guesses about the precise role of asymmetical stitches in the airflow, are providing the clearest picture to date of the knuckleball's bewildering dance.

Good pitching is almost always a business of shattering expectations—following fast pitches with slow, inside with outside. Only the knuckleball, however, sets up expectations, confounds them, renews them, and betrays them in the course of a single pitch. "I've seen knuckleballs break up and down on the same pitch," says Rick Cerone, who has caught (or tried to catch) plenty for major league teams. "I've seen knuckleballs break across and then break back."

In the days of such pioneers as Hoyt Wilhelm, who rode his knuckleball to the Hall of Fame, students of the pitch suggested that the lack of spin made the ball susceptible to every puff of wind. That idea has a grain of truth: a rapidly spinning ball does gain directional stability, like a gyroscope. But the knuckleball flutters in still air, too. Gusts alone do not explain its behavior.

A decade ago, a Tulane University engineer, Robert G. Watts, carried out the first detailed wind-tunnel experiments meant to gauge the influence of the seams—108 stitches of red, waxed thread that join two pieces of cowhide to form the baseball's cover. The stitches protrude into the airflow, providing a source of roughness greater than any scuff or nick. For most pitches, however, spin makes the location of the stitches irrelevant. "If it's spinning fast, then it looks to the air just like a rough sphere," Dr. Watts says, "and it doesn't make any difference how the strings are aligned." For a ball held stationary in a wind tunnel, experiments showed that the stitches could impart a sideways force. When aligned asymmetrically, so that the left and right sides of the ball present a different face to the air, the stitches disrupt the flow enough to create a pressure difference from side to side.

Sideways force alone does not explain the knuckleball, however. The ordinary curveball, breaking to the left or right, also relies on a lateral force, caused by its rapid spin. The sideways spin lowers the pressure on one side and raises it on the other. Similarly, a fastball with backspin gains lift, while a ball with topspin dives downward. The effect of spin is potent. A ball spinning at 1,800 revolutions per minute—a ball that will turn about fifteen times in its 60-foot, 6-inch journey to the plate—will feel a sideways force of more than an ounce. That will turn its path by about one and a half feet. And although the pressure is steady, the deflection is not. The sideways pressure is an accelerating force, so most of the curve comes during the last quarter of the journey, when the batter is presumably already in full stride. "The curvature is much greater as the ball gets closer to the plate," says Peter J. Brancazio, a Brooklyn College physicist and the author of *Sport Science*. A scuffed patch on the ball can also create sideways pressure, but how much is a matter of dispute. "I really wonder whether sandpapering one side of the ball can make as much of a difference as the players think," Dr. Brancazio says.

The knuckleball is another matter. It spins far too slowly to compete with a curveball on its own terms. The sideways force comes from the stitches instead. But as pitchers know and scientists have rediscovered, a knuckleball with zero spin is ineffectual—a slow roll, even if just a few degrees per second, gives the knuckleball its kinks. If the spin is just right, gradually changing the ball's orientation on its way across the diamond, the knuckleball accomplishes a midcourse correction. At first the seams' aerodynamic influence might be pushing the ball outside, away from the plate; then, suddenly, a small shift in the seams might reverse the force, causing the ball to plunge back across the inside corner. A piece of the puzzle comes from Joel W. Hollenberg of the Cooper Union School of Engineering, who is combining measurements on balls in a small wind tunnel with detailed three-dimensional computer simulations. His model calculates the effect of speed when the pitch is released and the angle of the release. It considers a variable release point, perhaps five and a half feet above the ground and three feet to the side of the pitching rubber. It needs to know the spin rate and the starting orientation of the stitches. And it builds in information about atmospheric conditions: wind, pressure, temperature, and humidity.

Dr. Hollenberg finds that tiny, almost imperceptible changes in some of these variables can drastically alter a trajectory. One typical pitch, thrown at seventy miles per hour, breaks about two feet inward when the stitches are pointed at an angle of sixty degrees. When the angle is reduced to fifty-five degrees, it breaks a bit farther inward. But when the angle is reduced to fifty degrees, the change is drastic—the pitch curves nearly four feet in the opposite direction. "The role of the stitches has been insufficiently appreciated," Dr. Hollenberg says. "It turns out there are still some very interesting things going on in that flow around the baseball."

Yet another effect revealed by his calculations adds to the complexity. At certain speeds, the changing orientation of the stitches causes the airflow to pull suddenly away from the leather surface, enlarging the wake behind the ball and sharply increasing the drag. That transition, a drag crisis, causes the ball to slow and plunge downward. So there is nothing random about the knuckleball's flightiness. Indeed, Dr. Hollenberg contends that he could turn his research into a practical offensive weapon—for pitchers, of course. He imagines setting up small computers "on location."

"We could use the program to look for the conditions that would take advantage of the known weaknesses of the batters," he says. But so far, he has not succeeded in attracting the interest of managers. The sensitivity of this unstable pitch to tiny variations in control may undermine Dr. Hollenberg's plan, and it may also explain why so few active players employ the knuckleball. In recent years, the major leagues had only four pitchers who relied almost totally on it for their livelihood.

Like knuckleballers, researchers toiling at the crossroads of science and sports are few but dedicated. They are well aware that their work strikes some colleagues as mundane if quirky, and they struggle for time on supercomputers and high-power wind tunnels. They know they are not contending for Nobel Prizes. More practical research in aerodynamics tends to focus on smoother surfaces and higher airspeeds. Even so, baseball researchers say their work is neither easy nor irrelevant. All kinds of physicists increasingly recognize that they must reckon with the surprising stubbornness of phenomena like the knuckleball, phenomena of everyday scales and everyday materials.

The problem of small instabilities governing large-scale behavior has become a serious issue for science. For baseball, it always has been—and never more so than in the waning days of summer, when, as any fan knows, a single pitch can have fateful consequences.   [JG]

INDEX

Abrahamson, James A., 333
acid rain, 269, 276
  Canadian research on, 245–49
Adelman, George, 133
Adler, Alfred, 138
Adler, Gerald, 151–52
adolescents:
  defense mechanisms in, 163
  lying by, 124–25
  passive smoking and, 185–86
  prejudice and, 127
adults:
  defense mechanisms in, 163
  lying by, 121, 124
  naps for, 135
  presence of empathy in, 159
Advanced X-Ray Astrophysics Facility, 33, 35
Africa, 62
  AIDS in, 191–92, 194
  evolution of humans in, 78–81
Ahearne, John F., 327
AIDS, 7, 190–98
  heterosexual transmission of, 190–94
  neurological symptoms of, 195, 197
  scavenger cells and, 195–98
  secondary infections associated with, 196
Akkadian, 97, 99–100
Alaska:
  assembly of, 61–62, 64–65
  North Slope of, 61, 64–65
  tundra of, 253
Alberta, fossilized dinosaur embryos discovered in, 71–72
alcoholism:
  deaths from, 183
  heritability of, 229–32
  memory and, 136–37
algae:
  acid rain and, 248–49
  phosphorus pollution and, 246–47
Allen, Lew, 58
Allport, Gordon, 127–28
Almagest (Ptolemy), 32
Alster, Bendt, 100
altruism:
  in children, 158
  during midlife, 161–65
Amazon basin, deforestation of, 266
Amoco Cadiz tanker accident, 250
Andean civilization, 88–93

Anderson, Carl, 25
Anderson, John, 40
Andrews, George, 359, 361–62
Andrews, Peter, 79, 81
animals:
  dieting and longevity in, 218–22
  electromagnetic fields and, 284–85
  experiments on retinoic acid in, 176–77
  flocking and schooling behavior of, 347–51
  liver transplants for, 209–10
  presence of empathy in, 159–60
  REM sleep in, 131–32
  in response to passive smoking, 183, 185
Annas, George J., 211
Anthony, David, 86
anthropologists, Maori heritage constructed by, 101–5
Anthropology as Cultural Critique (Marcus and Fischer), 105
antimatter, 5, 24–28
anti-missile defense, Brilliant Pebbles for, 331–34
antiprotons, detection and study of, 26
antiradiation garments, 330
anti-Semitic prejudices, 127
aorta rupture, 205–6
Aramaic, 97–99
Archimedes, 363–64
architecture, Peruvian, 88, 90–92
Arctic Ocean, 64–65
Argo Merchant tanker accident, 250
Aristotle, 351
Arnett, W. David, 47–49
arteriosclerosis, 185
arthritis, joint pain in damp weather from, 204–5
artificial hearts, 210
artificial life, 293–97
Arvizu, Dan E., 304–6
Asia, assembly of, 61–63
Asians, breast cancer in, 9
Askey, Richard, 359, 361–62
Aspero, Andean temple at, 91
aspirin, heart attacks and, 8
Assyrian dictionary project, 98–99
asteroids:
  in collisions with Mars, 51–53
  organic materials present on, 55, 57
Aswan High Dam, 241–43
atherosclerotic heart disease, 207